Social Media Mining and Social Network Analysis:

Emerging Research

Guandong Xu
University of Technology Sydney, Australia

Lin Li
Wuhan University of Technology, China

Managing Director: Lindsay Johnston
Editorial Director: Joel Gamon
Book Production Manager: Jennifer Yoder
Publishing Systems Analyst: Adrienne Freeland
Development Editor: Monica Speca
Assistant Acquisitions Editor: Kayla Wolfe
Typesetter: Christy Fic
Cover Design: Nick Newcomer

Published in the United States of America by
Information Science Reference (an imprint of IGI Global)
701 E. Chocolate Avenue
Hershey PA 17033
Tel: 717-533-8845
Fax: 717-533-8661
E-mail: cust@igi-global.com
Web site: http://www.igi-global.com

Library of Congress Cataloging-in-Publication Data

Social media mining and social network analysis: emerging research / Guandong Xu and Lin Li, editors.
p. cm.
Includes bibliographical references and index.
Summary: "This book highlights the advancements made in social network analysis and social web mining and their influence in the fields of computer science, information systems, sociology, organization science discipline and much more"--Provided by publisher.
ISBN 978-1-4666-2806-9 (hbk.) -- ISBN 978-1-4666-2807-6 (ebook) -- ISBN 978-1-4666-2808-3 (print & perpetual access) 1. Online social networks--Research. 2. Data mining. 3. Communication--Network analysis. I. Xu, Guandong. II. Li, Lin, 1977-
HM742.S6282 2013
006.7'54--dc23

2012033295

British Cataloguing in Publication Data
A Cataloguing in Publication record for this book is available from the British Library.

Table of Contents

Foreword xiii

Preface xv

Chapter 1
Extracting Sentiment Patterns from Syntactic Graphs 1
Alexander Pak, Université Paris-Sud, France
Patrick Paroubek, Université Paris-Sud, France

Chapter 2
Mobile Context Data Mining: Methods and Applications 19
Enhong Chen, University of Science and Technology of China, China
Tengfei Bao, University of Science and Technology of China, China
Huanhuan Cao, Nokia Research Center, China

Chapter 3
Clustering Algorithms for Tags 39
Yu Zong, West Anhui University, China & University of Science and Technology of China, China
Guandong Xu, University of Technology Sydney, Australia

Chapter 4
Grouping the Similar among the Disconnected Bloggers 54
Nitin Agarwal, University of Arkansas at Little Rock, USA
Debanjan Mahata, University of Arkansas at Little Rock, USA

Chapter 5
Topic Modeling for Web Community Discovery 72
Kulwadee Somboonviwat, King Mongkut's Institute of Technology Ladkrabang (KMITL), Thailand

Chapter 6
Analyzing Social Networks to Mine Important Friends 90
Carson K.-S. Leung, University of Manitoba, Canada
Irish J. M. Medina, University of Manitoba, Canada
Syed K. Tanbeer, University of Manitoba, Canada

Chapter 7
News Document Summarization Driven by User-Generated Content 105
Luca Cagliero, Politecnico di Torino, Italy
Alessandro Fiori, IRC@C: Institute for Cancer Research and Treatment at Candiolo, Italy

Chapter 8
Method for Modeling User Semantics and its Application to Service Navigation on the Web 127
Munehiko Sasajima, Osaka University, Japan
Yoshinobu Kitamura, Osaka University, Japan
Riichiro Mizoguchi, Osaka University, Japan

Chapter 9
Tag Cloud Reorganization: Finding Groups of Related Tags on Delicious...................................... 140
Alberto Pérez García-Plaza, UNED, Spain
Arkaitz Zubiaga, City University of New York, USA
Víctor Fresno, UNED, Spain
Raquel Martínez, UNED, Spain

Chapter 10
Global Community Extraction in Social Network Analysis.. 156
Xianchao Zhang, Dalian University of Technology, China
Liang Wang, Dalian University of Technology, China
Yueting Li, Dalian University of Technology, China
Wenxin Liang, Dalian University of Technology, China

Chapter 11
Local Community Extraction in Social Network Analysis.. 172
Xianchao Zhang, Dalian University of Technology, China
Liang Wang, Dalian University of Technology, China
Yueting Li, Dalian University of Technology, China
Wenxin Liang, Dalian University of Technology, China

Chapter 12
Supporting Social Interaction in Campus-Scale Environments by Embracing Mobile Social
Networking 182
Zhiwen Yu, Northwestern Polytechnical University, China
Yunji Liang, Northwestern Polytechnical University, China
Yue Yang, Northwestern Polytechnical University, China
Bin Guo, Northwestern Polytechnical University, China

Chapter 13
Recommending Related Microblogs.. 202
Lin Li, Wuhan University of Technology, China
Huifan Xiao, Wuhan University of Technology, China
Guandong Xu, University of Technology Sydney, Australia

Detailed Table of Contents

Sentiment analysis and opinion mining became one of key topics in research of social media and social networks. Polarity classification, i.e. determining whether a text expresses a positive attitude or a negative one, is a basic task of sentiment analysis. Based on traditional information retrieval techniques, such as topic detection, many researchers use a bag-of-words or an n-gram model to represent an analyzed text. Regardless of its simplicity, such a representation loses latent information contained in relations between words in a sentence. The authors consider this information to be important for sentiment analysis and thus propose a novel method for representing a text based on graphs extracted from sentence linguistic parse trees. The new method preserves the information of words relations and can replace a standard n-gram model. In this chapter, the authors give a description of their approach and present results of conducted experimental evaluations that prove the benefits of their text representation. In the authors' experiments, they work with English and French languages; however, their approach is generic and can be easily adapted to other languages.

The mobile devices, such as iPhone, iPad, and Android are becoming more popular than ever before. Many mobile-based intelligent applications and services are emerging, especially those location-based and context-aware services, e.g. Foursquare and Google Latitude. The mobile device is important since it can detect a user's rich context information with its in-device sensors, e.g. GPS, Cell ID, and accelerometer. With such data and suitable data mining methods better understanding of users is possible; smart and intelligent services thus can be provided. In this chapter, the authors introduce some mobile context mining applications and methods. To be specific, they first show some typical mobile context data types with a mobile phone which can be detected. Then, they briefly introduce mining methods that are related to two mostly used types of mobile context data, location, and accelerometer. In the following, we illustrate in detail two context data mining methods that process multiple types of context data and can deal with the more general problem of user understanding: how to mine users' behavior patterns and how to model users' significant contexts from the users' mobile context log. In each section, the authors show some state-of-the-art works.

With the development and application of social media, more and more user-generated contents are created. Tag data, a kind of typical user generated content, has attracted lots of interests of researchers. In general, tags are the freely chosen textual descriptions by users to label digital data sources in social tagging systems. Poor retrieval performance remains a major problem of most social tagging systems resulting from the severe difficulty of ambiguity, redundancy, and less semantic nature of tags. Clustering method is a useful tool to increase the ability of information retrieval in the aforementioned systems. In this chapter, the authors (1) review the background of state-of-the-art tagging clustering and the tag data description, (2) present five kinds of tag similarity measurements proposed by researchers, and (3) finally propose a new clustering algorithm for tags based on local information that is derived from Kernel function. This chapter aims to benefit both academic and industry communities who are interested in the techniques and applications of tagging clustering.

Social interactions are an essential ingredient of our lives. People convene groups and share views, opinions, thoughts, and perspectives. Similar tendencies for social behavior are observed in the World Wide Web. This inspires us to study and understand social interactions evolving in online social media, especially in the blogosphere. In this chapter, the authors study and analyze various interaction patterns in community and individual blogs. This would lead to better understanding of the implicit ties between these blogs to foster collaboration, improve personalization, predictive modeling, and enable tracking and monitoring. Tapping interactions among bloggers via link analysis has its limitations due to the sparse nature of the links among the blogs and an exponentially large search space. The authors present two methodologies to observe interaction within the blogs via observed events addressing the challenges with link analysis-based approaches by studying the opinion and sentiments of the bloggers towards the events and the entities associated with the events. The authors present two case studies: (1) "Saddam Hussein's Verdict" and (2) "The Death of Osama Bin Laden." Through these case studies, they leverage their proposed models and report their findings and observations. Although the models offer promising opportunities, there are a few limitations. The authors discuss these challenges and envisage future directions to make the model more robust.

The proliferation of the Web has led to the simultaneous explosive growth of both textual and link information. Many techniques have been developed to cope with this information explosion phenomenon. Early efforts include the development of non-Bayesian Web community discovery methods that exploit only link information to identify groups of topical coherent Web pages. Most non-Bayesian methods produce hard clustering results and cannot provide semantic interpretation. Recently, there has been growing interest in applying Bayesian-based approaches to discovering Web community. The Bayesian approaches for Web community discovery possess many good characteristics such as soft clustering results and ability to provide semantic interpretation of the extracted communities. This chapter presents a systematic survey and discussions of non-Bayesian and Bayesian-based approaches to the Web community discovery problem.

Chapter 6
Analyzing Social Networks to Mine Important Friends 90
Carson K.-S. Leung, University of Manitoba, Canada
Irish J. M. Medina, University of Manitoba, Canada
Syed K. Tanbeer, University of Manitoba, Canada

The emergence of Web-based communities and social networking sites has led to a vast volume of social media data, embedded in which are rich sets of meaningful knowledge about the social networks. Social media mining and social network analysis help to find a systematic method or process for examining social networks and for identifying, extracting, representing, and exploiting meaningful knowledge—such as interdependency relationships among social entities in the networks—from the social media. This chapter presents a system for analyzing the social networks to mine important groups of friends in the networks. Such a system uses a tree-based mining approach to discover important friend groups of each social entity and to discover friend groups that are important to social entities in the entire social network.

Chapter 7
News Document Summarization Driven by User-Generated Content 105
Luca Cagliero, Politecnico di Torino, Italy
Alessandro Fiori, IRC@C: Institute for Cancer Research and Treatment at Candiolo, Italy

The outstanding growth of the Internet has made available to analysts a huge and increasing amount of Web documents (e.g., news articles) and user-generated content (e.g., social network posts) coming from social networks and online communities that are worth considering together. On one hand, the need of novel and more effective approaches to summarize Web document collections makes the application of data mining techniques established in different research contexts more and more appealing. On the other hand, to generate appealing summaries the data mining and knowledge discovery process cannot disregard the major Web users' interests. This chapter presents a novel news document summarization system, namely NeDocS, that focuses on generating succinct, not redundant, yet appealing summaries by means of a data mining and knowledge discovery process driven by messages posted on social networks. NeDocS retrieves from the Web and summarizes news document collections by exploiting (1) frequent itemsets, i.e., recurrences that frequently occur in the analyzed data, to capture most significant correlations among terms and (2) a sentence relevance evaluator that takes into account term significance in a collection of social network posts ranging over the same news topics. This approach allows not disregarding sentences whose terms rarely occur in the news collection but are deemed relevant by Web users. To the best of our knowledge, the combined usage of frequent itemsets and user-generated content in news document summarization is an appealing research direction that has never been investigated so far.

The value of information accumulated on the Web is enhanced when it is provided to the user who faces a problematic situation that can be solved by the information. The authors have investigated a task-oriented menu that enables users to search for mobile Internet services not by category but by situation. Construction of the task-oriented menu is based on a user modeling method that supports descriptions of user activities, such as task execution and defeating obstacles encountered during the task, which in turn represents the users' situations and/or needs for certain information. They built task models of the mobile users that cover about 97% of the assumed situations of mobile Internet services. Then they reorganized "contexts" in the model and designed a menu hierarchy from the viewpoint of the task. The authors have linked the designed menu to the set of mobile Internet service sites included in the i-mode service operated by NTT docomo, consisting of 5016 services. Among them, 4817 services are properly connected to the menu. This chapter introduces a framework for a real scale task-oriented menu system for mobile service navigation with its relations to the SNS applications as knowledge resources.

Tag clouds have become an appealing way of navigating through Web pages on social tagging systems. Recent research has focused on finding relations among tags to improve visualization and access to Web documents from tag clouds. Reorganizing tag clouds according to tag relatedness has been suggested as an effective solution to ease navigation. Most of the approaches either rely on co-occurrences or rely on textual content to represent tags. In this chapter, the authors explore tag cloud reorganization based on both of them. They compare these clouds from a qualitative point of view, analyzing pros and cons of each approach. The authors show encouraging results suggesting that co-occurrences produce more compelling reorganization of tag clouds than textual content, being computationally less expensive.

Great efforts have been made in retrieving the structure of social networks, in which one of the most relevant features is community extraction. A community in social networks presents a group of people focusing on a common topic or interest. Extracting all communities in the whole network, one can easily classify and analyze a specified group of people, which yields amazing results. Global community extraction is due to this demand. In global community extraction (also global clustering), each person of the input network is assigned to a community in the output of the method. This chapter focuses on global community extraction in social network analysis, previous methods proposed by some outstanding researchers, future directions, and so on.

Chapter 11
Local Community Extraction in Social Network Analysis 172
Xianchao Zhang, Dalian University of Technology, China
Liang Wang, Dalian University of Technology, China
Yueting Li, Dalian University of Technology, China
Wenxin Liang, Dalian University of Technology, China

To identify global community structures in networks is a great challenge that requires complete information of graphs, which is infeasible for some large networks, e.g. large social networks. Recently, local algorithms have been proposed to extract communities for social networks in nearly linear time, which only require a small part of the graphs. In local community extraction, the community extracting assignments are only done for a certain subset of vertices, i.e., identifying one community at a time. Typically, local community detecting techniques randomly start from a vertex v and gradually merge neighboring vertices one-at-a-time by optimizing a measure metric. In this chapter, plenty of popular methods are presented that are designed to obtain a local community for a given graph.

Chapter 12
Supporting Social Interaction in Campus-Scale Environments by Embracing Mobile Social Networking 182
Zhiwen Yu, Northwestern Polytechnical University, China
Yunji Liang, Northwestern Polytechnical University, China
Yue Yang, Northwestern Polytechnical University, China
Bin Guo, Northwestern Polytechnical University, China

With the popularity of smart phones, the warm embrace of social networking services, and the perfection of wireless communication, mobile social networking has become a hot research topic. The characteristics of mobile devices and requirements of services in social environments pose challenges to the construction of a social platform. In this chapter, the authors elaborate a flexible system architecture based on the service-oriented specification to support social interaction in a university campus. For the client side, they designed a mobile middleware to collect social contexts such as proximity, acceleration, and cell phone logs, etc. The server backend aggregates such contexts, analyzes social connections among users, and provides social services to facilitate social interaction. A prototype of mobile social networking system is deployed on campus, and several applications are implemented to demonstrate the effectiveness of the proposed architecture. Experiments were carried out to evaluate the performance (in terms of response time and energy consumption) of our system. A user study was also conducted to investigate user acceptance of our prototype. The experimental results show that the proposed architecture provides real-time response to users. Furthermore, the user study demonstrates that the applications are useful to enhance social interaction in campus environments.

Chapter 13
Recommending Related Microblogs 202
Lin Li, Wuhan University of Technology, China
Huifan Xiao, Wuhan University of Technology, China
Guandong Xu, University of Technology Sydney, Australia

Computing similarity between short microblogs is an important step in microblog recommendation. In this chapter, the authors utilize three kinds of approaches—traditional term-based approach, WordNet-based semantic approach, and topic-based approach—to compute similarities between micro-blogs and recommend top related ones to users. They conduct experimental study on the effectiveness of the three

approaches in terms of precision. The results show that WordNet-based semantic similarity approach has a relatively higher precision than that of the traditional term-based approach, and the topic-based approach works poorest with 548 tweets as the dataset. In addition, the authors calculated the Kendall tau distance between two lists generated by any two approaches from WordNet, term, and topic approaches. Its average of all the 548 pair lists tells us the WordNet-based and term-based approach have generally high agreement in the ranking of related tweets, while the topic-based approach has a relatively high disaccord in the ranking of related tweets with the WordNet-based approach.

Chapter 14
On Group Extraction and Fusion for Tag-Based Social Recommendation 211
Guandong Xu, University of Technology Sydney, Australia
Yanhui Gu, University of Tokyo, Japan
Xun Yi, Victoria University, Australia

With the recent information explosion, social websites have become popular in many Web 2.0 applications where social annotation services allow users to annotate various resources with freely chosen words, i.e., tags, which can facilitate users' finding preferred resources. However, obtaining the proper relationship among user, resource, and tag is still a challenge in social annotation-based recommendation researches. In this chapter, the authors aim to utilize the affinity relationship between tags and resources and between tags and users to extract group information. The key idea is to obtain the implicit relationship groups among users, resources, and tags and then fuse them to generate recommendation. The authors experimentally demonstrate that their strategy outperforms the state-of-the-art algorithms that fail to consider the latent relationships among tagging data.

Compilation of References 224

About the Contributors 245

Index.. 252

Foreword

Social computing has emerged recently as a hot interdisciplinary research topic and reached exponential growth in various application areas enabling new digital contexts for almost any type of interaction between humans and computers, especially the Internet. Social computing refers to procedures that support the gathering, representation, processing, use, and dissemination of information that is distributed across social collectivities such as teams, communities, organizations, and markets. Among the social computing domain, social networking and social media are deemed as two most popular and booming information services from the perspectives of not only academia but also the market. These new applications and services create a large volume of new heterogonous and homogeneous data sources, which bring in new research questions and challenges to the database, data management, and data mining research communities. Although an increasing number of conferences and workshops are held every year to discuss and exchange the latest progress and findings of commonly interested topics, such as social network analysis and social media mining, there are still many open research issues attracting numerous researchers and developers to tackle the difficulties and limitations of existing systems. This newly released edited book is undoubtedly a timely and valuable volume filling the gap between research and application.

Dr. Xu and Dr. Li have been working on these research themes since 2010 after successfully organizing the First International Workshop on Social Network Analysis and Social Media Mining over the Web. Knowing from many years' collaboration, I trust they indeed are serious and active researchers whose research profiles are evidenced by many quality and important publications, some joint works with me. At this point, I am very glad to see the completion of the book, which makes important contribution to the related research communities.

I believe this is a valuable and practical handbook on social network analysis and social media mining, and it will provide a useful referential book for both academia and industry.

Yanchun Zhang
Victoria University, Australia

Yanchun Zhang *is a Professor and the Director of Centre for Applied Informatics at Victoria University. He received the National "Thousand Talent Program" Award from China in 2010, and is currently a Director on the Australia-China Joint Lab on Social Computing and E-Health, a joint initiative from Graduate University of Chinese Academy of Science and Victoria University. He obtained a PhD degree in Computer Science from the University of Queensland in 1991. He has been active in areas of database and information systems, distributed databases, Web and Internet technologies, Web information systems, Web data management, Web mining, Web search, Web services, and e-research. He has published over 220 research* *papers in international journals and conference proceedings and authored/edited 12 books. His research has been supported by a number of Australian Research Council (ARC) linkage projects and discovery project grants. He was the winner of 2005 Victoria University Vice-Chancellor's Medal for Excellence in Research, and 2011 Victoria University Vice-Chancellor's Peak Award for Research and Research Training in Research Supervision, respectively. He was a member of Australian Research Council (ARC) College of Experts (2008-2010). He is currently a steering committee member of the ARC Research Network in Enterprise Information Infrastructure (EII). He held Honorary Professor positions at several universities/institutions in China, including Chinese Academy of Sciences, Wuhan University, Xiamen University, Northerneast University, Hubei University, Hebei Polytechnic University, and Hebei Normal University, and Visiting Professor position at Nagoya University in Japan (2006-20077). He is the Editor-In-Chief of World Wide Web Journal (Springer) and Health Information Science and Systems Journal (BioMed Central), and the Editor of the Web Information Systems Engineering and Internet Technologies Book Series from Springer. He is Chairman of the Web Information Systems Engineering (WISE) Society. He is the Australian representative of the International Federation of Information Processing (IFIP)'s Working Group WG6.4 on Internet Applications Engineering.*

Preface

BACKGROUND

The emergence of Web-based communities and hosted services such as social networking sites, wikis, and folksonomies brings the tremendous freedom of Web autonomy and facilitates collaboration and knowledge sharing between users. Along with the interaction between users and computers, social media is rapidly becoming an important part of our digital experience, ranging from digital textual information to diverse multimedia forms. These aspects and characteristics constitute the core of second generation Web.

A prominent challenge lies in modeling and mining this vast pool of data to extract, represent, and exploit meaningful knowledge, and to leverage structures and dynamics of emerging social networks residing in social media. Social networks and social media mining combines data mining with social computing as a promising direction and offers unique opportunities for developing novel algorithms and tools ranging from text and content mining to link mining.

We have successfully organized the first international workshop on social network analysis and social media mining over the Web and are continuing the second and third international workshops in conjunction with the DASFAA2010, DASFAA2011, and DASFAA2012 conferences. The accepted papers cover important research topics and novel applications of Social Web computing. About 50 people participate in the workshop every year and discuss the topics submissions. The overall goal of the workshop is to bring together academia, researchers, and industrial practitioners from computer science, information systems, statistics, sociology, behavior science, and organization science discipline, and provide a forum for recent advances in the field of social networks and social media from the perspectives of data management and mining.

Upon the past successful experience of organizing the workshop and the inspiration of rapid growth of research interests and outcomes available on these emerging areas, we compile the extension of these quality publications into a single volume along with the most recent results from a broad range of contributors to reflect the research advances in Social Web and social network analysis. It is believed such a compilation will provide a substantial summarization and review to those who are interested in related topics from the perspectives of algorithms and applications.

SUMMARY OF CHAPTERS

The coverage of each chapter presented is particularly summarized as follows:

Chapter 1 is about sentiment analysis and opinion mining. It utilizes graphs extracted from sentence linguistic parse trees to preserve the information of word relations, which can replace a standard n-gram model. This chapter gives a description of the proposed approach and presents results of conducted experimental evaluations that prove the benefit of this new text representation.

Chapter 2 introduces some mobile context mining applications and methods. To be specific, it first shows some typical mobile context data types with a mobile phone, which can be detected. Then, it briefly introduces mining methods that relate to two mostly used types of mobile context data, location and accelerometer. It illustrates in detail two context data mining methods that process multiple types of context data and can deal with the more general problem of user understanding: how to mine a user's behavior pattern and how to model a user's significant context from the user's mobile context log.

Chapter 3 first reviews the background of state-of-the-art tagging clustering and the tag data description. Then it presents five kinds of tag similarity measurements proposed by researchers. Finally, a new clustering algorithm is proposed for tags based on local information, which is derived from kernel function. This chapter aims to benefit both academic and industry communities who are interested in the techniques and applications of tagging clustering.

Chapter 4 studies and understands social interactions evolving in online social media, especially in the blogosphere. This chapter analyzes various interaction patterns in community and individual blogs to better understand the implicit ties between these blogs to foster collaboration and improve personalization, predictive modeling, and enable tracking and monitoring. Two methodologies are presented to observe interaction within the blogs via observed events addressing the challenges with link analysis-based approaches by studying the opinion and sentiments of the bloggers towards the events and the entities associated with the events.

Chapter 5 addresses the development of non-Bayesian Web community discovery methods that exploit only link information to identify groups of topical coherent Web pages. Most non-Bayesian methods produce hard clustering results and cannot provide semantic interpretation. Recently, there have been growing interests in applying Bayesian-based approaches to discovering Web community. The Bayesian approaches for Web community discovery possess many good characteristics such as soft clustering results, and the ability to provide semantic interpretation of the extracted communities. This chapter presents a systematic survey and discussions of non-Bayesian and Bayesian-based approaches to the Web community discovery problem.

Chapter 6 analyzes social networks to mine important friends among many users in social media. In particular, this chapter theoretically investigates, from a frequent pattern mining prospective, the research problem of distinguishing close friends who post messages on your wall from those acquaintances. Theoretical concepts of both ego-centric and socio-centric groups of friends are also implemented and materialized into a tree-based social media mining system that discovers important ego-centric groups of friends of any individual social media user as well as important socio-centric groups of friends in the social networks. A step-by-step illustrative example is given to demonstrate how relevant social network information is capturing in a tree structure, from which important groups of friends can be mined.

Chapter 7 presents a novel news document summarization system, namely NeDocS, that focuses on generating succinct, not redundant, yet appealing summaries by means of a data mining and knowledge discovery process driven by messages posted on social networks. NeDocS retrieves from the Web and summarizes news document collections by exploiting frequent itemsets, i.e., recurrences that frequently occur in the analyzed data, to capture most significant correlations among terms and a sentence relevance evaluator that takes into account term significance in a collection of social network posts ranging over the same news topics.

Chapter 8 investigates a task-oriented menu, which enables users to search for mobile Internet services not by category but by situation of the users. Construction of the task-oriented menu is based on a user modeling method that supports descriptions of user activities, such as task execution and defeating obstacles encountered during the task, which in turn represents users' situations and/or needs for certain information. This chapter introduces a framework for a real scale task-oriented menu system for mobile service navigation with its relations to the SNS applications as knowledge resources.

Chapter 9 discusses tag clouds, which have become an appealing way of navigating through Web pages on social tagging systems. Recent research has focused on finding relations among tags to improve visualization and access to Web documents from tag clouds. Reorganizing tag clouds according to tag relatedness has been suggested as an effective solution to ease navigation. Most of the approaches either rely on co-occurrences or rely on textual content to represent tags. This chapter explores tag cloud reorganization based on both of them. These clouds are compared from a qualitative point of view, analyzing the pros and cons of each approach. Encouraging results are shown, suggesting that co-occurrences produce more compelling reorganization of tag clouds than textual content, being computationally less expensive.

Chapter 10 discusses retrieving the structure of social network, in which one of the most relevant features is community extraction. A community in social networks presents a group of people focusing on a common topic or interest. By extracting all communities in the whole network, a specified group of people can be easily classified and analyzed, which yields amazing results. Global community extraction is due to this demand. This chapter focuses on global community extraction in social network analysis, previous methods proposed by some outstanding researchers, future directions, and so on.

Chapter 11 talks about local community extraction. The community extracting assignments are only done for a certain subset of vertices, i.e., identifying one community at a time. Typically, local community detecting techniques randomly start from a vertex, v, and gradually merge neighboring vertices one-at-a-time by optimizing a measure metric. In this chapter, plenty of popular methods are presented that are designed to obtain a local community for a given graph.

Chapter 12 elaborates a flexible system architecture based on the service-oriented specification to support social interaction on a university campus. A prototype of mobile social networking system is deployed on campus, and several applications are implemented to demonstrate the effectiveness of the proposed architecture. Experiments were carried out to evaluate the performance (in terms of response time and energy consumption) of this system. A user study was also conducted to investigate user acceptance of our prototype.

Chapter 13 describes how to compute similarity between short microblogs, which is an important step in microblog recommendation. Three kinds of approaches, i.e., traditional term-based approach, WordNet-based semantic approach, and topic-based approach are utilized to compute similarities between micro-blogs and recommend top related ones to users. An experimental study is conducted on the effectiveness of the three approaches in terms of precision.

Chapter 14 reports an interesting work about extracting group information from user annotation behavior for social recommender systems. With the recent information explosion, social websites have become popular in many Web 2.0 applications where social annotation services allow users to annotate various resources with freely chosen words, i.e., tags, which can facilitate users' finding preferred resources. However, obtaining the proper relationship among user, resource, and tag is still a challenge in social annotation-based recommendation researches. This chapter aims to utilize the affinity relationship between tags and resources and between tags and users to extract group information. The key idea is to obtain the implicit relationship groups among users, resources, and tags, and then fuse them to generate recommendation. It experimentally demonstrates that the proposed strategy outperforms the state-of-the-art algorithms that fail to consider the latent relationships among tagging data in terms of recommendation precision.

AUDIENCE OF THIS BOOK

This book is designed as a reference book for both academic and industry communities interested in the techniques and applications of Social Web, Web data management, Web mining, socially collaborative systems, as well as Web community and social network analysis for either in-depth academic research or industrial development in related areas.

This edited volume will not only combine the most recent research progresses and applications in social Web mining and social network analysis as a handbook to academia and researchers in computer science, information systems, statistics, sociology, behavior science, and organization science discipline, but also provide a compilation for disseminating and exchanging recent advances in the field of social networks analysis and social Web mining from the perspective of developmental practice for industrial practitioners.

Chapter 1
Extracting Sentiment Patterns from Syntactic Graphs

Alexander Pak
Université Paris-Sud, France

Patrick Paroubek
Université Paris-Sud, France

ABSTRACT

Sentiment analysis and opinion mining became one of key topics in research of social media and social networks. Polarity classification, i.e. determining whether a text expresses a positive attitude or a negative one, is a basic task of sentiment analysis. Based on traditional information retrieval techniques, such as topic detection, many researchers use a bag-of-words or an n-gram model to represent an analyzed text. Regardless of its simplicity, such a representation loses latent information contained in relations between words in a sentence. The authors consider this information to be important for sentiment analysis and thus propose a novel method for representing a text based on graphs extracted from sentence linguistic parse trees. The new method preserves the information of words relations and can replace a standard n-gram model. In this chapter, the authors give a description of their approach and present results of conducted experimental evaluations that prove the benefits of their text representation. In the authors' experiments, they work with English and French languages; however, their approach is generic and can be easily adapted to other languages.

INTRODUCTION

The increase of interest in sentiment analysis is associated with the appearance of Web-blogs and social networks, where users post and share information about their likes/dislikes, preferences, and lifestyle. Many websites provide an opportunity for users to leave their opinion about a certain object or a topic. For example, the users of IMDb[1] website can write a review on a movie they have watched and rate it on a 5-star scale. As a result, given a large number of reviews and rating scores, the IMDb reflects general opinions of Internet users on movies. Many other related Web-resources, such as cinema schedule websites, use the information from the IMDb to provide in-

DOI: 10.4018/978-1-4666-2806-9.ch001

formation about the movies including the average rating. Thus, IMDb reviews influence the choice of other users, who will have a tendency to select movies with higher ratings.

Another example is social networks. It is popular among users of Twitter[2] or Facebook[3] to post messages that are visible to their friends with an opinion on different consumer goods, such as electronic products and gadgets. Jansen (2009) called Twitter as "electronic word of mouth." The companies who produce or sell those products are interested in monitoring the current trend and to analyze people's interest. Such information can influence their marketing strategy or bring changes in the product design to meet the customers' needs.

Therefore, there is a need in algorithms and methods of automatic collection and processing of opinionated texts. Such methods are expected to classify the texts by their polarity (positive or negative), estimate the sentiment and determine the opinion target and the holder, where the target is the object or a subject of the opinion and the holder is usually the author of the text, but not limited to (Toprak, et al., 2010).

Bag-of-words is one of the first models of text representation, which is nowadays often used in sentiment analysis. In this approach, text is represented usually as a set of unigrams (or bigrams) disregarding their order and relations within the text. Common machine learning techniques such as Naive Bayes or SVM are then used to perform the sentiment classification of the given text. Although the accuracy of such approaches can be quite high, especially when using advanced feature selection techniques and additional lexicons of opinionated texts. We think that this model should be improved or replaced by the one that can identify more complex sentiment expressions rather than only simple ones such as *good movie* or *bad acting*.

One of the problems of bag-of-words representation is the information loss when representing a text as a collection of unrelated terms. However, these relations are often very important and may change the degree and the polarity of a sentiment expressed in the text. We illustrate this problem with a simple example. Let us consider a simple phrase: "This book is *bad*." The sentiment of this phrase is obviously negative and a standard classifier based on unigrams model will easily classify this sentence correctly provided a good training dataset. Now let us make the sentence a little more complex: "This book is *not bad*." In this case, a simple unigram model will probably fail. However, a bigram model will still work, capturing *not bad* as a term with a positive polarity. If we make the sentence more complex: "This books is *surprisingly not that bad*," both unigram and bigram models will fail. To make them work, a more sophisticated treatment of negations is needed.

Other than handling negations, the n-gram model has problems with capturing long dependencies. A bigram model will capture "I like" as a positive pattern in a sentence such as "*I like* fish," but not in "*I* really *like* fish." If we advance the task and move to a more refined polarity classification, i.e. identifying not only the polarity of a text (positive or negative), but also the degree of the polarity (low/high or even more precise), the n-gram model cannot provide sufficient information.

In order to solve the problems of the n-gram model, we propose to use a dependency parse tree of a sentence to generate a text representation. A dependency tree is a graphical representation of a sentence where nodes correspond to words of the sentence and edges represent syntactic relations between them such as *object*, *subject*, *modifier* etc. Figure 1 depicts a dependency parse tree of a sentence "I do not like fish very much."

Such a representation of the sentence suits very well sentiment analysis purposes and also opinion mining:

- From the given tree, we can easily identify negations: (not) → neg → (like).
- We can find intensity markers: (very) → advmod → (much) → advmod → (like).

Figure 1. A dependency parse tree of a sentence "I do not like fish very much." Nodes represent words, edges represent relations between words.

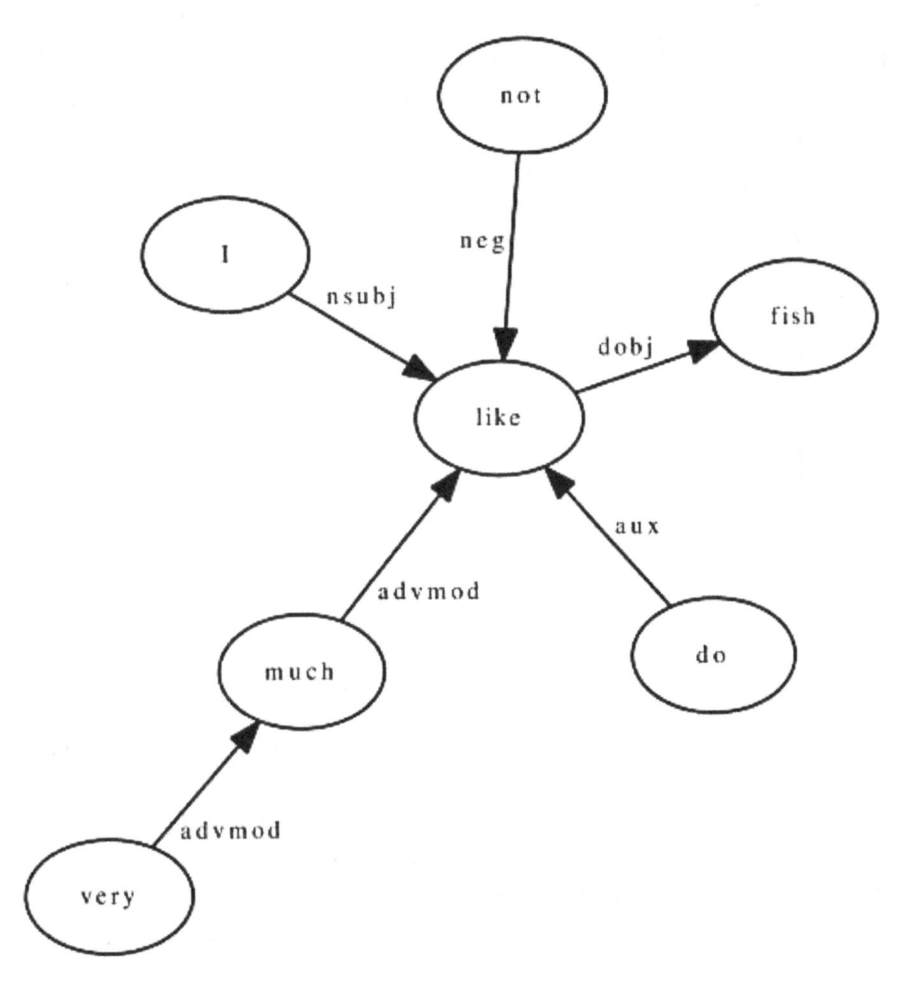

- Opinion holder: (I) → nsubj → (like); and opinion target: (like) → dobj → (fish).

In our approach, we use subgraphs from a sentence dependency tree to represent a given text. Similarly to n-grams, we define the size of a subgraph, which is equal to the number of edges it contains. Thus, a subgraphs of size 1 contains 1 edge and 2 nodes, a subgraph of size 2 contains 2 edges and 3 nodes, etc. For example, a sentence "I like fish much" can be represented by subgraphs of size 2 as depicted in Figure 2.

In the next section, we discuss prior research on sentiment analysis and give a brief overview of related works.

SENTIMENT ANALYSIS AND OPINION MINING

With the population of blogs and social networks, opinion mining and sentiment analysis became a field of interest for many researchers. A very broad overview of the existing work was presented in the work of Pang and Lee (2008). In their survey, the authors describe existing techniques

Figure 2. A representation of a sentence "I like fish much" with subgraphs of size 2

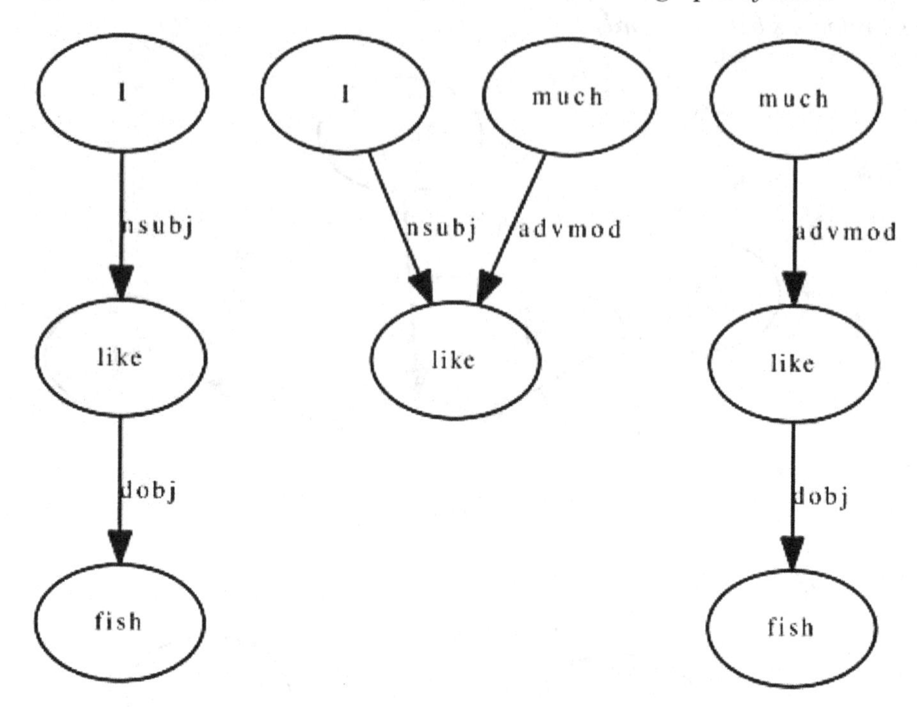

and approaches for opinion-oriented information retrieval.

Opinion and Sentiment Analysis (OSA) models vary greatly in their orientation. They may be either oriented toward discovering expression of opinion based on more or less rational considerations, judgment, or appreciations, or oriented toward the modeling and representation of the expression of the sentiment/emotions that one entertains about an object or an issue. They vary also greatly in the number of dimensions that they use to represent opinion or sentiments and in the more or less coarse granularity of their semantics.

The more of less subjective characterization of an opinion expression rendered through the combination of three attributes: opinion polarity, opinion intensity, and source engagement (with respect to the opinion expression). Listing the attributes of all the models we have encountered and putting them into relation yielded a graph that is too complex to be easily usable because of the numerous edges. Therefore, to organize the presentation of the different models, we decided to sort the different attributes according to an arbitrary order of importance, supposed to represent the distance from the center of our preoccupation, i.e. expression of opinion or sentiment. From the most important to the least, we have thus: polarity, intensity, target, information (the more or less factual character of the expression of opinion), engagement, and source. For instance, an OSA model having only the attribute polarity would be judged to be nearer the center of our preoccupation than a model with would have both polarity and target. With this attribute ordering, the twenty different OSA models organize themselves into a quasi-linear sort. From the most generic to the most specific model, we have identified six levels in the hierarchy of models.

Level 1 of our hierarchy lists authors who have not proposed any attribute in particular, but have addressed the subject of opinion and sentiment in language. They are associated in our taxonomy to the most generic (top) attribute

OSA model. Level 2 shows authors who do not have any polarity in their model and level 3 those who did not address Intensity, and so on. Then we have used the same methodology at each level to refine our hierarchy. At level 1, we find the models of Quirk et al. (1985), Kamps et al. (2004), and Berthard et al. (2004). They have defined other attributes of opinion expression, like polarity, intensity, target, information, etc. Quirk et al. (1985) have introduced the notion of private state, which regroups all the expressions of subjectivity like emotions, opinions, attitudes, evaluations, etc. This notion is also present in the model of Wiebe et al. (2005), Pang and Lee (2008). The models of Dave et al. (2003), Turney (2002), Harb et al. (2008), Somasundaran et al. (2008), Kim and Hovy (2006), and Stoyanov et al. (2004) are located at level 2. The models of Mullen and Collier (2004), Stoyanov et al. (2004) and Yu and Hatzivassiloglou (2003) were considered more specific than those of level 2 because they stressed the importance of target and source for opinion mining. The work of Yannik-Mathieu (1991) is characterized by a categorization of verbs expressing feelings. The model of Martin and White (2005) deals with evaluative aspects. The authors have mentioned three subtypes of evaluation, characterized by their respective attributes, which are attitude, engagement, and graduation. Attitude refers to values returned by judgment from one or more sources and can be associated to emotional responses. Its three subtypes are judgment, affect, and appreciation. Engagement explicit the position, the implication of the source with respect to its expression of opinion. It is one of the main characteristics of subjectivity. Martin and White (2005) have introduced the concept of graduation, which is further declined using force and focus. It expresses the strength of the opinion expression, so we merged this concept with intensity. Models of Choi et al. (2005) and Riloff et al. (2003) are also at the same level as Riloff et al. (2006), Turney and Littman (2003), and Yannik-Mathieu (1991). At last, Pang and Lee (2008) and Wiebe et al. (2005) propose the most complete models, which regroup together all the attributes.

Although opinion mining and sentiment analysis are often considered together in research and some authors even use these two terms as synonyms, we believe that it is necessary to make a distinction between two concepts: *opinions* and *sentiments*. In order to draw the line between these concepts, we propose the following definitions:

- Opinion is a statement of the personal position or beliefs regarding an event, an object, or a subject (opinion target).
- Sentiment is the author's emotional state that may be caused by an event, an object, or a subject (sentiment target).

To illustrate these differences, we show the examples in Table 1. In order to simplify the examples, we have limited sentiment and opinion values only to *positive* and *negative*. However, we believe that opinions have a more complex structure.

From the given examples, we conclude the following statements:

1. An opinion always has a target, while a sentiment may not have any cause.
2. An opinion may be associated with a sentiment, but there are cases with standalone opinions and standalone sentiments.
3. Opinions and sentiments may not agree on their polarity.

To formalize our model, we represent a sentiment as a triple (Holder, Polarity, Target), where the Holder has an emotional state with the Polarity (positive or negative) and the Target is optional. An opinion is represented by a triple (Holder, Claim, Target), where the Holder has the Claim about the Target. The representation of an opinion is similar to the one of a sentiment, but the Claim replaces the Polarity and the Target is a required element.

Table 1. Example of sentences with expressed sentiments and opinions

Sentence	Sentiment	Opinion	O. Target	S. Target
I am happy, cooking and listening to music on my iPod	positive	-	-	-
I am happy with the quality of my new iPod	positive	-	-	iPod
I think iPod is the best mp3 player	-	positive	iPod	-
I am happy that I have bought an iPod, it is the best mp3 player	positive	positive	iPod	purchase of iPod
I am happy that I have bought an iPod, although its price is too high	positive	negative	price	purchase of iPod
I read the new iPod press-release	-	-	-	-

From our examples, in the sentence "I think iPod is the best mp3 player," the Claim of the opinion is "<the Target> is the best mp3 player" and the Target is "iPod." In the sentence "I am happy that I have bought an iPod, although its price is too high," the Claim is "the price of <the Target> is too high" and the Target is "iPod." Our representation is similar to opinion quadruple (Topic, Holder, Claim, Sentiment) in the work of Kim and Hovy (2004), where Topic corresponds to Target in our model and in our model, we separate sentiments from opinions.

While sentiments are always polar, it is not the same case for opinions. An opinion claim often may be simplified and replaced with polarity. However, in the example "I believe the price of iPod will be $399," there is no positive or negative polarity. Nevertheless, in some context, it might have a positive or a negative meaning (e.g. "Apple's products are always overpriced. I believe the price of iPod will be $399."). Balahur et al. (2010) reported that it is necessary to distinguish between three perspectives: the author, the reader, and the text itself, as from the different perspectives there may be a different point of view on the information presented in the text.

Like/dislike statements deserve a special attention, because they combine sentiments and opinions. For example, a statement "I like iPod" expresses both a positive sentiment and a positive opinion. In this case, the positive sentiment towards the target implies that the sentiment holder has a positive claim about the target. In our research, we limit our scope to sentiments and polar opinions such as in like/dislike statements. We assume that the sentiment (opinion) holder is the author of a text and the polarity may take a positive or a negative value.

RELATED WORK

An early work by Pang et al. (2002) on polarity classification using bag-of-words model and machine learning reported 82.7% accuracy. This approach was tested on a movie review dataset collected from IMDb, which was later used by other researchers thus producing comparable results. The authors reported that using unigram features with binary weights yielded the highest accuracy.

In the follow up work, Pang and Lee (2004) augmented the classification framework with a additional preprocessing step. During this step, sentences from an analyzed text are being classified as subjective or objective. The authors translated subjectivity classification into a graph-partitioning problem and used the min-cut max-flow theorem to solve it. Finally, the sentences labeled as "subjective" are extracted and passed to a general polarity classifier (bag-of-words model with SVM). The reported statistically significant improvement of the classification accuracy was from 82.8% to 86.4%.

Whitelaw et al. (2005) used appraisal theory to produce additional features to be used in the classification. The authors built a taxonomy of appraisal and used it to identify appraisal groups within a text, such as "extremely boring" or "not that very good." For each appraisal group, a frame with 5 slots is filled: Attitude, Orientation, Force, Focus, and Polarity. Combinations of the first 3 of these slots were used to generate a feature vector. When backed up with a bag-of-words based classifier, the proposed method yielded 90.2% accuracy and 78.3% standalone.

Martineau and Finin (2009) took a different approach to increase the accuracy of sentiment classification. Instead of adding supplementary features or text preprocessing steps, they focused on the words weighting scheme. The authors presented delta tf-idf weight function which computes the difference of a word's tf-idf score in a positive and a negative training set. They claimed that the proposed technique boosts the importance of words unevenly distributed between the positive and the negative classes, thus these words should contribute more in the classification. Evaluation experiments on three different datasets showed statistically significant improvement of the classification accuracy. They achieved 88.1% accuracy on the movie dataset.

Paltoglou and Thelwall (2010) performed a thorough study on different weighting schemes and the impact on the sentiment analysis systems' performance. In their study, the authors have tested different variations of the classic tf-idf scheme on three datasets: movie reviews, product reviews, and the blog dataset (Macdonald & Ounis, 2006). The best results were yielded by a variation of smoothed delta tf-idf. In the experimental setup of leave-one-out cross validation, the polarity classification accuracy on the movie and the product review datasets was 95-96% depending on the scoring function variant.

Matsumoto et al. (2005) focused on a problem of the bag-of-word model, the information loss when representing a text by a non-related terms, thus losing the information contained in word order and syntactic relations between words in a sentence. To solve this problem, the authors proposed new features: word subsequences and dependency subtrees. Word subsequences were defined as a sequence of words obtained from a sentence by removing zero or more words. Dependency subtrees were obtained by extracting a part of a dependency tree, a sentence representation where nodes represent words and edges represent syntactic relations between words. Efficient mining algorithms were then used to find frequent subsequences and subtrees in the dataset. The combination of the proposed features with traditional n-gram features yielded 92.9% classification accuracy on the movie dataset.

Sentence dependency tree has been widely used in the sentiment analysis domain. A recent research by Arora et al. (2010) noted the problems of the standard bag-of-words text representation. The authors suggested an algorithm to extract subgraph features using genetic programming. However, the obtained features were not used to replace the standard n-gram model, but rather as a complementary set of features. Another recent research by Nakagawa et al. (2010) used dependency parse tree to obtain features that were used to train a CRF classifier for sentiment polarity detection. Zhuang et al. (2006) used dependency tree to extract feature-opinion pairs, where the first member of the pair is a feature term (such as *movie*) and the second is an opinionated term (such as *masterpiece*). The dependency tree was used to establish relations between feature words and opinion keywords. In the work of Chaumartin (2007), dependency parse tree was used to normalize headlines to a grammatically correct form for further sentiment tagging. Meena et al. (2007) used dependency tree to analyze a sentence construction along with WordNet[4] to perform sentence level sentiment classification.

OUR METHOD

Subgraph Representation

To represent a text using our method, first we need to obtain a dependency parse tree of each sentence. In our research, we have worked with English and French texts. The following examples will be in English; however, the corresponding steps are similar in French. Stanford Parser (de Marneffe, et al., 2006) was used to parse English, Xerox Incremental Parser (XIP) (Aït-Mokhtar, et al., 2002) was applied to French texts.

We use the output of typed dependencies of Stanford Parser to obtain a dependency tree of a sentence. For the sentence, "I do not like fish very much," the parser produces a list of dependencies from which we can reconstruct a dependency tree:

- nsubj(like-4, I-1)
- aux(like-4, do-2)
- neg(like-4, not-3)
- dobj(like-4, fish-5)
- advmod(much-7, very-6)
- advmod(like-4, much-7)

We want to obtain a tree where each node has a finite meaning. In our example, nodes *not* and *do,* do not have final meaning, and the node *like* contains only a partial meaning (it lacks the negation). Thus, after we obtain the dependency tree, we combine certain nodes as follows:

- Nodes linked with the negation relation *neg*, such that the edge (neg) → not → (like) becomes a single node (not like).
- Auxiliary verbs with the main verb *aux*, such that the edge (do) → aux → (like) becomes a single node (do like).
- Nodes connected with *cop* edge (copulae).

We prune the following edges to avoid unnecessary information:

- Determiners, such as (a) → det → (book) becomes (book).
- Possessives, such as (my) → poss → (book) becomes (book).
- Noun modifiers, such as (dog) → nn → (food) becomes (food).

Applying the rules above, the resulting tree for our example would look as in Figure 3.

Finally, the sentence is represented by a set of all possible subgraphs of a size S, where S is equal to the number of edges of subgraphs. In our experiments, we used S = 1 and S = 2.

Most of sentiment expressions have similar grammatical structures. For example, in the following expressions: "I like fish" and "I like movies," only the object differs, while the rest of the construction remains the same. We would like to train our system to recognize these expressions. For this reason we have added wildcard nodes in subgraphs (Arora, et al., 2010), which can match any word in a sentence.

For each subgraph, we obtain a permutation of subgraphs containing various number (from 0 to S − 1) of wildcard nodes. To do that, we replace each node of a subgraph by a wildcard node except for verbs, adjectives and adverbs as they usually express sentiments. We also forbid the creation of two adjacent wildcard nodes. Examples of the obtained wildcard subgraphs for a graph (I) → nsubj → (like) → dobj → (fish) are depicted in Figure 4.

Feature Construction

We represent a given text *T* as a feature vector $T = \{w_1, w_2, \cdots, w_K\}$, where w_i is a weight of a subgraph *i* in text *T; K* is the number of subgraphs in *T*. Similarly to the previous studies on sentiment classification, we experiment with the following types of weighting schemes:

Figure 3. The dependency tree after combining nodes and pruning edges

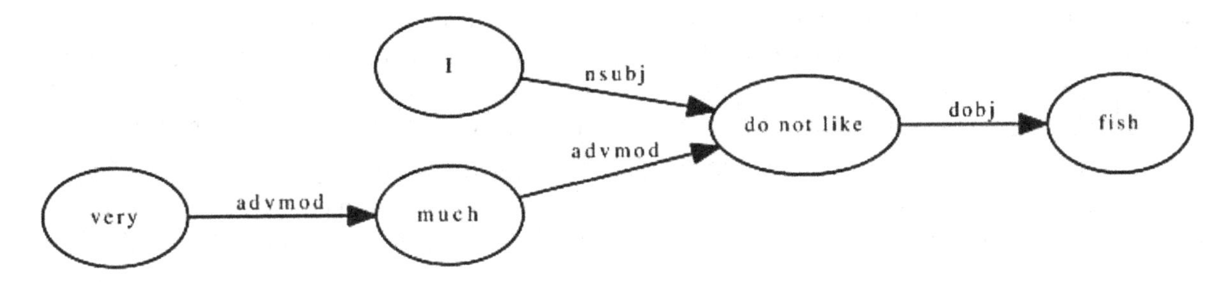

Figure 4. Obtained wildcard subgraphs for a sentence "I like fish"

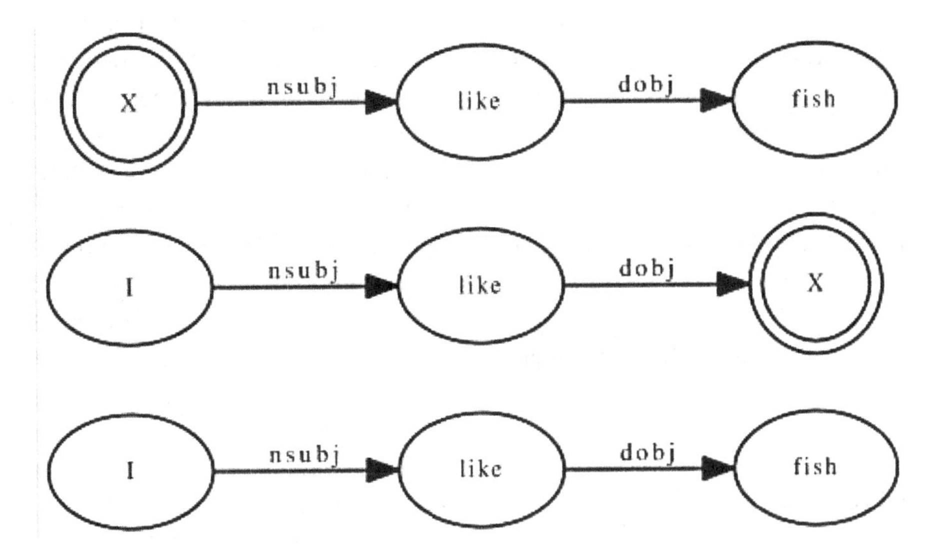

1. Binary, represents whether the graph is presented in the text (Pang, et al., 2002): $w_i = 1\ \forall i \in T$.
2. Frequency count: $w_i = tf_i$ where tf_i (term frequency) is the number times a subgraph i occurs in T.
3. Smoothed delta TFIDF (Martineau and Finin, 2009): $w_i = tf_i \cdot \Delta idf_i$ where $\Delta idf_i = \log [(N_1 \cdot df_2 + 0.5)/(N_2 \cdot df_1 + 0.5)]$ N_1 and N_2 are total numbers of documents of class 1 and 2, df_1 and df_2 are class frequencies of a graph i (i.e. numbers of documents of classes 1 and 2 in which the graph is occurred). In our case, classes 1 and 2 are positive and negative documents.

Discounting Scheme

Prior research on sentiment analysis showed that review authors tend to express sentiments in the last part of the text (Becker, et al., 2010). Thus, we decided to capture the position of the sentence in which the subgraph occurs in relation to the whole text. We introduce two strategies on how to add position information when constructing the feature vector. In the first strategy, we divided text into quantiles (3, 4, and 5) and treated subgraphs from different quantiles as independent features. In the second strategy, we added a discounting scheme for term frequency counting:

$$tf_i = \Sigma \forall_{pi} \in_T f(p_i)$$

where {p} is a set of sentence indexes where subgraph i occurs and f is a discounting function. p_i is equal to the index of the sentence in which the subgraph occurs divided by the total number of sentences. Thus if the subgraph occurs in the first sentence, $p_i = 0$, and if it occurs in the last sentence, $p_i = 1$. As for the discount function, we have tried the following:

1. Uniform, evenly increases weights of the sentences in the end of the review: $f(p) = p$
2. Sigmoid, gives more weight to the sentences in the end of the review: $f(p) = 1 / (1 + e^{-10p+5})$
3. Cosine square, gives more weight to the sentences in the beginning and in the end of the review: $f(p) = \cos^2(\pi x)$
4. Sine square, gives more weight to the sentences in the middle part of the review: $f(p) = \sin^2(\pi x)$

The graphs of the discount functions are given in Figure 5.

EXPERIMENTS AND RESULTS

Data and Evaluation Setup

To evaluate our classifier, we used two datasets: one in English and the other one in French. For English, we used the movie-review dataset, which is often used in sentiment analysis research[5] and initially was used in the work of Pang et al. (2002). The dataset contains 2000 written movie reviews mined from the IMDb website. The dataset is evenly split into a positive and a negative set (i.e. 1000 texts in each set). Each document contains a raw text without any HTML formatting. We use our subgraph text representation model with subgraphs of size 1 and also present the results with a size 2. We depict top-5 subgraphs (selected by Δidf score) from positive and negative set in Figure 6 and Figure 7, respectively, to give an idea of the extracted features.

We apply 10-fold cross validation and measure the classification average accuracy. This way we can compare our results with Pang et al. (2002) that we have chosen as the baseline. Pang et al. have obtained the best accuracy of 82.7% using SVM classifier on unigram model with binary features. We used an open source implementation of SVM classifier from the LIBLINEAR package (Fan, et al., 2008) with default parameters and a linear kernel.

For French, we use the video game review dataset from the DOXA project[6], which aims at building a domain independent industrial platform for opinion mining. The corpus is made of video game reviews collected from 8 dedicated sites[7]. The corpus and the sentiment annotations are described in the work of Paroubek et al. (2010). The annotations synthesize the sentiments expressed by the authors of the reviews at the document level and at the paragraph level (an arbitrary unit of roughly 200 words). For a given annotation unit (a whole document or a paragraph) DOXA annotation provide information about the semantic category of the sentiment drawn from an ontology of 17 semantic categories, inspired by the work of Mathieu (1991), split into 3 broad classes concerning respectively: affect, intellectual appreciation and a combination of both. Valency information and intensity are represented together by means of a of a six-value scale: neutral (no sentiment expressed), strong negative, weak-medium-negative, mixed (both positive and negative sentiment expressed together), weak-medium-positive, strong-positive. For the target of the sentiment, the annotation are taken from a domain specific ontology. Note that when several sentiments are expressed together about several targets in the same annotation unit, specific annotations (link), mapping sentiments to targets can be used to express any kind of mapping: one-to-one, one-to-many, many-to-one and many-to-many. The default mapping between sentiment and targets is many-to-many and there cannot be more than 5 sentiments and 5 targets

Figure 5. Discount functions (from the left to the right, top to bottom): uniform, sigmoid, cosine square, sine square

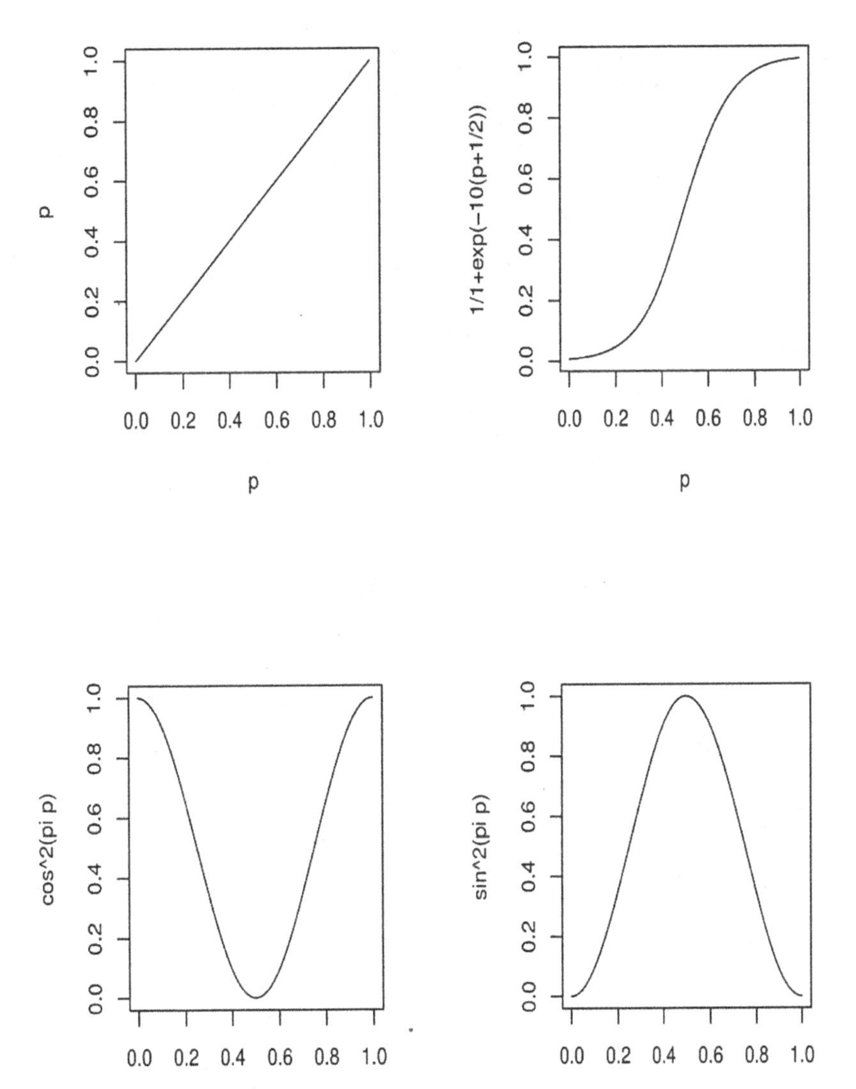

assigned to an annotation unit. An example of the annotations is given in Table 2.

We take all the documents with positive polarity (strong positive and weak-medium positive) all the documents with negative polarity (strong negative and weak-medium negative) and assign them to a positive and a negative class respectively. We do not use documents marked as neutral or mixed. This way, we obtain 387 positive documents and 250 negative documents. We further divide the positive set into a training set and a test set by taking all the documents that have been annotated by two annotators and assigning them to the test set. The remaining documents are moved to the training set. The same procedure is performed for the negative documents. Table 3 summarizes the dataset statistics.

Figure 6. Positive subgraphs extracted from the movie-review database

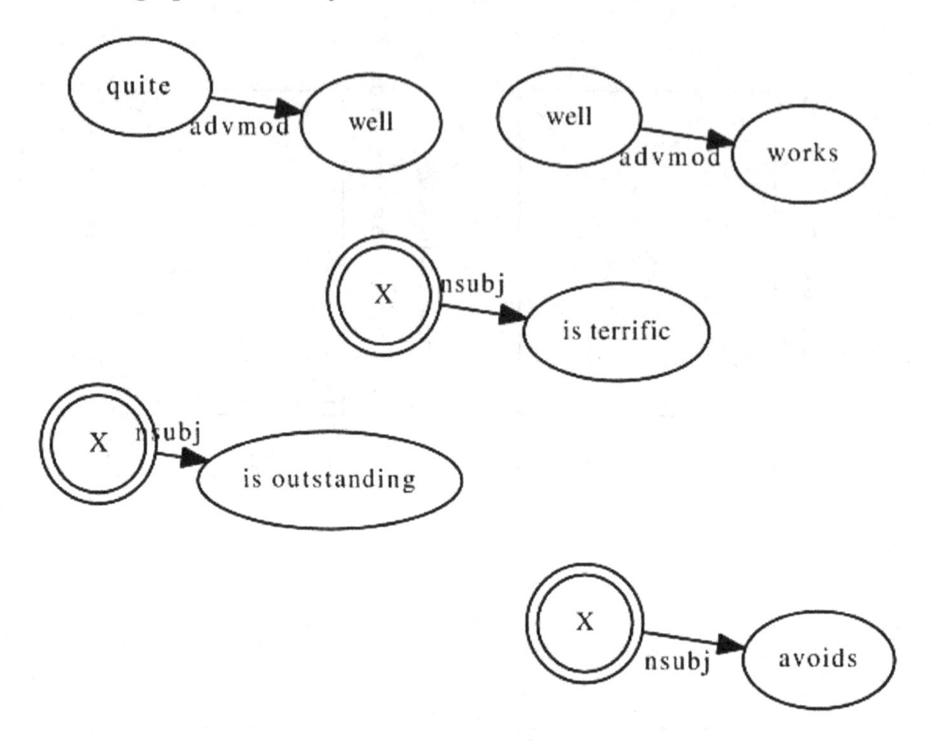

Figure 7. Negative subgraphs extracted from the movie-review database

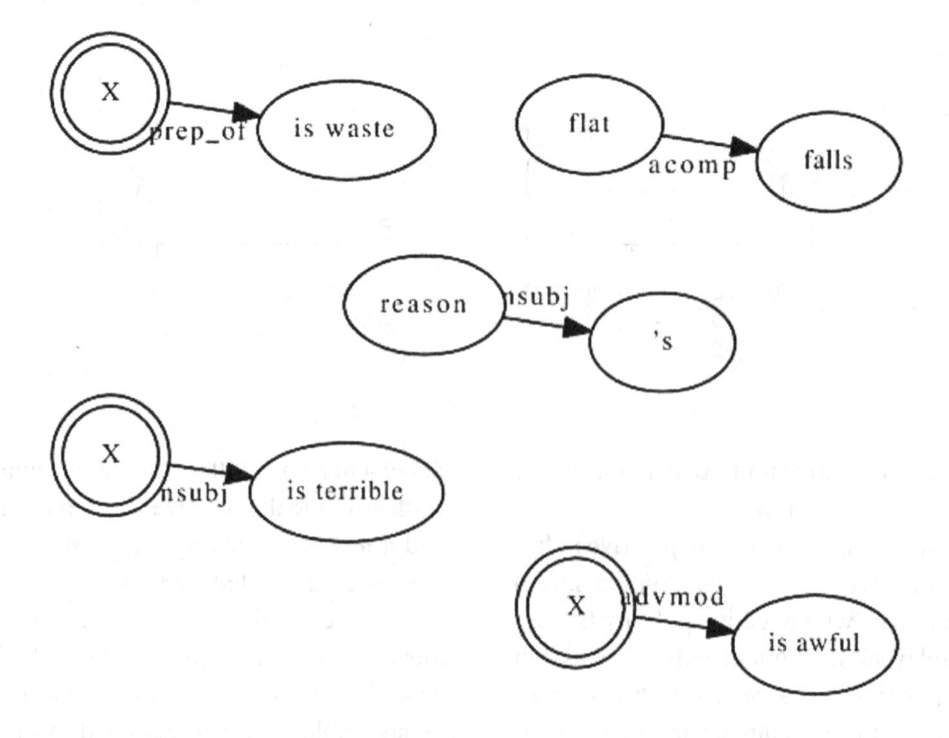

Table 2. DOXA macro/meso annotation of opinion

Anotation Unit	même façon: un individu rapide, un qui vole, un costaud. Même si vous changez de formation, la façon de se diriger dans les niveaux sera la même. Il est donc dommage de devoir se retaper le tutorial pour les 4 équipes. J'apprends donc qu'une couleur correspond à une fonction (rouge= force, jaune=vol, bleu=vitesse). Il est donc nécessaire d'utiliser les différentes compétences afin de terminer les niveaux. Par moment, plusieurs solutions seront possibles, à vous d'opter pour celle qui vous semble la mieux appropriée. Le tutorial terminé, je peux enfin en *(same manner: someone fast, another who flies, a though guy. Even if you change team, the way to move between levels will be the same. It is too bad to have to redo the whole tutorial for the four teams. So I learn that functions are color coded (red= strength, yellow=flight, blue=speed). You need to use the different capabilities to finish a level. Sometimes several solutions will be possible, it's up to you to choose the most appropriate. Once I finished the tutorial, I can at last)*
Attribute	**Value**
semantic category	*Displeasure-Disappointment*
polarity-intensity	*strong-negative*
topic	*GameTutorialMode*
link	*[[Displeasure-Disappointment], GameTutorialMode]*

As a baseline, we trained an SVM classifier with n-gram features and delta TFIDF weighting scheme. Negations were handled by attaching a negation particle to the preceding and the following word (Pak & Paroubek, 2010), when generating n-grams.

Results

Weighting Scheme

First, we present results obtained using the English dataset. We examined which weighting scheme works the best for our subgraph representation model. We have compared binary, frequency and delta TFIDF schemes. The results of the evaluation are presented in Table 4.

From the presented results, we can observe that the delta TFIDF scheme yields the highest accuracy of 85.1%. This result outperforms the baseline on 2.4%. Binary weighting scheme (81.9%) performs better than frequency counting (80.7%). Similar results were obtained in previous researches (Pang, et al., 2002; Paltoglou & Thelwall, 2010). Thus, for the further experiments we use the delta TFIDF weighting scheme.

Table 3. Number of documents in each set

Class	Train	Total
Positive	334	53
Negative	197	35
Total	531	88

Discounting Scheme

Next, we have examined whether capturing a subgraph position within the text improves the classification results. We have compared uniform, sigmoid, cosine and sine discounting functions as well as the quantile strategy using 3, 4, and 5 quantiles. The evaluation results are presented in Table 5.

Table 4. Accuracy and precision comparison for unigram, bigram, trigram models and our proposed subgraph representation model

Weighting Scheme	Ave. acc (%)	Δ
Baseline	82.7	0
Binary	81.9	-0.6
Frequency (tf)	80.7	-2
Delta TFIDF	85.1	2.4

Table 5. Reported accuracy of polarity classification using different discounting schemes

Discounting Scheme	Ave. acc (%)	Δ
No discounting	85.1	0
Uniform	84.4	-0.6
Sigmoid	84.45	-0.55
Cosine square	82.55	-2.55
Sine square	82.5	-2.6
3 quantiles	83.75	-1.35
4 quantiles	82.65	-2.45
5 quantiles	81.5	-3.6

We have obtained similar results as in Pang et al. (2002) and did not confirm the importance of the last part of the review (Becker & Aharonson, 2010). Capturing position using quantiles and discount functions does not improve the classification accuracy. Using sigmoid function yields the best performance (84.45%) as compared to other schemes and it is better than the baseline accuracy, while the accuracy is slightly lower than using no discounting scheme. Perhaps, a more refined notion than the subgraphs positions is needed to obtain better results. Our explanation is that authors use their own style of writing and therefore express their opinions in different parts of reviews.

We have also tried to index the dataset using subgraphs of size 2. However, the obtained results were much lower. The best achieved accuracy was 76.5%.

French Dataset

The results of the evaluation using the French dataset are presented in Table 6. Unigram, bigram, and trigram correspond to the baseline n-gram models respectively. Subgraph-1, subgraph-2 and subgraph-3 correspond to the proposed subgraph model with size 1, 2, and 3, respectively.

As we can observe from the table, the best accuracy is achieved by the subgraph representation model with subgraph size 1 (78.41%). The best accuracy of the n-gram model is yielded by the unigram model (73.86%). The performance of the n-gram models degrades as the order of n-gram increases. The same happens with the subgraph model accuracy decreases if we increase the size of subgraphs. Our explanation for the degradation of the performance is the lack of training data to learn a more complex model. Because in our data we have more positive opinionated documents, the average classification precision of the positive documents is higher than the classification precision of the negative ones. We have also tried to combine unigram, bigram, and trigram models to achieve a better accuracy, but it did not bring significant results. The same method was applied to the subgraph model combining subgraphs of a different size but it did not improve the results either.

Table 6. Reported accuracy and precision of polarity classification using unigram, bigram, trigram models, and our proposed subgraph representation model on the French dataset

Model	Ave. acc (%)	Ave. prec (%)	$Prec_{pos}$ (%)	$Prec_{neg}$ (%)
Unigram	73.86	69.57	90.57	48.57
Bigram	72.73	69.11	86.79	51.43
Trigram	64.77	60.08	83.02	37.14
Subgraph-1	78.41	74.8	92.45	57.14
Subgraph-2	64.77	61.05	79.25	42.86
Subgraph-3	60.23	59.22	64.15	54.29

CONCLUSION

With the popularity of blogs and social networks, opinion mining and sentiment analysis became a field of interest for many researches. An early work on supervised sentiment classification using n-gram model gave promising results and motivated many researches to use this model. However, n-gram text representation cannot capture complicated sentiment expressions and is hard to scale for bigger problems, such as identifying the polarity intensity or opinion holder/target tagging. A new text model is needed to improve the performance of sentiment analysis and opinion mining systems.

In our research, we developed a new text representation based on sentence parsing dependency tree. We represent a text as a collection of subgraphs, where nodes are words (or a wildcard) and edges represent relations between them. Such a representation allows to fill the information loss occurred when representing a text as a collection of n-grams without relation information.

We have tested our model on the movie-review dataset, which was often used in sentiment analysis community. An SVM classifier with features based on the extracted subgraphs yields a better performance than traditional system based on the unigram model. The highest accuracy we obtained is 85.1%. However, we think it can be improved by using advanced techniques such as feature selection algorithms or utilizing additional sentiment lexicons.

We have examined different weighting schemes for feature construction and concluded that the best performance is achieved using subgraphs of size 1 with the delta TFIDF scheme. We have also tried to note the position of the subgraphs within the given text by utilizing different discount functions and dividing texts into quantiles. However, it did not produce any significant improvement.

In the future work, we plan to add a machine-learning algorithm instead of hand-written rules for combining and pruning nodes during the preprocessing stage on the dependency tree. We also plan to obtain longer sentiment patterns by extracting subgraphs with more nodes. We believe that would improve our text representation model.

ACKNOWLEDGMENT

This work has been done thanks to the financial support of the project DOXA and CAP-DIGITAL.

REFERENCES

Aït-Mokhtar, S., Chanod, J.-P., & Roux, C. (2002). Robustness beyond shallowness: Incremental deep parsing. *Natural Language Engineering*, *8*, 121–144. doi:10.1017/S1351324902002887

Arora, S., Mayfield, E., Penstein-Rosé, C., & Nyberg, E. (2010). Sentiment classification using automatically extracted subgraph features. In *Proceedings of the NAACL HLT 2010 Workshop on Computational Approaches to Analysis and Generation of Emotion in Text, CAAGET 2010*, (pp. 131-139). Morristown, NJ: Association for Computational Linguistics.

Balahur, A., Steinberger, R., Kabadjov, M., Zavarella, V., van der Goot, E., & Halkia, M. … Belyaeva, J. (2010). Sentiment analysis in the news. In *Proceedings of the Seventh Conference on International Language Resources and Evaluation (LREC 2010)*. Valletta, Malta: LREC.

Becker, I., & Aharonson, V. (2010). Last but definitely not least: On the role of the last sentence in automatic polarity-classification. In *Proceedings of the ACL 2010 Conference Short Papers, ACLShort 2010*, (pp. 331–335). Morristown, NJ: Association for Computational Linguistics.

Berthard, S., Yu, H., Thornton, A., Hativassiloglou, V., & Jurafsky, D. (2004). Automatic extraction of opinion propositions and their holders. In *Proceedings of AAAI Spring Symposium on Exploring Attitude and Affect in Text*. AAAI.

Chaumartin, F.-R. (2007). Upar7: A knowledge-based system for headline sentiment tagging. In *Proceedings of the 4th International Workshop on Semantic Evaluations, SemEval 2007*, (pp. 422-425). Morristown, NJ: Association for Computational Linguistics.

Choi, Y., Cardie, C., Riloff, E., & Patwardhan, S. (2005). Identifying sources of opinions with conditional random fields and extraction patterns. In *Proceedings of HLT/EMNLP*. HLT/EMNLP.

Dave, K., Lawrence, S., & Pennock, D. M. (2003). Mining the peanut gallery: Opinion extraction and semanctic classification of product reviews. In *Proceedings of the 12th International World Wide Web Conference*. Budapest, Hungary: IEEE.

de Marneffe, M. C., Maccartney, B., & Manning, C. D. (2006). Generating typed dependency parses from phrase structure parses. In *Proceedings of LREC*. LREC.

Fan, R. E., Chang, K.-W., Hsieh, C.-J., Wang, X.-R., & Lin, C.-J. (2008). Liblinear: A library for large linear classification. *Journal of Machine Learning Research, 9*, 1871–1874.

Harb, A., Planitié, M., Poncelet, P., Roche, M., & Trousset, F. (2008). *Détection d'opinions: Apprenons les bons adjectifs*. Paper presented at Actes de l'Atelier Fouille des Données d'Opinions, conjointement Conférence INFORSID 08. Fontainebleau, France.

Jansen, B. J., Zhang, M., Sobel, K., & Chowdury, A. (2009). Twitter power: Tweets as electronic word of mouth. *Journal of the American Society for Information Science and Technology, 60*, 2169–2188. doi:10.1002/asi.21149

Kamps, J., Marx, M., Mokken, R. J., & de Rijke, M. (2004). Using wordnet to measure semantic orientation of adjectives. [). LREC.]. *Proceedings of LREC, 4*, 174–181.

Kim, S.-M., & Hovy, E. (2004). Determining the sentiment of opinions. In *Proceedings of the 20th International Conference on Computational Linguistics*, (p. 1367). Morristown, NJ: Association for Computational Linguistics.

Macdonald, C., & Ounis, I. (2006). *The TREC Blog06 collection: Creating and analysing a blog test collection. Technical Report*. Glasgow, UK: University of Glasgow.

Martin, J. R., & White, P. R. R. (2005). *The language of evaluation: Appraisal in English*. New York, NY: Palgrave Macmillan.

Martineau, J., & Finin, T. (2009). Delta TFIDF: An improved feature space for sentiment analysis. In *Proceedings of the Third AAAI Internatonal Conference on Weblogs and Social Media*. San Jose, CA: AAAI Press.

Matsumoto, S., Takamura, H., & Okumura, M. (2005). Sentiment classification using word subsequences and dependency sub-trees. In *Proceedings of the Pacific-Asia Conference on Knowledge Discovery and Data Mining*. IEEE.

Meena, A., & Prabhakar, T. V. (2007). Sentence level sentiment analysis in the presence of conjuncts using linguistic analysis. *Lecture Notes in Computer Science, 4425*, 573–580. doi:10.1007/978-3-540-71496-5_53

Nakagawa, T., Inui, K., & Kurohashi, S. (2010). Dependency tree-based sentiment classification using crfs with hidden variables. In *Proceedings of the Human Language Technologies: The 2010 Annual Conference of the North American Chapter of the Association for Computational Linguistics, HLT 2010*, (pp. 786-794). Morristown, NJ: Association for Computational Linguistics.

Pak, A., & Paroubek, P. (2010). Twitter as a corpus for sentiment analysis and opinion mining. In *Proceedings of LREC*. LREC.

Paltoglou, G., & Thelwall, M. (2010). A study of information retrieval weighting schemes for sentiment analysis. In *Proceedings of the 48th Annual Meeting of the Association for Computational Linguistics, ACL 2010*. Morristown, NJ: Association for Computational Linguistics.

Pang, B., & Lee, L. (2004). A sentimental education: Sentiment analysis using subjectivity summarization based on minimum cuts. In *Proceedings of the ACL 2004*, (pp. 271–278). ACL.

Pang, B., & Lee, L. (2008). Opinion mining and sentiment analysis. *Foundational Trends in Information Retrieval*, *2*(1-2), 1–135. doi:10.1561/1500000011

Pang, B., Lee, L., & Vaithyanathan, S. (2002). Thumbs up? Sentiment classification using machine learning techniques. In *Proceedings of the ACL 2002 Conference on Empirical Methods in Natural Language Processing, EMNLP 2002*, (vol. 10, pp. 79-86). Morristown, NJ: Association for Computational Linguistics.

Paroubek, P., Pak, A., & Mostefa, D. (2010). Annotations for opinion mining evaluation in the industrial context of the DOXA project. In *Proceedings of LREC*. LREC.

Quirk, R., Leech, G., & Startvik, J. (1985). *A comprehensive grammar of the English language*. New York, NY: Longman.

Riloff, E., Patwardhan, S., & Wiebe, J. (2006). Feature subsumption for opinion analysis. In *Proceedings of EMNLP*. EMNLP.

Riloff, E., Wiebe, J., & Wilson, T. (2003). Learning subjective noun using extraction pattern bootstrapping. In *Proceedings of the 7th Conference on Natural Language Learning*, (pp. 25–32). Edmonton, Canada: ACL.

Somasundaran, S., Ruppenhofer, J., & Wiebe, J. (2008). Discourse level opinion relations: An annotation study. In *Proceedings of the 9th SIGdial Workshop on Discourse and Dialogue*, (pp. 129–137). Association for Computational Linguistics.

Stoyanov, V., Cardie, C., Littman, D., & Wiebe, J. (2004). Evaluating an opinion annotation scheme using a new multiperspective question and answer corpus. In *Proceedings of the AAAI Spring Symposium on Exploring Attitude and Affect in Text*. AAAI.

Toprak, C., Jakob, N., & Gurevych, I. (2010). Sentence and expression level annotation of opinions in user-generated discourse. In *Proceedings of the 48th Annual Meeting of the Association for Computational Linguistics, ACL 2010*. Stroudsburg, PA: Association for Computational Linguistics.

Turney, P. (2002). Thumbs up or thumbs down? Semantic orientation applied to unsupervised classification of reviews. In *Proceedings of 40th Meeting of the Association for Computational Linguistics*, (pp. 417–424). ACL.

Whitelaw, C., Garg, N., & Argamon, S. (2005). Using appraisal groups for sentiment analysis. In *Proceedings of the 14th ACM International Conference on Information and Knowledge Management, CIKM 2005*, (pp. 625–631). New York, NY: ACM.

Wiebe, J., Wilson, T., & Cardie, C. (2005). *Annotating expressions of opinions and emotions in language*. Dordrecht, The Netherlands: Kluwer Academic Publishers. doi:10.1007/s10579-005-7880-9

Yannik-Mathieu, Y. (1991). *Les verbes de sentiment – De l'analyse linguistique au traitement automatique*. Paris, France: CNRS Editions.

Yu, H., & Hatzivassiloglou, V. (2003). Towards answering opinion questions: Separating facts from opinions and identifying the polarity of opinion sentences. In *Proceedings of EMNLP*, (pp. 129–136). Sapporo, Japan: EMNLP.

Zhuang, L., Jing, F., & Zhu, X.-Y. (2006). Movie review mining and summarization. In *Proceedings of the 15th ACM International Conference on Information and Knowledge Management, CIKM 2006*, (pp. 43–50). New York, NY: ACM.

ENDNOTES

[1] The Internet Movie Database: http://imdb.com

[2] Twitter: http://twitter.com

[3] Facebook: http://facebook.com

[4] A large lexical database of English: http://wordnet.princeton.edu/

[5] http://www.cs.cornell.edu/people/pabo/movie-review-data/otherexperiments.html

[6] DOXA is a project (DGE no 08-2-93-0888) supported by the numeric competitiveness center CAP DIGITAL of Île-de-France region which aims among other things at defining and implementing an OSA semantic model for opinion mining in an industrial context. See http://www.projet-doxa.fr

[7] www.ecrans.fr, www.gamehope.com, www.gamepro.fr, www.jeuxactu.com, www.jeuxvideo.com, www.jeuxvideo.fr, www.play3-live.com

Chapter 2
Mobile Context Data Mining:
Methods and Applications

Enhong Chen
University of Science and Technology of China, China

Tengfei Bao
University of Science and Technology of China, China

Huanhuan Cao
Nokia Research Center, China

ABSTRACT

The mobile devices, such as iPhone, iPad, and Android are becoming more popular than ever before. Many mobile-based intelligent applications and services are emerging, especially those location-based and context-aware services, e.g. Foursquare and Google Latitude. The mobile device is important since it can detect a user's rich context information with its in-device sensors, e.g. GPS, Cell ID, and accelerometer. With such data and suitable data mining methods better understanding of users is possible; smart and intelligent services thus can be provided. In this chapter, the authors introduce some mobile context mining applications and methods. To be specific, they first show some typical mobile context data types with a mobile phone which can be detected. Then, they briefly introduce mining methods that are related to two mostly used types of mobile context data, location, and accelerometer. In the following, we illustrate in detail two context data mining methods that process multiple types of context data and can deal with the more general problem of user understanding: how to mine users' behavior patterns and how to model users' significant contexts from the users' mobile context log. In each section, the authors show some state-of-the-art works.

DOI: 10.4018/978-1-4666-2806-9.ch002

1. INTRODUCTION

In recent years, the mobile device market is continuously increasing in a very fast speed. Users are more likely to use their mobile devices to play games, listen to music, and browse Internet anywhere and anytime. Mobile device is becoming important than ever before. A special feature of mobile device is its personality to user. Unlike traditional devices, e.g., PC, mobile device can detect user's rich mobile contexts, such as, user location and physical activity. More importantly, with the mobile context data, user understanding is becoming possible, e.g., mining habits of user in usual days, mining significant locations that user has visited, mining the contexts that the user likes to play games or listen to music, and inferring transportation mode when user is on the way, and so on. When we can understand user better, corresponding intelligent services would be provided naturally. Especially, in recent days, location based services are becoming popular. For example, users vote their favorite places with the mobile devices, which can detect location and publish the voting results through the social network. This has benefits both for those locations that users advertise for them and for users to let them have many funny things. Such application has attracted billions of users to engage in it.

Mobile devices are usually equipped with many sensors. Users may be familiar with two types of sensors, GPS sensor and Accelerometer sensor. With the GPS sensor, user can query where they are and with the Accelerometer sensor, user's movement and activity can be detected, this function is needed by many games. Besides these, there are also other sensors, like WiFi, Light, and Temperate sensors, they all play part in the mobile context sensor system and detect the user's context in different perspectives. Not at least, the camera and microphone can also be looked as sensors by which we can get the user's surrounding pictures and sounds. Table 1 shows the common sensors that the state-of-the-art mobile phones have. In addition, Cell ID which represents the location of cell site can also be used to reflect user's location, but with low accuracy.

Table 1. Mobile phone sensors

Sensor Name	Description
GPS	Detect location in out-door, have high accuracy.
WiFi	Detect network connection, can also detect location.
Accelerometer	Detect movement and gravity change.
Compass	Detect the direction change.
Bluetooth	Detect nearby Bluetooth devices.
Light	Detect light change.
Proximity	Detect the presence of nearby objects without physical contact.
Barometer	Detect change of atmospheric pressure.
Gyroscope	Detect orientations.
Camera	Detect nearby pictures.
Microphone	Detect nearby sounds.

With all these sensors, we can exactly know what the user's context looks like. Especially, if we collect all of the mobile context information historically, the user's habit will be mined and estimated, and then the mobile device will do things automatically and behave intelligently like a smart assistant. For example, to manage user's content, activity and give recommendations what user should to do next. Within all these sensors, location-related and accelerometer sensors are important since they reflect user's behaviors mostly. From the location history data which can be collected by GPS or Cell ID, user's significant places and routes can be mined. In addition, from the accelerometer data, user's current behavior status such as walking, driving can be known. Such information can be used, e.g., to provide more relevant Ads given the user's significant places, or to recommend places that the user may like to go to or to give suggestions when the user is in driving.

In the rest of this chapter, we will briefly introduce some location and accelerometer data mining methods in Section 2 and Section 3, these two types of data are the most important data in the whole mobile context sensor system, which records user's trajectory and physical activity. Then, we show methods and models to mine multiple types of mobile context data in together to reflect user's behavior habits in Section 4 and 5. At last, we conclude this chapter and point out the future works in Section 6.

2. MINING LOCATION AND TRAJECTORY DATA

Location data is very important to understand user. Through the location data, we can know the user's trajectory, the user's significant places where he/she frequently visited, the user's significant routines, such as the routines from user's home to office and likewise. This knowledge gives us opportunity to develop those location-based services. There are a lot of works mining knowledge from user's location data collected by mobile device (mostly from mobile phones). We will briefly introduce some basic tasks of mining location data in this section.

GPS sensor is a simple and accurate way to know the user's current location. However, the GPS sensor has its drawbacks. One drawback is about its power consumption, which means that GPS sensor cannot be always opened given the limited power. Thus, it is impossible to monitor user's movement with the GPS sensor on a mobile phone in all time. Another drawback is the GPS sensor need to communicate with satellites in an open place thus when users are indoor the GPS sensor will be useless. However, the fact is that users spend most of their time in the building on weekdays. Such drawbacks make the GPS sensor unsuitable as location data collector. However, as the widely used location data recorder, there are many GPS data to be mined, thus we will introduce some common tasks of GPS data mining in this section.

As mentioned before, Cell ID can be taken as another type of location sensor, which represents the cellular site when the mobile phone is connected. Mobile phones always need to connect to cellular sites for communication. Usually, a cellular site can cover area of several kilometers and all mobile phones in this area can communicate through it. Thus, the location of mobile phones can be taken as the location of their connected cellular site, which is represented by a Cell ID. The drawback of Cell ID is the accuracy, since a cellular site usually covers a large area. For those applications that need to collect user's historical data but do not require to be very accurate, using the Cell ID data to replace GPS sensor data is a proper way. After introducing GPS data mining methods, we will also introduce some typical examples which utilize Cell ID data.

2.1. Mining Collective Intelligences from GPS Data

GPS data is hard to collect directly, especially those related to personal trajectories, thus it is difficult to mine personal knowledge from mobile device GPS data. However, some services can still collect many location data from GPS-equipped mobile devices, such as, some map applications, that data are mostly collected when a user is in traveling, e.g. to a new city or a new place. From these user trajectory data, collective intelligence can be mined, e.g., social locations or trajectories.

Data Preprocessing

Usually, we can collect user's location data from GPS sensor when people turn on it. However, people can arbitrarily turn the GPS sensor on or off, thus the collected GPS data is not continuous and usually very sparse. We need to carefully preprocess it. An example of GPS log is like Table 2. In each timestamp, the latitude, longitude, altitude, and speed information can be recorded.

Table 2. A toy example of GPS log

Timestamp	Latitude	Longitude	Altitude	Speed
t_1	lat_1	lng_1	alt_1	s_1
t_2	lat_2	lng_2	alt_2	s_2
t_3	lat_3	lng_3	alt_3	s_3
...	...	...	...	...

The interval of timestamp depends on the sampling rate of GPS sensor. Usually we only focus on Latitude and Longitude attributes. To process such GPS log, firstly extracting stay places is useful to overcome the data spareness problem (Zheng, Liu, Wang, & Xie, 2008; Zheng, Zhang, Xie, & Ma, 2009), here stay place means those place that user have stayed for a long time. Stay place extraction is the procedure to determine those time ranges and GPS points in GPS log when user is in a fixed place. It can be divided into such steps: firstly determining two thresholds which limit the time lasting and place area, then merging those points in GPS log which are in the same time range and the same area that do not exceed the threshold, and at last, the center of the merged GPS points are computed to represent the stay place. When processing a new GPS point, if there exists a stay place and the distance between them is lower than the distance threshold, this new point can be taken as that stay place. There are also other data preprocessing steps such as to deal with the missing data problem, e.g., using adjacent GPS points to fill missing points (Adams, Phung, & Venkatesh, 2006).

Extracting Social Locations and Social Routines

After extracting stay places, those co-visited places of users can be determined. We can define social location as those places where multiple users usually visited. There are some reasons that social location is important, first given the social locations, many location-aware services can be provided, such as recommendation or advertisement; second, social location is the basic unit to compare different users, after mining user behavior patterns on these social locations, user similarity computing will be very useful in many applications, such as collaborative filtering for recommendation.

After extracting stay places from different single user's GPS log, we need to merge them together with a clustering method, then take the clusters as social locations. There are two commonly used approaches to cluster the stay places. One is hierarchy-based, which clusters the stay places in hierarchy levels, and thus the cluster results can be shown in different scale. Another type is flat-based, which just clusters high-density points to a cluster. Both approaches do not need to specify the number of clusters, since it is not practical to determine the number of social locations in advance. The stay places are all 2-Dimensional float data type, thus density-based clustering methods are very proper, e.g., DBSCAN. For the flat-based clustering approach (Adams, et al., 2006), DBSCAN method can be used directly. And for hierarchy-based (Zheng, et al., 2008, 2009), OPTICS method (Ankerst, Breunig, Kriegel, & Sander, 1999) can be used.

Besides extracting social locations, there are also works on extracting social trajectories, for example, trajectories that connect several famous locations (Zheng, et al., 2009).

Applications

Given the social locations, many interesting applications had emerged. For example, Adams et al. (2006) proposed a new type of browser to manage user's content such as photos and videos according to social locations and user interactions, compared with traditional content management methods, e.g., keyword or time based, to index the content with context is obviously easier to understand and can let users memorize what happens easily when recording those content; Zheng et al. (2009) proposed a user-location interest inference model, which was directly extended from HITS model and can rank the users and locations, thus the highly ranked users can be named as famous tourists and highly ranked locations can be recommended to new tourists. Li et al. (2008) computed user similarity based on user's history location data. User similarity computing is useful in collaborative filtering for recommendation. The basic steps are just like what we have mentioned, i.e. extracting stay places, social contexts and trajectories, and then computing user similarity based on their co-occurred locations and trajectories.

Location Recommendation: Location and trajectory recommendation is the most commonly applications leveraged from location data mining. Since it can reflect the wisdom of crowds. And, it is an automation process to utilize the trajectory data. When doing location recommendation, not only the popularity of location need to be considered, Zheng et al. (2009) proposed that user's experience is also an important factor. It is useful to design a model to consider both factors. They use the HITS (Kleinberg, 1999) inference model, which calculates user's experience and location's popularity in the same framework.

In this model, user and location are represented as hubs and authorities of HITS model. The computing of user's experience score depends on several factors, e.g., the number of places that the user visited, the visited places' popularity. The computing of the location's popularity also needs to take into account of factors like the number of users who have visited the place, and the experience of these users. In a word, the location and user have a reinforcement relationship, and the HITS model can properly reflect it. Such an inference model ranks the users and locations, and provide high ranked nodes as recommendations. Based on this HITS-based inference model and clustered social locations, social trajectories can also be ranked, thus can be recommended to user as travel routes.

Learning Transportation Mode: Another application of mining GPS data is to infer user's current transportation mode. The task is to judge whether the user is in driving, walking or in other transportation modes based on the given input of user's current GPS data. It is a typical supervised modeling work that needs training examples when user is in different transportation modes. To deal with the training data, some pre-processing steps are needed, that includes segmentation of different transportation mode from GPS log, feature extraction, classification model selection, and training. Usually, some typical features include maximum velocity, minimum velocity, average velocity, accelerator, last time transportation mode and so on. The classification model can be chosen from SVM, decision tree, naive bayes or sequential based classification model like CRF. Zheng et al. (2008) showed a framework of how to train transportation mode classifier with GPS data. Besides inferring transportation mode from user's GPS data, in following sections, we will also introduce inferring transportation mode with the accelerometer data.

2.2. Mining Significant Locations and Routines from Cell ID Log

Though Cell ID data can be collected easily, there is still much uncertainty. The issue is that cell tower usually covers a large area, and thus it is hard to determine the mobile phone's real location accurately. In another, the true physical location of cell site is also hard to get for some privacy and security reasons. Since cell sites usually have overlap in coverage, most previous works (Eagle, Clauset, & Quinn, 2009; Laasonen, Raento, & Toivonen, 2004; Meneses & Moreira, 2006) take the cell IDs and their transitions as a network, and take the movement of mobile phone as the walk in the network. For example, in a Cell ID log, if Cell ID A and Cell ID B are occurred adjacently, which means the user in the coverage of both cell site A and cell site B, thus an edge between A and B can be constructed. With the Cell ID network, the true locations of cell sites can be neglected.

Significant Location Extraction

Different from mining GPS data, it is easy to get a single user's full historical Cell ID data. Thus, it is possible to extract the personal significant locations from Cell ID log. Significant locations are those places where a specific user frequently visited and stayed for a long time. To do this we need to note some issues. First, the area of cell site coverage can be very large, up to several kilometers in diameter, especially in those sparsely populated areas, in another, areas covered by cell sites mostly overlapped where several cell IDs can be received by the same mobile phone even when is in stationary and at last, usually there is no Cell ID to physical location map thus it is hard to evaluate the mining results.

As mentioned earlier, to represent Cell IDs by a network, an intuitive idea is to partition the cell ID network into different clusters so that each cluster represents as a significant place. For example, Eagle et al. (2009) showed that three methods can be used to segment the cell ID network, which include normalized cut-based, community-based, and threshold group-based methods respectively. These methods are usually used to segment images and social networks. Take the community-based method as an example. It tries to partition the network to communities, where the communities have the property that the nodes in the same community have more connections and the nodes in different community have few connections. Such property can be measured with a formula called Q-modularity function.

The problem of partitioning network is to find a solution having the largest Q value. As we know, it is an NP-hard problem and researchers proposed lots of greedy algorithms to find out an approximate solution for it. We first let each node be a community; then according to the Q value, we iteratively merge the pair of communities, which can lead the Q value to have the largest increase until the maximum Q value is achieved. The advantage of this method is that it does not require the number of communities.

The drawback of the partition-based methods is that it did not consider the time difference of each node that user has stayed. Actually, most of cell IDs in the network are just noises and occurred only when user is moving, e.g., driving a car. Thus, we need to filter those noisy cell IDs and only keep the significant location related cell IDs, e.g., when user is in a significant location and the phone received cell IDs. According to the difference of time spending between different locations, we can design a simple rule to both cluster cell IDs and remove noise IDs. That is, given the connection network and the time spent on each node, we only keep those cell IDs that have lower time spent compared to its neighbor nodes and take the connected subgraphs as the significant locations.

There are also other approaches to extract significant locations from Cell IDs. Yang (2009) proposed a method to compute temporal correlation of cell IDs at first, and then merged correlated cell IDs. In addition, Laasonen et al. (2004) proposed to cluster cell IDs to locations and bases, where locations and bases are a set of cell IDs that satisfy several conditions such as subgraph diameter cannot exceed 2 or time spent have to be higher than a threshold. Bayir, Demirbas, and Eagle (2009) proposed to distinguish whether the user is in stationary or moving before constructing the cell ID network. The intuitive idea is when users start to move or is on the move, the cell ID changing is fast and can be computed as indicator, which is called as Mobility index. It thus can be used to discover places when the mobility index becomes lower. When mobility index is lower, the cell IDs can be clustered as the locations, and thus can be compared when user enters this location next time. They also propose to use time spent in each location to compute the significance.

Route Prediction

Given the GPS data, it is not possible to make prediction about one user's next destination since the lack of enough personal historical data. However, with abundant Cell ID data, it is possible to predict user's route by mining user's trajectory patterns. Laasonen (2005a, 2005b) proposed to compute the association of different locations and use association confidence to predict next location, this approach seems to lack a significant factor of periodic pattern that user mostly have though they have considered the time. Eagle et al. (2009) proposed an X-Factor model, which combines the user's significant location and time factors including day, week and a latent variable called abnormal behavior to indicate whether the user of this day is normal or abnormal to a bayes net model, then infer user's destination in a probabilistic framework.

3. MINING ACCELEROMETER DATA

Besides that location data can reveal the user's high scale movements, accelerometer sensor can detect the user's low-level movements (Randell & Muller, 2000). In this section, we introduce some examples showing how to use machine-learning methods to mine user's patterns from accelerometer data.

Accelerometer data mining is usually related to Activity recognition, which means recognizing user's high-level goals accurately from low-level accelerometer data, for example, to infer user's transportation mode, or user's movement goal, etc. Nowadays, most mobile phones contain triaxial accelerometer sensors, which return real valued estimate of acceleration along the x-, y-, z-axes. The velocity and displacement can also be estimated. Usually we can take accelerometer data as a 3-dimensional float data.

3.1. Learning Transportation Mode from Accelerometer Data

As previously introduced, we can use GPS data to infer user's transportation modes; however, the limitation is that users usually did not open their GPS sensor. Compared with GPS sensor, accelerometer sensor has low power consuming and usually is opened. Thus, it is more proper to utilize accelerometer data to infer transportation mode. Users' accelerometer data have different values in different modes; we can carefully design features to train a transportation mode classification model.

Since most mobile phones have only one accelerometer which is different from many research works that use multiple accelerometers to detect user's activity, the design of features and data preprocessing is important here. Firstly, the position and direction of accelerometer sensor should not have constraints and thus need to transform the 3-axis data to another type, which does not depend on direction. Secondly, since the accelerometer data is continuous thus we need to segment and extract training examples. Thirdly, feature design needs to consider most properties of different mode.

To cope with the problem of not being constrained by the accelerometer positions, it is a good solution that computing the magnitude of 3-axis (Longstaff, Reddy, & Estrin, 2010; Nham, Siangliulue, & Yeung, 2009; Reddy, Burke, Estrin, Hansen, & Srivastava, 2008). In another, Fourier transformation is also useful. To segment the training example, a short time range is proper, since some real applications require real-time processing. The features usually include energy, mean, and variance of signals as we have mentioned previously when using GPS data.

Nham et al. (2009) proposed to use GDM and SVM as the classification model, and their results show SVM is a little better. Reddy et al. (2008) experiment with more classification models including K-Nearest Neighbor, Naive Bayes, C4.5 Decision Trees, SVM, continuous Hidden Markov Model, a two-stage system involving a Decision Tree and a discrete Hidden Markov Model. In their results, the sequential model like HMM, the two-stage system with Decision Tree and HMM got the best results.

3.2. Activity Recognition

Transportation mode is just a subset of user activities. If we have enough labels, such as climbing up or down stairs, vacuuming, brushing teeth, and corresponding training examples, these activities can also be inferred by training classification models. A typical activity recognition problem can be described as following steps:

1. Collecting Accelerometer Data.
2. Processing Raw Accelerometer Data.
3. Constructing Feature vectors.
4. Training Classification Models.
5. Inferring new accelerometer data with the trained classifier.

In step 2, processing raw accelerometer data is to segment the data log given the activity labels; and in step 3, common features include magnitude, energy, mean, variances (Randell & Muller, 2000; Ravi, Mysore, & Littman, 2005; Xi, Bin, & Aarts, 2009). Along this line, Xi, et al. (2009) experiment with the accelerometer placed on the waist to detect more user activities. They also found that Bayes classifier has the advantage of scalability, requiring little effort in classifier retraining and performing well. To reduce the feature vector dimension, they used principal component analysis to remove the correlations among features. Ravi et al. (2005) proposed to recognize activities like brushing teeth, and they also showed that the combined classifiers with voting strategies can boost the recognition results. Usually, normal activities like transportation modes can be easily recognized since the abundance of training data. However, due to the lack of training examples, or the problem of unbalanced classes, all classifiers perform poorly when classifying abnormal activities. Hu, Zhang, Yin, Zheng, and Yang (2010) proposed to use HDP-HMM to estimate the number of activities given the accelerometer data, then incorporate a Fisher Kernel into the One Class SVM model to detect abnormal activities.

4. MINING PERSONALIZED BEHAVIOR PATTERN

Methods mentioned in previous sections just deal with single type data, such as location data or accelerometer data. It is obvious that single type data can only reflect one aspect of user context and may not reveal many user habits. From this section, we show mining methods that can deal with the whole types of context data. We begin with a work of mining personalized behavior patterns, which is an adaption of association rule mining between the mobile context and user behaviors.

Users spend a fair amount of time with their mobile devices in everyday, like playing games, listening to music or browsing Web pages. Meanwhile, the rich user interaction information can be captured by the mobile device and can be used to understand user habits. Such understanding of user habits may bring a great business values, such as targeted advertising and personalized recommendations. Usually, some user interactions are context-aware, that is, the occurrences of these user interactions are influenced by the contexts of users. For example, some users would like to listen to music with their smart phones when taking a bus to the workplace but rarely do the same thing on other contexts. Therefore, the associations between user interaction records and the corresponding contexts can be mined to characterize user habits, which are refereed as behavior patterns. In this section, we will format this problem and show an algorithm to solve it (Cao, Baọ, Yang, Chen, & Tian, 2009, 2010). Unlike previous sections, we find there exists very few similar works of mining the whole types of mobile context data in literature, and the procedures of such works is more complex than single-type context data mining algorithms; thus, we would like to illustrate them in detail here.

4.1. Problem Statement

We first define some related notations as follows to easy understand the problem of behavior pattern mining.

Definition 1 (Context): Given a contextual feature set $F = \{f_1, f_2, ..., f_{K'}\}$, a context C_i is a group of contextual feature-value pairs, i.e., $C_i = \{(x_1 : v_1), (x_2,: v_2), ..., (x_l, v_l)\}$, where $x_n \in F$ and v_n is the value for $x_n (1 \leq n \leq l)$. A context with l contextual feature-value pairs is called a l-context.

Definition 2 (Interaction Record): An interaction record is an item in the interaction set $\Gamma = \{I_1, I_2, ..., I_Q\}$, where $I_n (1 \leq n \leq Q)$ denotes a kind of user interaction.

Definition 3 (Context Record, Context Log): A context record $r =< Tid, C_i, I >$ is a triple of a timestamp Tid, a context C_i, and a user interaction record I. A context log $R = r_1 r_2 \cdots r_{N_R}$ is a group of context records ordered by timestamps.

We here define context as set of contextual feature-value pairs, where contextual feature denotes a type of context data, such as day period, location, and audio level. For example, *{(Holiday?: No),(Time: AM8:00-9:00)}* is a context with two contextual feature-value pairs. A context record captures the most detailed available context and occurrence of a user interaction during a time interval. To be noticed that here "available" means that a context record may miss the values of some contextual features though the set of context-features, whose values should be collected is predefined. Moreover, interaction records can be empty (denoted as "Null") because user interaction do not always happen. A toy context log example is shown in Table 3.

Context log includes the history context data and user interaction records, and thus can be data source for mining behavior patterns. A challenge here is the unbalanced occurrences of contexts and interaction records that traditional association rule mining approach did not consider since we only want to mine the association between context and user interaction. For example, suppose that Sam usually listens to music when taking a bus during workdays' AM8:00-9:00. When the context *{(Holiday?: No),(Time: AM8:00-9:00),(Transportation: On vehicle)}* appears, Sam usually listens to music but the exact time points when the interaction happens are uncertain. Consequently, the occurrences of the interaction *Listening to music* are very sparse compared with the occurrences of the context *{(Holiday?: No),(Time: AM8:00-9:00),(Transportation: On vehicle)}*, which causes the traditional association rule mining approach can hardly discover the behavior pattern *{(Holiday?: No),(Time: AM8:00-9:00),(Transportation: On vehicle)}* $\Longrightarrow$ *Listening to music*.

The observation from context logs is that if a user interaction is influenced by the context, the corresponding interaction record I usually co-occurs within the time ranges when continuously appears. Therefore, an intuitive idea is not only consider the co-occurrences of contexts and interaction records in separate context records but also consider their co-occurrences in the whole time ranges of contexts.

To be specific, let us take context logs as time ordered sequences of context records and calculate the support of a context by taking into account its time ranges of appearances. For a candidate behavior pattern, $C_i \Rightarrow I$ the support (denoted as $Sup(C_i \Rightarrow I)$) is still calculated by counting the context records where $C_i \cup I$ occurs. But for a context C_i, the support (denoted as $Sup(C_{i)})$ is

Table 3. A toy context log

Timestamp	Context	Interaction
t1	(Holiday?: No),(Time: AM8:00-9:00), (Transportation: On vehicle)	Null
t2	(Holiday?: No),(Time: AM8:00-9:00), (Transportation: On vehicle)	Listening music
t3	(Holiday?: No),(Time: AM8:00-9:00), (Transportation: On vehicle)	Null
……		
t22	(Holiday?: No),(Time: AM8:00-9:00), (Transportation: On vehicle)	Null
t23	(Holiday?: No),(Time: AM8:00-9:00), (Transportation: On vehicle)	Null
t24	(Holiday?: No),(Time: AM8:00-9:00), (Transportation: On vehicle)	Listening music
……		
t38	(Holiday?: No),(Time: AM8:00-9:00)	Null
t39	(Holiday?: No),(Time: AM8:00-9:00)	Null
……		
t43	(Holiday?: No),(Time: AM8:00-9:00), (Transportation: On vehicle)	Null
t44	(Holiday?: No),(Time: AM8:00-9:00), (Transportation: On vehicle)	Null
……		
t58	(Holiday?: No),(Time: AM8:00-9:00), (Transportation: On vehicle)	Listening music
t59	(Holiday?: No),(Time: AM8:00-9:00), (Transportation: On vehicle)	Null

calculated separately in two different cases. If C_i continuously appears in several adjacent context records and all of these context records do not contain any non-empty interaction record, we will count C_i once. Otherwise, we will count C_i by the number of interactions in these context records. In this way we ensure that $Sup(C_i)$ is always not smaller than $\sum_I Sup(C_i \Rightarrow I)$.

Take the toy context log in Table 3 for example, with the new definition of support, the support of context *{(Holiday?: No),(Time: AM8:00-9:00),(Transportation: On vehicle)}* is 4, and the support of behavior *{(Holiday?: No),(Time: AM8:00-9:00),(Transportation: On vehicle)}* $\Longrightarrow$ *Listening to music* is still 3, so the corresponding confidence is 3/4=0.75. Such a confidence is sufficiently large to distinguish the pattern from noisy behaviors.

The formal problem statement of behavior pattern mining is as follows.

Definition 4 (Match Context): A context record $r = < Tid, C_i, I >$ matches a context C_j if C_j is a sub-context of C_i. For simplicity, we can use $r \perp C_i$ to denote that r matches C_i.

Definition 5 (Context Range): Given a context C_i, and a context log $R = r_1 r_2 \cdots r_{N_R}$, we say that $R_i^n = r_i r_{i+1} \cdots r_{i+n}$ is a context range of C_i if 1) $\forall_{1 \le j \le n}(r_{i+j} \perp C_i)$; 2) Both $(r_{i-1} \perp C_i)$ and $(r_{i+n+1} \perp C_i)$ are false.

For example, consider the context log in Table 3, the context ranges of the context *{(Holiday?: No),(Time: AM8:00-9:00), (Transportation: On vehicle)}* are the context records with timestamps t_1, t_2, t_3, the context records with timestamps t_{22}, t_{23}, t_{24}, the context records with timestamps t_{43}, t_{44}, and the context records with timestamps t_{58}, t_{59}, respectively.

Definition 6 (Support, Confidence): Given a context C_i, an interaction record I, and a context log R, the support of C_i w.r.t. I (denoted as $Sup(C_i \Rightarrow I)$) is $\sum_m Count_m(I)$, where

$Count_m(I)$ denotes the number of context records which contain I in the m-th context range of C_i in R.

Given an interaction set Γ, the support of C_i (denoted as $Sup(C_i)$) is $\sum_{I \in \Gamma} Sup(C_i \Rightarrow I) + N_0$, where N_0 denotes the number of context ranges of C_i that do not contain non-empty interaction records. Moreover, the confidence of C_i w.r.t. I (denoted as $Conf(C_i \Rightarrow I)$) is $\frac{Sup(C_i \Rightarrow I)}{Sup(C_i)}$

Definition 7 (Promising Context, Behavior Pattern): Given a context C_i, a context log R, two user defined parameters min_sup and min_conf, if $\exists_I Sup(C_i \Rightarrow I) \geq \min\_sup$, C_i is called a promising context. Moreover, if $Sup(C_i \Rightarrow I) \geq \min\_sup$ and $Conf(C_i \Rightarrow I) \geq \min\_conf$, $C_i \Rightarrow I$ is called a behavior pattern.

Generally, a behavior pattern can be considered meaningless if the corresponding confidence is less than a threshold. The parameter min_sup can be determined by comprehensively considering min_conf and the distribution of contextual feature-value pairs.

4.2. Algorithm for Mining Behavior Patterns

Traditional association rule mining algorithms divide the mining procedure into two stages. In the first stage, all frequent item sets are found from the transaction database. In the second stage, the rules are generated from the frequent item sets and their confidences are calculated. However, in behavior pattern mining we only need to consider the association between the context and user interaction, thus the two stages can be integrated since the upper bound of the number of behavior patterns is linear to the number of promising contexts.

A naive algorithm is to enumerate all contexts appearing in the context log as candidate promising contexts and then counts their corresponding supports and confidences w.r.t. each interaction. However, this algorithm is inefficient since its time complexity is $O(N_R \cdot \prod_{k=1}^{K} N_k)$, where N_R indicates the number of context records and N_k indicates the number of appeared values for the contextual feature f_k in the context log. Wisely, we can reduce the number of candidate promising contexts $\prod_{k=1}^{K} N_k$ to a much smaller number by leveraging the *Apriori-property* of behavior patterns. That is, given a context C_i and an interaction *I*, if $Sup(C_i \Rightarrow I) > a$, for any sub-context of C_i denoted as C_j (i.e., $\forall_{p_i \in C_i^p} \in C_i$, where p_i denotes a contextual feature-value pair), $Sup(C_j \Rightarrow I) > a$. Along this line, we propose an algorithm for behavior pattern mining based on the framework of Apriori-all algorithm (Agrawal & Srikant, 1994), which is named GCPM (Generating Candidate promising contexts for behavior Pattern Mining). The main idea of GCPM is to generate candidate promising *l*+1-contexts by *joining* promising *l*-contexts.

Definition 8 (Join Context): Given two contexts $C_i = \{(x_1 : v_1), (x_2, v_2), ..., (x_l : v_l)\}$ and $C_j = \{(y_1 : u_1), (y_2 : u_2), ..., (y_l : u_l)\}$, if \$ $\forall_{1<n\leq l}(x_n = y_{n-1} \wedge u_n = u_{n-1})$, we say that C_i and C_j can join. The joined context of C_i and C_j is denoted as $C_i \cdot C_j = \{(x_1 : v_1), (x_2 : v_2), ..., (x_l : v_l), (y_l : u_l)\}$.

For example, given C_i={(*a*:2),(*c*:4),(*d*:7)} and C_j={(*c*:4),(*d*:7),(*e*:1)}, C_i can join with C_j, and $C_i \cdot C_j$ = {(*a*:2),(*c*:4), (*d*:7),(*e*:1)}. In contrast, given C_h = {(*c*:3),(*d*:7),(*e*:1)}, C_i and C_h cannot join because they have different values for contextual feature *c*.

Given a context log *R* and a user interaction set $\Gamma = \{I_1, I_2, ..., I_Q\}$, GCPM firstly finds all appearing 1-contexts as $\Lambda^1 = \{C_i^1\}$ and set *l* to be 1. Then, given a set of candidate promising l-contexts $\wedge^l$, the following procedure is executed iteratively until no candidate promising contexts can be generated.

1. Scan R and find all promising l-contexts from $\wedge^l$ and the corresponding behavior patterns.
2. Generate $\wedge^{l+1}$ by joining candidate promising l-contexts C_i^l and C_j^l where $\exists_{I\in\Gamma} Sup(C_i^l, I) \geq min\_sup \wedge Sup(C_j^l, I) \geq min\_sup$
3. If $\wedge^{l+1} = \varphi$, the algorithm terminates. Otherwise l++ and go to step 1).

Through iterative joining stages, GCPM largely reduces the number of candidate promising contexts. The time complexity of GCPM is $O(N_R \cdot \sum_l |\wedge^l|$. Furthermore, since the main cost of GCPM comes from counting the supports of candidate promising contexts w.r.t. each interaction, and to improve the efficiency of this stage, a novel data structure called *CH-Tree (Context Hash Tree)* for quickly updating the supports of candidate promising contexts can be used (Cao, et al., 2009).

Table 4 shows some behavior patterns mined from two real context logs D_A and D_B which are collected from two volunteers. From these behavior patterns we can infer some habits of the corresponding users. For example, the first pattern of D_A implies that user who provides D_A usually listens to music during workdays' AM8:00-9:00 when taking a vehicle (Speed=High).

5. MINING SIGNIFICANT PERSONALIZED CONTEXTS

Besides those contexts that are related to user interactions, we also want to know the significant contexts that users usually take, such as the contexts when the user is in the office, or user's typical habits of playing basketball in weekend afternoon. In this section, we introduce an unsupervised approach to mine significant personalized context of mobile users. The basic assumption is that each user has several typical contexts such as working, driving, or shopping, etc., user spent most of time in those contexts. In each context, the context data distribution is different from others. We can learn the context data distribution of each context. When given the new context data, the learned context model can be used to infer the current context.

There are few prior works of modeling user context with supervised models, like we mentioned in previous section, e.g., transportation mode learning and activity recognition. The drawback of such approaches is that they need user predefine the context and provide training examples that are hard to collect. In addition, the predefined or supervised model can only recognize user defined context and cannot process those contexts that

Table 4. Examples of mined behavior patterns

	behavior pattern	Sup	Conf
DA	Context: {(Holiday?:No),(Period:Morning), (Time: AM8:00-9:00),(Speed: High)}=>Interaction: Listening music	14	0.64
	Context: {(Holiday?: No),(Period: Afternoon), (Profile: General),(Movement:No),(Coordinate: (39.8554 116.4097))} => Interaction:Message session	15	0.68
DB	Context: {(Holiday?: Yes)(Time: AM7:00-8:00), (Battery: High),(Ring: Silent)}=>Interaction: Listening visible radio	5	0.62
	Context: {(Day: Thursday),(Period: Noon), (Time: AM12:00-PM13:00),(Profile: Silent)}=>Interaction: Accessing Web	4	1

are hard to predefine, e.g., the user's significant personalized contexts. Therefore, it is very attractive to exploit unsupervised techniques for mobile context mining and modeling when the domain knowledge is not available.

For example, suppose Table 5 shows a part of the context log of Ada, we can see that in a workday and during the time at AM8:00-AM9:00, Ada is in vehicle and often uses the music application, which might imply the context is driving a car to work place. In the following, if several adjacent context records in a context log are mutually similar, we say that they make up a context session. The context records in the same context session may capture the similar context information, and if two contextual feature-value pairs usually co-occur in same context sessions, they may also represent the same context. The unsupervised approach can automatically discover these highly related contextual feature-value pairs by taking advantage of their co-occurrences. Once a group of highly related contextual feature-value pairs is found, users can assign them meaningful context tags for context-aware recommendations. For example, if we discovered that the contextual feature-value pairs (Is a holiday?: Yes), (Time range: AM10:00-11:00), and (Movement: Moving) are highly related, Ada will be encouraged to tag these contextual feature-value pairs with an explicit context label "Go shopping" and define the services she wants on that context, such as playing some music or recommending the information of fashion dress. This kind of semi-automatic context-aware configuration is more convenient than a manual alternative that let Ada define the contextual feature-value pairs of "Go shopping" by herself.

Along this line, a two-stage unsupervised approach for learning the personalized contexts of mobile users can be proposed (Bao, Cao, Chen,

Table 5. A toy context log

Time	Context record
t_1	{(Holiday?: No),(Time: AM8-9),(Speed: High),(Audio: Low),(Application: Music)}
t_2	{(Holiday?: No),(Time: AM8-9),(Speed: High),(Audio: Middle)}
t_3	{(Holiday?: No),(Time: AM8-9),(Speed: High),(Audio: Middle)}
t_{38}	{(Holiday?: No),(Time: AM10-11),(Movement: No),(Audio: Low),(Inactive: Long)}
t_{39}	{(Holiday?: No),(Time: AM10-11), (Movement: No),(Audio: Low), (Inactive: Long)}
t_{40}	{(Holiday?: No),(Time: AM10-11), (Movement: No),(Audio: Low), (Inactive: Long)}
t_{58}	{(Holiday?: Yes),(Time: AM10-11), (Movement: No),(Cell ID: 2552), (Audio: Middle)}
t_{59}	{(Holiday?: Yes),(Time: AM10-11), (Movement: No),(Cell ID: 2552), (Audio: High)}
t_{60}	{(Holiday?: Yes),(Time: AM10-11), (Movement: No),(Cell ID: 2552), (Audio: Middle)}

Tian, & Xiong, 2011; Tengfei, Happia, Enhong, Jilei, & Hui, 2010). In the first stage, we segment the context log into context sessions where context records in the same context session are very similar with an adaptive segmentation method. In the second stage, we can group those contextual feature-value pairs that are usually co-occurred in the same context session with topic model methods. By using the topic models, we learn the context in the form of probabilistic distributions of context data. In the following, we will illustrate these two steps in detail.

5.1. Extracting Context Sessions

Given a context log $R = r_1 r_2 \dots r_n$, extracting context sessions from R is a procedure of segmenting R into N segments $S = \{s_1, s_2, \dots, s_N\}$, where s_i ($1 \le i \le N$) denotes a context session which consists of a group of adjacent and similar context records, and S is called a N-segmentation of R. There are two challenges: First, it is hard to estimate the number of context sessions (i.e. N). Second, it is also difficult to define a unified similarity threshold to determine where the original context log should be segmented for each different user. Therefore, we need an adaptive segmentation method, which can automatically choose the right number of context sessions and the segmentation points.

Here, we refer to a minimum entropy approach, which was originally proposed to segment pixels of an image (Hermes & Buhmann, 2003) and can be easily extended to segment context logs. The basic idea is to transform the objective of finding the optimized segmentation to finding the minimum conditional entropy of the context logs given the segmentation. In other words, we represent the user's activity with entropy, when she is in stable contexts, the entropy will be low. To be specific, if we measure the similarity between two adjacent context records through the probability that they are assigned into the same context session by a random segmentation, the objective becomes seeking the segmentation $S^* = \arg\max_S L(R \mid S)$, and it is equal to seeking the maximum $\log L(R|S)$. If we assume that 1) for each context record r, the probability to be assigned into a given context session s is independent, and 2) for each context feature-value pair p of a given context record r, the probability to be assigned into a given context session s is independent, $\log L(R|S)$ can be expressed as follows.

$$\begin{aligned}\log L(R \mid S) &= \sum_s \sum_{r_s} \log P(r_s \mid s) = \sum_s \sum_p n_{s,p} \log P(p \mid s) \\ &= N_p \sum_s \sum_p P(p,s) \log P(p \mid s) = -N_p \cdot H(p \mid s), \quad (1)\end{aligned}$$

where s denotes a context session in S, r_s denotes a context record in s, $n_{s,p}$ indicates the occurrence number of the feature-value pair p in s, and we use $\frac{n_{s,p}}{N_p}$ to estimate $P(p,s)$, where N_p denotes the number of all feature-value pairs in R. $H(p|s)$ denotes the conditional entropy of all contextual feature-value pairs given all context sessions. Therefore, the original problem is transformed to $S^* = \arg\max_S H(p \mid s)$.

We use greedy optimization to solve this problem (Hermes & Buhmann, 2003). To be specific, to search an N-segmentation with the minimum entropy, we first find a N+1-segmentation. Then we try to merge each pair of adjacent context sessions and find a N-segmentation with the minimum entropy. In this bottom up method, we initialize N equal to n, and decrease it. When the growth rate of the local minimum entropy becomes not that obvious, we stop the algorithm to make a balance between the complexity of the segmentation and the corresponding local minimum entropy.

5.2. Learning Significant Personalized Contexts

In the following, we illustrate the ways of learning personalized contexts using topic models. Intuitively, if we take contextual feature-value pairs as words, take context sessions of the given user as documents, and take latent contexts as topics, we can take advantage of topic models to learn personalized contexts for each user from his/her context sessions. However, we cannot directly apply topic models to mobile context modeling because the occurrences of the contextual features and the corresponding values. Contextual feature-value pairs are dependent on different factors: The contextual features are dependent on some external conditions, such as the availability of the corresponding signal; while the contextual values are dependent on the latent contexts and the corresponding contextual features. To discriminate the generation of contextual features and that of contextual values, we extend two of the existing topic models for fitting mobile context modeling.

Single-Context-Based Context Model: If we assume that one context session reflects one latent context, we can extend a typical singe-context-based topic model named the Mixture Unigram (MU) model (Nigam, McCallum, Thrun, & Mitchell, 2000) to the Mixture Unigram Context (MUC) model. MUC assumes that a context session in one user's log is generated by a prior contextual feature distribution and a prior context distribution together. To be specific, given K contexts and F contextual features, the MUC model assumes that a context session s is generated as follows. Firstly, a global prior context distribution θ is generated from a prior Dirichlet distribution α. Secondly, a prior contextual feature distribution π_s is generated from a prior Dirichlet distribution γ. Then, a context c_s is generated from θ. Finally, a contextual feature $f_{s,i}$ is generated from π_s, and the value of $f_{s,i}$ denoted as $v_{s,i}$ is generated from the distribution $\phi_{c_s,f_s,i}$. Moreover, there are totally $K \times F$ conditional distributions of contextual feature-value pairs $\{\phi_{k,f}\}$ which follow a Dirichlet distribution β. Figure 1 shows the graphical representation of the MUC model.

The likelihood of MUC is too complex to be inferred directly, and the approximate method

Figure 1. The graphical representation of the MUC model

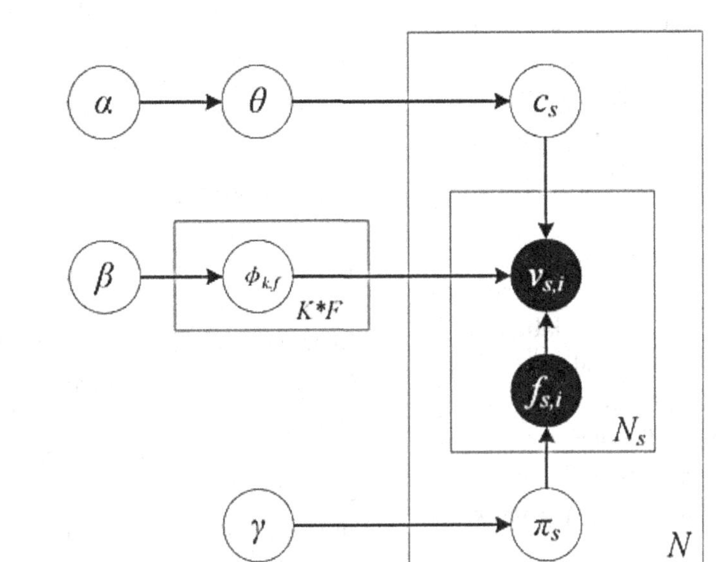

based on Gibbs sampling can be adopted (Heinrich, 2008; Resnik & Hardisty, 2010). In each iteration, the probability of Gibbs sampler assigns a context session s, denoted as c_s, to the latent context k is defined as follows.

$$P(c_s = k \mid C_{\neg s}, S) \propto P(v_s \mid c_s = k, C_{\neg s}, F, V_{\neg s}) P(c_s = k \mid C_{\neg s})$$

and if indicating the token (s,i) as m, we have:

$$P(c_s = k \mid C_{\neg s}, F, V_{\neg s}) = \prod_{i=1}^{N_s} \frac{n_{\neg s,k,f_m,v_m} + \beta_{v_m}}{\sum_v n_{\neg s,k,f_m,v_m} + \sum_{v \in V_{fm}} \beta_v}$$

$$P(c_s = k \mid C_{\neg s}) = \frac{n_{\neg s,k} + \alpha_k}{N - 1 + \sum_{k'}^{K} \alpha_{k'}}$$

where $\neg s$ means removing s from S (the given user's context session set), $C_{\neg s}$ denotes the context labels of other context sessions expect for s, V and F denote all contextual values and all contextual features in S, respectively, and v_s denotes all contextual values in s. $n_{\neg s,k,f,v}$ indicates the frequency that the contextual feature-value pair $(f{:}v)$ is labeled with the k-th context in all context sessions expect for s, v_f denotes the set of contextual values for the contextual feature f, and $n_{\neg s,k}$ indicates the number of context sessions with the k-th context expect for s. In all, the general idea is that in each iteration the probability of setting c_s equal to k not only depends on the probability of generating v_s from the k-th context but also the current probability of the k-th context in S.

After several rounds of Gibbs sampling, eventually each context session of given user's context log will be assigned a final context label. For a given user, we can derive the personalized contexts from the labeled context sessions by estimating the probability distribution of contextual feature-value pairs generated by a particular context. To be specific, the probability that a contextual feature-value pair $p_m = (f_m : v_m)$ is generated by the context c_k is estimated as

$$P(p_m \mid c_k) = P(v_m \mid c_k, f_m) P(f_m),$$

$$P(v_m \mid c_k, f_k) = \frac{n_{k,f_m,v_m} + \beta_{v_m}}{\sum_v n_{k,f_m,v} + \sum_{v \in V_{fm}} \beta_v} \quad (2)$$

$$P(f_m) = \frac{\sum_{k'}^{K} \sum_v n_{k',f_m,v} + \gamma f_m}{\sum_f \sum_{k'}^{K} \sum_v n_{k',f_m,v} + \sum_f \gamma_f}$$

Multiple-Context-Based Context Model: In practice, it is more general to assume that one context session may reflect multiple latent contexts. To this end, we propose a multiple-context-based context model (LDAC) which is extended from the Latent Dirichlet Allocation (LDA) model (Blei, Ng, & Jordan, 2003).

The graphical representation of the LDAC model is illustrated in Figure 2. From Figure 2 we can see that different feature-value pairs in the same context session s can now be extracted from different latent contexts $c_{s,i}$. Similar to MUC, the Gibbs sampling approach can also be adopted for estimating the parameters in LDAC, and the personalized contexts of mobile users can be also derived from the labeled contextual feature-value pairs according to Equation 2. Besides these, in both MUC and LDAC, we use perplexity as the measure to determine the number of latent contexts for each mobile user (Blei, et al., 2003).

5.3. A Case Study

Table 6 and Table 7 show the contexts of a user learned by MUC and LDAC respectively. After contacted with the volunteer who owns the data, we know he has a typical personalized context that *he usually plays basketball in weekends' afternoon (PM14:00-17:00)*. Then we manually check the learnt contexts of MUC and LDAC, and find that both of them discover a group of contextual feature-value pairs corresponding to

Figure 2. The graphical representation of the LDAC model

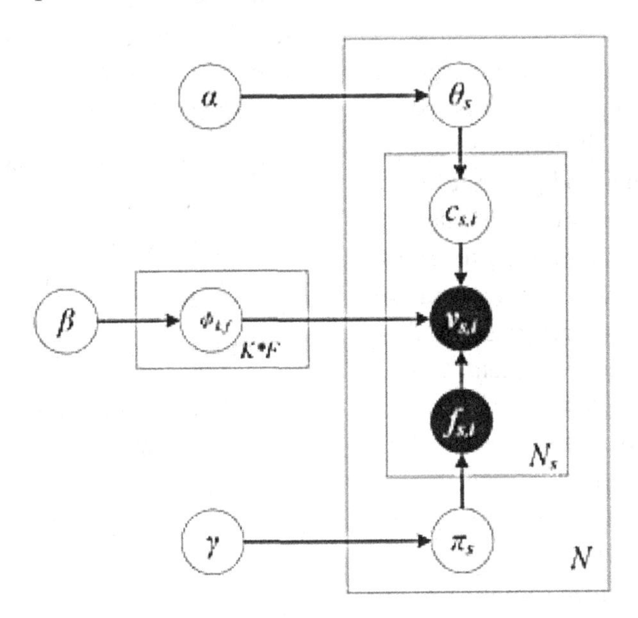

that context. For simplicity, we denote the context learnt by MUC as c_a and the context learnt by LDAC as c_b and show them in Table 6 and Table 7. We can observe that most of the contextual feature-value pairs of c_a are reasonable, such as (*Day name: Saturday*), (*Day period: Afternoon*), (*Location: Basketball area*). However, it also contains two noisy contextual feature-value pairs, namely, (*Time range: PM13:00-14:00*) and (*Time range: PM17:00-18:00*), and misses some more relevant contextual feature-value pairs such as (*Time range: PM14:00-15:00*) and (*Time range: PM15:00-16:00*). Table 6 lists all contextual feature-value pairs in c_b. From this figure, we can see that all listed contextual feature-value pairs are sensible to represent the user context. As expected, both MUC and LDAC work very well, and compared to MUC, LDAC model gets a better performance.

Table 6. Context C_a learnt by MUC

(Is a holiday?: Yes)
(Day name: Saturday)
(Day period: Afternoon)
(Time range: PM13:00-14:00)
(Time range: PM16:00-17:00)
(Time range: PM17:00-18:00)
(Location: Basketball area)
(Area ID: 21761)
(Cell ID: 10066)
(Profile: Outdoor)
(Movement: Not moving)
(Battery level: High(50%-80%))
(Battery level: Full(>80%))
(Inactive time: Middle(5-30 minutes))

Table 7. Context C_b learnt by LDAC

(Is a holiday?: Yes)
(Day name: Saturday)
(Day name: Sunday)
(Day period: Afternoon)
(Time range: PM14:00-15:00)
(Time range: PM15:00-16:00)
(Time range: PM16:00-17:00)
(Location: Basketball area)
(Area ID: 21761)
(Cell ID: 10066)
(Profile: Outdoor)
(Movement: Not moving)
(Battery level: Full(>80%))
(Inactive time: Middle(5-30 minutes))

6. CONCLUSION

In this chapter, we have introduced several mobile context data mining examples. The first part is about location data mining. GPS sensor is equipped by many of mobile phones and generates many user location data every day, from these data social locations can be mined. However, since most GPS data is not continuously recorded and thus do not include users' full trajectories, it is hard to find personal knowledge from them. Cell ID data is easy to collect though with low accuracy. We introduce some works to mine personal significant locations from Cell ID log. The second part is about accelerometer data mining. According to accelerometer data, we can infer user transportation modes or other activity patterns by training and applying classification models. The third part is on how to mine user personal behavior patterns from context log. We illustrate an adaptive association rule mining approach to mine the relationships between context and user interactions. The fourth part is an unsupervised approach to modeling personal context of mobile users. The intuitive idea is using probability distributions to model user contexts.

Given the limited space, we do not introduce more mobile context data mining works. However, there are more and more user mobile context data generated every day, and thus there are always needs to understand user by mining user context data. We believe more context-aware services will emerge in the future, and mobile data mining approach will play an important role in understanding users' behavior.

REFERENCES

Adams, B., Phung, D., & Venkatesh, S. (2006). Extraction of a social context and application to personal multimedia exploration. In *Proceedings of ACM Multimedia 2006*, (pp 987-996). ACM Press.

Agrawal, R., & Srikant, R. (1994). Fast algorithms for mining association rules in large databases. In *Proceedings of the 20th International Conference on Very Large Data Bases*. IEEE.

Ankerst, M., Breunig, M. M., Kriegel, H. P., & Sander, R. (1999). OPTICS: Ordering points to identify the clustering structure. In *Proceedings of the 1999 ACM SIGMOD International Conference on Management of Data*. Philadelphia, PA: ACM Press.

Bao, T., Cao, H., Chen, E., Tian, J., & Xiong, H. (2011). An unsupervised approach to modeling personalized contexts of mobile users. Retrieved from http://dm.ustc.edu.cn/docs/2010/mobileuser.pdf

Bayir, M. A., Demirbas, M., & Eagle, N. (2009). Discovering spatiotemporal mobility profiles of cellphone users. In *Proceedings of the World of Wireless, Mobile and Multimedia Networks & Workshops, 2009, WoWMoM 2009*. IEEE Press.

Blei, D. M., Ng, A. Y., & Jordan, M. I. (2003). Latent dirichlet allocation. *Journal of Machine Learning Research*, *3*, 993–1022.

Cao, H., Bao, T., Yang, Q., Chen, E., & Tian, J. (2009). An effective approach for mining mobile user habits. In *Proceedings of the 19th ACM International Conference on Information and Knowledge Management*, (pp. 1677-1680). ACM Press.

Cao, H., Bao, T., Yang, Q., Chen, E., & Tian, J. (2010). An effective approach for mining mobile user habits. In *Proceedings of the 19th ACM International Conference on Information and Knowledge Management*. Toronto, Canada: ACM Press.

Eagle, N., Clauset, A., & Quinn, J. A. (2009). *Location segmentation, inference and prediction for anticipatory computing*. Retrieved from http://reality.media.mit.edu/pdfs/anticipatory.pdf

Heinrich, G. (2008). *Parameter estimation for text analysis*. Leipzig, Germany: University of Leipzig.

Hermes, L., & Buhmann, J. M. (2003). A minimum entropy approach to adaptive image polygonization. *IEEE Transactions on Image Processing, 12*(10), 1243–1258. doi:10.1109/TIP.2003.817240

Hu, D. H., Zhang, X. X., Yin, J., Zheng, V. W., & Yang, Q. (2010). Abnormal activity recognition based on HDP-HMM models. Retrived from http://www.cse.ust.hk/~vincentz/ijcai09_abnormalAR.pdf

Kleinberg, J. M. (1999). Authoritative sources in a hyperlinked environment. *Journal of the ACM, 46*(5), 604–632. doi:10.1145/324133.324140

Laasonen, K. (2005). Clustering and prediction of mobile user routes from cellular data. *Knowledge Discovery in Databases, 3721*, 569–576.

Laasonen, K. (2005). *Route prediction from cellular data.* Paper presented at the Workshop on Context-Awareness for Proactive Systems (CAPS). New York, NY.

Laasonen, K., Raento, M., & Toivonen, H. (2004). Adaptive on-device location recognition. *Pervasive Computing, 3001*, 287–304. doi:10.1007/978-3-540-24646-6_21

Li, Q., Zheng, Y., Xie, X., Chen, Y., Liu, W., & Ma, W. Y. (2008). Mining user similarity based on location history. In *Proceedings of the 16th ACM SIGSPATIAL International Conference on Advances in Geographic Information Systems.* Irvine, CA: ACM Press.

Longstaff, B., Reddy, S., & Estrin, D. (2010). *Improving activity classification for health applications on mobile devices using active and semi-supervised learning.* Paper presented at the Pervasive Computing Technologies for Healthcare (PervasiveHealth), 2010 4th International Conference on NO PERMISSIONS. New York, NY.

Meneses, F., & Moreira, A. (2006) *Using GSM CellID positioning for place discovering.* Paper presented at the Pervasive Health Conference and Workshops. New York, NY.

Nham, B., Siangliulue, S., & Yeung, S. (2009). *Predicting mode of transport from iPhone accelerometer data.* Palo Alto, CA: Stanford University.

Nigam, K., McCallum, A. K., Thrun, S., & Mitchell, T. (2000). Text classification from labeled and unlabeled documents using EM. *Machine Learning, 39*(2), 103–134. doi:10.1023/A:1007692713085

Randell, C., & Muller, H. (2000). *Context awareness by analyzing accelerometer data.* Paper presented at the Fourth International Symposium on the Wearable Computers. New York, NY.

Ravi, N. D. N., Mysore, P., & Littman, M. L. (2005). *Activity recognition from accelerometer data.* Paper presented at the Proceedings of the IAAI 2005. New York, NY.

Reddy, S., Burke, J., Estrin, D., Hansen, M., & Srivastava, M. (2008). Determining transportation mode on mobile phones. In *Proceedings of the Wearable Computers, 2008.* IEEE Press.

Resnik, P., & Hardisty, E. (2010). *Gibbs sampling for the uninitiated.* University Park, MD: University of Maryland.

Tengfei, B., Happia, C., Enhong, C., Jilei, T., & Hui, X. (2010). An unsupervised approach to modeling personalized contexts of mobile users. In *Proceedings of the Data Mining (ICDM), 2010.* IEEE Press.

Xi, L., Bin, Y., & Aarts, R. M. (2009). *Single-accelerometer-based daily physical activity classification. In Proceedings of the Engineering in Medicine and Biology Society, 2009.* IEEE Press.

Yang, G. (2009). Discovering significant places from mobile phones: A mass market solution. *Mobile Entity Localization and Tracking in GPS-less Environments, 5801*, 34–49. doi:10.1007/978-3-642-04385-7_3

Zheng, Y., Liu, L., Wang, L., & Xie, X. (2008). Learning transportation mode from raw GPS data for geographic applications on the web. In *Proceedings of the 17th International Conference on World Wide Web*. Beijing, China. IEEE.

Zheng, Y., Zhang, L., Xie, X., & Ma, W.-Y. (2009). Mining interesting locations and travel sequences from GPS trajectories. In *Proceedings of the 18th International Conference on World Wide Web*. Madrid, Spain: IEEE.

Chapter 3
Clustering Algorithms for Tags

Yu Zong
West Anhui University, China & University of Science and Technology of China, China

Guandong Xu
University of Technology Sydney, Australia

ABSTRACT

With the development and application of social media, more and more user-generated contents are created. Tag data, a kind of typical user generated content, has attracted lots of interests of researchers. In general, tags are the freely chosen textual descriptions by users to label digital data sources in social tagging systems. Poor retrieval performance remains a major problem of most social tagging systems resulting from the severe difficulty of ambiguity, redundancy, and less semantic nature of tags. Clustering method is a useful tool to increase the ability of information retrieval in the aforementioned systems. In this chapter, the authors (1) review the background of state-of-the-art tagging clustering and the tag data description, (2) present five kinds of tag similarity measurements proposed by researchers, and (3) finally propose a new clustering algorithm for tags based on local information that is derived from Kernel function. This chapter aims to benefit both academic and industry communities who are interested in the techniques and applications of tagging clustering.

INTRODUCTION

With the proliferation of social media, lots of User Generated Content (UGC) have been brought and UGC becomes one of the main prevailing Web trends (Baeza-Yates, 2009). Various types of data, e.g., text, photo, music, and video, are generated and viewed.

As a typical type of UGC, social tag (also known as collaborative tag or social annotation) has obtained significant development and it also has become popular for their revolutionary ways of organizing online resources. Tags are simple, uncontrolled and ad-hoc labels that are assigned by users to describe or annotate any kind of resource. Since the distribution or types of contents are diverse and change dynamically, tagging is especially suitable for online copra.

Many social tagging sites have been established, such as Del.icio.us (http://delicious.com)

DOI: 10.4018/978-1-4666-2806-9.ch003

for Web pages bookmarking, Flickr (http://www.flickr.com) for photo sharing, CiteULike (http://citeulike.org) for academic publishing management, Youtube (http://youtube.com) for video sharing, etc. The former discussed tagging sites are only focused on special topics. For the common request, however, some general sites have also provided the features for users. Users are allowed to add freely uncontrolled tags for products in Amazon. Twitter (http://twitter.com), a kind of micro-blog, allows users to annotate their short tweets with a type of hash tags for content annotation. Recently, lots of Weibo sites, such as, Sina Weibo (Weibo.com), Yahoo Weibo (http://itwwt.com/tag), etc., have been established in China for users to annotate their topics.

The low technical barrier of tag based recommender system and easy usage of tagging have attracted a large amount of research interest. The user-contributed tags are not only an effective way to facilitate personal organization but also provide a possibility for users to search for information or discover new things. However, the ambiguity, redundancy and less semantic nature are the major problems suffering all tagging systems. For example, for one same resource, different users will use their own textual description to annotate, resulting in the tagging behavior much ambiguous and redundant. In order to deal with these difficulties, recently clustering method has been introduced into tag-based recommender system to find meaningful information conveyed by tags. As the user tagging behaviors can be modeled as data record with triple attributes, i.e. user, resource, and tag, clustering on tag data could be conducted on these three attributes respectively. The effectiveness of clustering of tag data is the ability of aggregating tags into topic domains. In this chapter, we (1) briefly discuss research background and related work on tagging clustering, (2) introduce the form of tag data and various tag similarity measurements, (3) propose a clustering algorithm named Local Information Passing Clustering algorithm (LIPC). Especially, in LIPC, We first estimate a KNN neighbor directed graph G of tags, the Kernel density of each tag in its neighborhood is calculated at the same time; we then use Local coverage and Local Kernel to capture the local information of each tag; thirdly, we define two operators, that is, I and O, to pass the local information on G; then a tag priority is generated when I and O are converged; at last, we use the tag priority values to find out the clusters of tags by using Depth First Search (DFS) on G. Experimental results demonstrate the efficiency and the improved outcome of tag clusters by using the proposed method.

BACKGROUND

Recently tag has been widely used in recommender systems for many applications (Durao, 2010; Jaschk, 2007; Tso-Sutter, 2008). The common usage of tags in these systems is to add them as an additional feature to re-model users or resources over the tag vector space, and in turn, making tag-expanded collaborative filtering recommendation or personalized recommendation. However, as the tags are of syntactic nature, in a free style and do not reflect sufficient semantics, the problems of redundancy, ambiguity and less semantics of tags are often incurred in all kinds of social tagging recommender systems. For example, for one resource, different users will use their own words to describe their feeling of likeness, such as "favorite, preference, like" or even the plural form of "favorites"; and another obstacle is that not all users are willing to annotate the tags, resulting in the severe problem of sparseness, redundancy and ambiguity of tags. In order to deal with these difficulties, recently clustering method has been introduced into social tagging recommender systems to find meaningful information conveyed by tag aggregates. The aim of tagging clustering is to reveal the coherence of tags from the perspective of how resources are annotated and how users annotate in the tagging behaviors. Undoubtedly, the tag cluster form is able to deliver user tagging interest or resource topic information in a more concise and semantic way, which, in some extent, to handle the problems of tag sparseness and redundancy, in turn facilitating the tag-based recommender systems. Thus, this demand mainly motivates the research of tagging clustering in social annotation systems.

In the context of tagging clustering, most of the researches on tagging clustering are directly using the traditional clustering algorithms such as K-means (Gemmell, 2008) or Hierarchical Agglomerative Clustering (Shepitsen, 2008) on tag data, which possess the inherent drawbacks, such as the sensitivity of initial values and high computational cost etc. In Hayes (2007), topic relevant partitions are created by clustering resources rather than tags. By clustering resources, it improves recommendations by distinguishing between alternative meanings of query. While in Chen (2010), clusters of resources are shown to improve recommendation by categorizing the resources into topic domains. A framework named Semantic Tagging clustering Search which is able to cope with the syntactic and semantic tag variations is proposed in van Dam (2010). Lehwark et al. use Emergent-Self-Organizing-Maps (ESOM) and U-Map techniques to visualize and cluster tag data and discover emergent structures in collections of music (Lehwark, 2008). Self-organizing maps are artificial neural networks that map high-dimensional data points to nodes in a low-dimensional grid, that is, the output layer. Usually the grid is two-dimensional and can be viewed as a graphical map, where similar data points are placed close together or at the same point on the map. In order to deal with the high dimensionality of tag data set, Marco et al. use a Self Organizing Map (SOM) to cluster tagged bookmarks which were taken from the website http://delicious.com/about/ (Marco, 2009). State-of-the-art methods suffice for simple search, but they often fail to handle more complicated or noisy Web page structures due to a key limitation. Miao et al. propose a new method for record extraction that captures a list of objects in a more robust way based on a holistic analysis of a Web page (Miao, 2009). In Giannakidou (2008), a co-clustering approach is employed, that exploits joint groups of related tags and social data resources, in which both social and semantic aspects of tags are considered simultaneously. Li et al. were motivated by the key observation that in a social network, human users tend to use descriptive tags to annotate the contents that they are interested, and in particular, they proposed an Internet Social Interest Discovery system (ISID) to discover the common user interest and cluster the users and their saved URLs by different interest topics (Li, 2008). Ramage et al. explore the usage of tags in K-means clustering by using an extended vector space model which included tags as well as page text (Ramage, 2009). In order to reduce the set of different tags which to be considered by a part-of-speech tagger, Felipe et al. propose a clustering algorithm by using the hidden Markov model (Felipe, 2006). Practically, the authors (1) train the HMM by considering information not only from the source language but also from the target language; (2) obtain taggers involved in machine translation system; (3) generated the tag clusters by executing a bottom-up agglomerative clustering algorithm. Due to the complexity of social tag data, recently researchers focus on spectral clustering that has been proven effective in addressing complex data. However, existing spectral clustering algorithms work with 2-way relationships. To overcome this problem, Karydis et al. develop a novel data-modeling scheme and a tag-aware spectral clustering procedure that uses tensors to store the multi-graph structures that the personalized similarity (Karydis, 2009). By taking the idea that the tag semantics is the key for deep understanding the correlation of objects in their mind, Jiang et al., propose the concept of core-tag and the model of core-tag clusters. In order to find out the core-cluster, a core-tagging clustering algorithm CET Clustering based on ensemble method is designed (Jiang, 2009).

Tag Data Description

In a social tagging system or tagging services, the users are easily allowed to organize, share and retrieve online resources with tags. A user can assign a tag to a resource according to his/her background, or other user's tag. Figure 1 gives an example about tagging on Del.icio.us tagging system. In each bookmark, the upper left corner shows the title of the Web page and the lower right corner gives the tags assigned to that page.

Though tag data is similar to rating data, there also have two major differences between them: (1) tag data does not contain users' explicit preference information on resources, but ranking data has; (2) tag data involves users, tags and resources, however, rating data only contain users and resources.

The users of social tagging systems have created large amounts of tag data, which have attracted much attention from the research community. There, in general, have two types of tags in the social tagging system. The first tag type is the triple tags or machine tag, which uses a special syntax to define extra semantic information about the tag and make it easier or more meaningful for interpretation by a computer program. For example, "geo: long=48.900000" is a tag for geographical longitude coordinate whose value is 48.900000. Another tag type is called as Hashtags, which were commonly used as short message in social tagging system, such as Twitter or identi.ca. Figure 2 gives an example of Hashtag. A person can search for the string #freiheitstattangst and this tagged word will appear in the search engine results. In addition, these Hashtags show up in a number of trending topics websites, including Twitter's own front page.

In fact, the usage of tagging also has its advantages and disadvantages, as follows concluded.

The advantages of tagging are:

Figure 1. An example of tagging in Del.icio.us

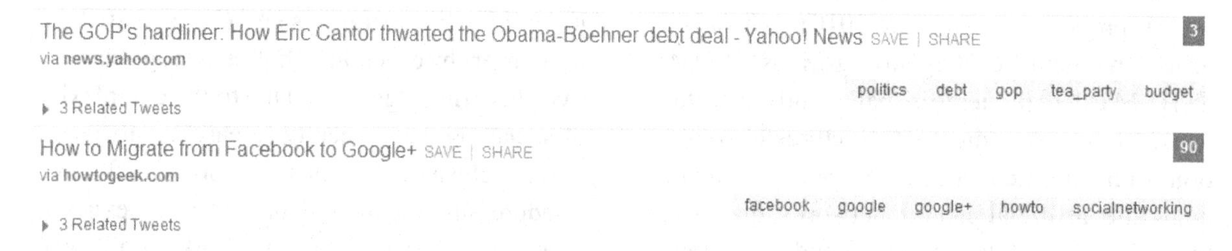

Figure 2. An example of hashtag in Twitter

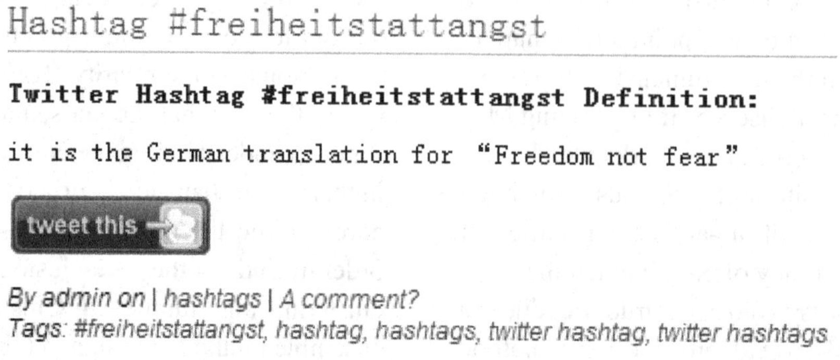

Hashtag #freiheitstattangst

Twitter Hashtag #freiheitstattangst Definition:

it is the German translation for "Freedom not fear"

tweet this

By admin on | hashtags | A comment?
Tags: #freiheitstattangst, hashtag, hashtags, twitter hashtag, twitter hashtags

- **Flexibility:** The flexibility of tagging allows users to classify their collections of items or resources in the ways that they find useful.
- **Simplicity:** In a typical tagging system, there is no explicit information about the meaning of each tag. A user is allowed to apply new tags to an item or resources or to use older tags.
- **Share Property:** Users in a tagging system can use tags to share the interesting items or resource with others.

The disadvantages of tagging are:

- **Ambiguity:** Because users can freely choose tags, the resulting metadata can include homonyms (e.g., the same tags with different meanings) and synonyms (e.g., different tags for the same concept). For example, the tag "orange" may refer to the fruit or the color.
- **Spamming:** Tagging systems open to the public are also open to tag spam. In tagging system, people can apply excessive number of tags or unrelated tags to an item or results for getting more attraction.

In spite of there have two types of tag in tagging system, however, they have the common information, that is tag data always contains three parts: users, U, tags, T, and resources, R. Formally, social tag data can be viewed as a set of triples (Guan, Bu, Mei, Chen, & Wang, 2009; Guan, Wang, & Bu, 2010). Each triple (u,t,r) represents a user u annotates a tag t to a resource r. A social tagging system can described as $D = < U,T,R,A^N >$, where exists a set of users, U; a set of tags, T; a set of resources, R; and a set of annotations, A^N. The annotations are represented as a set of triples contains a user, tag and resource defined as: $A^N \subseteq < u,t,r >: u \in U, t \in T, r \in R$. Therefore a social tagging system can be viewed as a tripartite hyper-graph (Mika, 2007) with users, tags and resources represented as nodes and the annotations represented as hyper-edges connecting one user, tag and resource. Figure 3 is an example of this hyper-graph about tag data. From Figure 3, we can find that u_1 has assigned t_1 to r_1, t_3 to r_2 and r_4, as well as t_5 to r_3 and r_4. Resource r_1 has been annotated by tag t_1, t_2 and t_4.

Tag Similarity Measurements

Given a set of data objects, the goal of a clustering algorithm is to depart them into a set of clusters so that objects in the same cluster are close together, while objects in different clusters are far apart. The similarity between objects takes an important role in clustering. In tagging clustering situation, there have three different roles—tags, resources, and users, need to be considered. According to the aim of tagging clustering, several different similarity measurements are defined as follows to capture the similar objects in tag data set.

Lexical Similarity

The tag data set obtained from a tagging system can be enriched by using online lexical resources. In this way, tags can be replaced by concepts and homonyms, and then, Lexical Similarity can be

Figure 3. An example of hyper-graph of tagging system

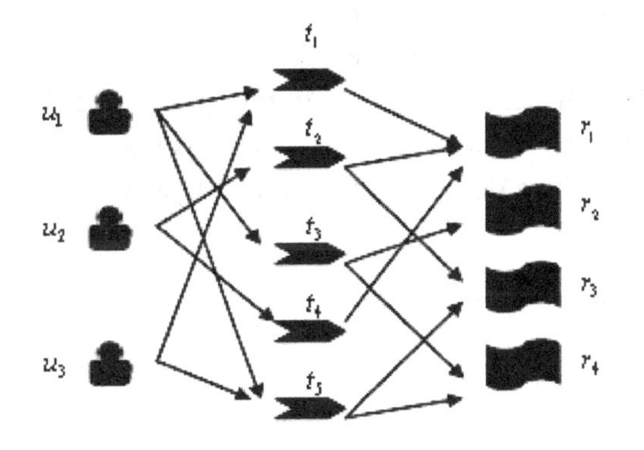

used to capture the similarity of tags. Lexical Similarity measurement use an external thesaurus or semantic lexicon such as Wordnet (Katharina, 2007) or Wikipeda (Heep, 2006) to obtain the relationship between terms which were used as tags. This measurement, in one hand, suffers when new concepts and words are used, or when spell checking on misspelled tag fails. On the other hand, the tag relationships are less likely to reflect the way that tags are used by a particular group of users, as the meaning and association between tags is likely to vary between users interested in different topics. For example, the tag "cluster" has two specific meanings in Computer Science, which are both different to general English (Edwin, 2008).

Document-Term Similarity

It is an interesting research task that we group resources, which share tags into clusters and then compare the similarity of all resources within a cluster. This task based on a hypothesis that a cluster of resources shared a tag should be more similar than a randomly constructed set of resources. In order to describe the similarity between resources, researchers introduce Document-term similarity, which stems from document-similarity in text clustering to capture similar tags. The similarity between tags is calculated from the textual similarity of documents. Brooks (2006) first created a term-frequency vector for each tag, and then use cosine similarity to compare pairs of tags, where cosine similarity is defined as follows.

Given two tags $t_i = \{tr_{i1}, tr_{i2}, .., tr_{im}\}$ and $t_j = \{tr_{j1}, tr_{j2}, .., tr_{jm}\}$, where tr_{il} and tr_{jl}, $l=1,...,m$ is the TF-IDF score of tags occurred in a resource. The cosine similarity between t_i and t_j is represented by using a dot product and magnitude as:

$$Similarity\left(t_i, t_j\right) = \frac{t_i \cdot t_j}{\|t_i\| \cdot \|t_j\|}$$

Co-Occurrence Similarity

In tagging system, users allowed to annotate interesting resource by using any selected words (tags). Essentially, similarity between tags could be capture by the co-occurrence resources. In this situation, we (1) construct a tag vector which is consisted by a set of resources which labeled by the same tag, e.g. $t_i = \{r_1, r_2, .., r_m\}$, where r_l, $l=1, ..., m$ is a resource; (2) calculate the Co-occurrence similarity between tags as follows.

$$Similarity(t_i, t_j) = \frac{|t_i \cap t_j|}{|t_i \cup t_j|}$$

Social Similarity

In tagging system, each resource can be represented by the set of tags which have been used for its annotation, e.g. $r_i = \{\cup t_x\},\ \forall t_x \in T : f(r_i, t_x) = 1$, where $f(r_i, t_x) = 1$ indicates resource r_i is labeled by tag t_x. Thus, finding the relation between a resource and an attribute indicates capturing the similarity between resources's tags and the attributes. Giannakidou et al. propose a similarity measurement named Social Similarity to capture the similarity between tags (Giannakidou, 2008). See Box 1.

Normalized Levenshtein Similarity

Syntactic variations detection is done by creating a set $T' \subset P(T)$, where $P(T)$ represents the power set of T. Each element of T' represents a cluster of tags where each tag occurs only in one element or cluster, i.e. if $X, Y \in T'$, $X \neq Y$, and $a \in X$, $b \in Y$, then this implies $a \neq b$, where X,Y indicate two different cluster and a, b denote two tags respectively. Then we denote by m' the objective function that indicates a label for each $X \in T'$, $m' : T' \rightarrow L$. Furthermore, for each $l \in L$ such that $l \in X$, i.e., l is one of the

Box 1.

$$Similarity(t_x, t_y) = SoS(t_x, t_y) = \frac{\sum_{i=1}^{n} r_i : (u_w, r_i, t_x) \in A \quad and \quad (u_z, r_i, t_y) \in A}{\max(\sum_{i=1}^{n} r_i : (u_w, r_i, t_x) \in A, \sum_{i=1}^{n} (u_z, r_i, t_y) \in A)}$$

where $u_w, u_z \in U, r_i \in R$.

tags in cluster *X*. In order to calculate the similarity between tag t_i and t_j, van Dam, et al. (van Dam, 2010) defined a new similarity measurement named normalized Levenshtein similarity.

$$Similarity(t_i, t_j) = \frac{lev_{ij}}{\max(length(t_i), length(t_j))}$$

where lev_{ij} is the Levenshtein similarity (Levenshtein, 1966). The normalized Levenshtein similarity addresses the string lengths.

Local Information Passing Clustering Algorithm for Tags

In the context of conventional tagging clustering, the first step is to define the tag similarity by using the similarity measurements which are described above; and then to find out the clustering structure from tag data by executing clustering algorithm; eventually to make use this structure for further applications such as forming recommended information. In this case, the quality of the clustering result has critical effect for the tag based recommender system. Most of the researches on tagging clustering are directly use the traditional clustering algorithms on tag data. These clustering algorithms often focus on local aspect of tag data and cannot capture the global information of tagging. However, various tags used in tag data apparently possess different significance in tag groups due to the semantic or domain topic tendency of tags, for example, "image" should locate close to the center of one tag cluster of "photograph," while some other tags possessing the broad and diverse topic relatedness to any other tag clusters, for example, "blue," are always scattered around the outer brim of tag aggregate clouds.

Bearing this observation in our mind, the basic idea of this chapter is originated from the latent significance of each tag in tagging activities to creating tag clusters. Particularly, we propose a clustering algorithm named Local Information Passing Clustering algorithm (LIPC). In LIPC, we first construct a KNN neighbor directed graph *G* of tags, the kernel density of each tag in its neighborhood is calculated at the same time; We then use Local coverage and Local Kernel to capture the local information of each tag; thirdly, we define two operators, that is, *I* and *O*, to pass the local information on *G*; then a tag priority is generated when *I* and *O* operators are converged; at last, we use the tag priority to find out the clusters of tags. Experimental results demonstrate the efficiency and the improved outcome of tag clusters by using the proposed method.

TagVector Model and Tag Similarity

Li et al. analyze the bookmark data set and find a phenomenon that the distribution of URLs, Users, and Tags follows power law distribution. This phenomenon indicates that most URLs are only bookmarked once and most users only bookmark one URL, in the same way, most tags are only annotated on one URL (Li, 2008). Recently, an experiment on detecting the pair-wise relationship between tags and resources and between tags and users has shown that only a small part of resources are annotated frequently by many tags, whereas as a large number of resources are annotated once, and that the same observation of power law distribution also exists in the relationship between tags and users. Most of applications on tags are using tags to describe resources or users, that is, a resource or user is defined as a tags vector. In this model, thus the tag vector is in a very high dimension due to the free style of tag texts. In addition, most of tags are redundant and ambiguous, in turn; bring in a difficulty of similarity calculation. Therefore, tagging clustering is able to capture the topic domains of tags, which is expected to partially handle the above problems. In a tagging system, resources are mostly fixed and unique. Tag can be described by a set of resources which the tag has been assigned to it by users. In this way, a tag vector was constructed by using resources as dimensions, e.g., $t_i = (r_1, ..., r_m)$. The similarity between any two tags is defined as Definition 1.

Definition 1: Given two tags $t_i = (r_{i1}, ..., r_{im})$ and $t_j = (r_{j1}, ..., r_{jm})$, the similarity $Similarity(t_i, t_j)$ is defined as the ratio of co-occurred resource.

$$Similarity(t_i, t_j) = \frac{t_i \cdot t_j}{|| t_i || \cdot || t_j ||}$$

A Working Example of Tagging Clustering

In the above discussed tag vector model, the tag is usually in a very high dimension due to the free style of tag texts, which results in the problem of redundancy and ambiguity of tags, in turn; bring in the difficulty in tag computing such as the similarity calculation in terms of tag vector. Therefore, tagging clustering is often employed to capture the topical aggregates of tags, which is to capture the structural semantics of tags. In real applications, we start from the tripartite graph of social annotations to compose a resource-tag matrix by accumulating the frequency of each tag along users. In this expression, each tag is described by a set of resources to which this tag has been assigned, i.e., $t_i = (w_{i1}, ..., w_{im})$, where w_{ik} denotes the weight on resource dimension r_k of tag t_i, which could be determined by the occurrence frequency. In this manner, the similarity between any two tags is defined as Definition 1.

Upon the mutual tag, similarity is determined; various clustering algorithms could be applied to partition the tags. Figure 4 gives a simple working example to show how five tags are assigned to two groups by using various clustering strategies (in red or black dashed circles). The clustering results in Figure 4 show that different clustering algorithm will capture totally different clustering result. So the design of clustering algorithm for tags is an important task in tagging clustering.

Local Information Passing Clustering Algorithm

In real world, how should we know other peoples whom we did not know before? The recommendations from our friends are commonly used way. In Web world, users are always using tags to appraise a resource and other users can accept the resource according to the annotated tags. This behavior of Web could be regarded as the copy of real world, that is, the social network. Similarly, the tags could be regarded as the recommendation information. If we assume that the most similar K tags are the K friends of one tag, we can use the behavior of social network recommender system in the real world to simulate the tag's recommendation. However, these recommendation information are always locally, so we need to define operators to pass these information to all the tags. In this section, we first use the KNN neighbor method to find out the K nearest neighbors of one tag and then construct a KNN directed graph *G*. Local information is defined by using the kernel density estimator method. In order to pass local information, we define two operators *I* and *O* to

Figure 4. A working example of tagging clustering

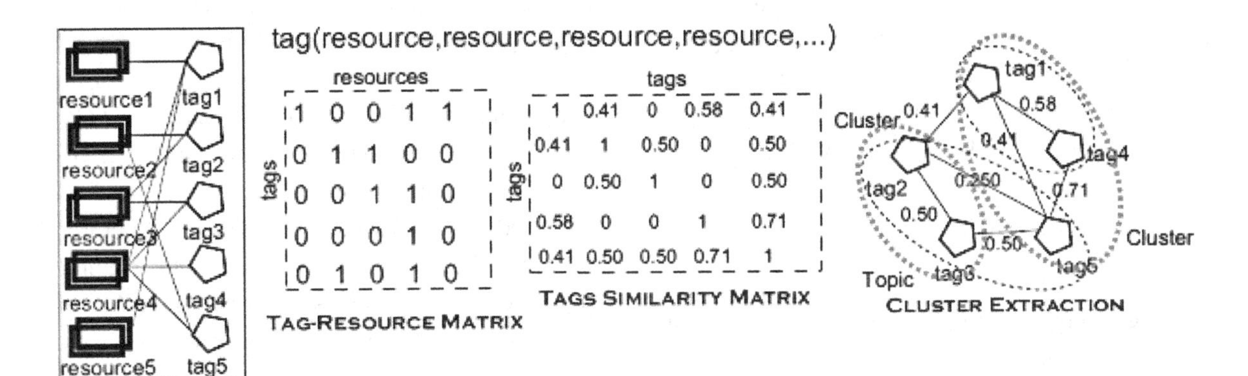

transit the local information to all the tags and the priority of each tag is generated. The purpose of this chapter is to find the groups with similar tags, so we devise a clustering algorithm based on tag priority to generate tag clusters.

KNN Directed Graph and Local Information

According to Definition 1, a similarity matrix S could be constructed for a given tag data set. From S, we can find KNN neighbors of each tag and then a KNN directed graph G could be created. Graph $G=<V,\{E\}>$, where V is the tag set and $\{E\}$ is the directed edge set between tags, $<p,q>\in\{E\}$ denotes that tag q is a KNN neighbor of tag p.

Figure 5 shows an example of a part of graph G. On one hand, the black node is tag p, and five heavy dark nodes with black line are the KNN-neighbors of p and there have arches between p and them. On the other hand, p is the KNN-neighbor of four nodes which are denoted by light line circles and there have arches between these nodes and p. In this way a KNN directed graph G could be constructed and the adjacency matrix A of G is defined as Definition 2.

Definition 2: Given a KNN directed graph G, the adjacency matrix is defined as A, where $A(p,q)=1$, if the directed arch $<p,q>$ exists, and $A(p,q)=0$, otherwise.

The KNN Kernel estimate method (Liu, 2007) has mainly been used in capturing local character and density distribution. In this chapter, we first use it to calculate the KNN kernel density of each node, and then define two important indexes named Local Coverage *(LC)* and Local Kernel *(LK)* to capture the local information of each node.

Definition 3: Given a node $p\in G$ and its KNN neighborhood $N(p)$, the Local Coverage of p is defined by the KNN kernel density of its neighbors:

$$LC(p)=\sum\nolimits_{q\in N(p)} f(q),$$

where $f(q)$ denotes the KNN kernel density value of node q.

$LC(p)$ is defined as the sum of KNN kernel density of p's KNN neighbors, the higher value

Figure 5. An example of a part of directed graph G

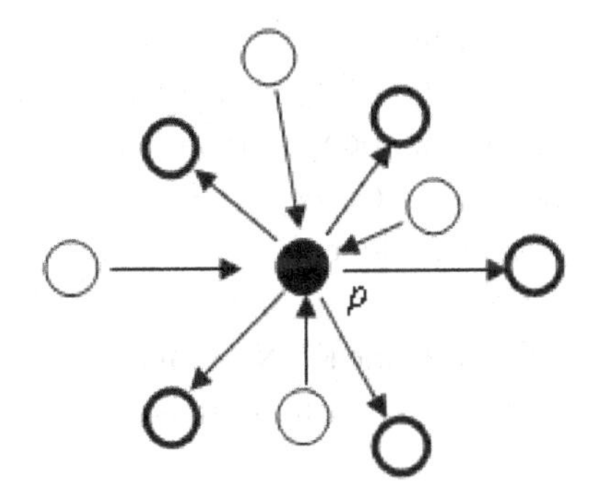

of *LC(p)* indicates more nodes with higher KNN kernel density in its neighborhood, and the probability of it locating in a high density area is higher.

Definition 4: Given a node $p \in G$ and a node set Q which contains p, the Local Kernel of p is defined as the KNN kernel density of Q:

$$LK(p) = \sum\nolimits_{q \in Q} f(q).$$

According to Definition 4, we can find that *LK(p)* is the sum of KNN kernel density of nodes which directed to p. The higher value of *LK(p)* indicates more nodes with higher KNN kernel density connected to p and p has a higher dependability to represent a local center.

LIPC Algorithm Framework

LC(p) and *LK(p)* capture the local information of each node in G. In this section we define a passing method to pass the local information to all the nodes and generate the authority and coverage of all the nodes based on *LC(p)* and *LK(p)*. We define I and O operators to transit the local information *LC(p)* and *LK(p)* respectively, as shown in Definition 5.

Definition 5: Assume vector LC_i indicates the local coverage value of all nodes in ith iteration and vector LK_i indicates the local kernel value of all nodes in ith iteration respectively. Then the I and O operators are defined as:

$$I: LC_i = A^T \times LK_{i-1},\ O: LK_i = A \times LC_{i-1}.$$

The function of operator I is to pass the LK information of nodes which are directed to p, while, the function of operator O is passing the LC information of node p. After the convergence of operation I and O, we use LK_i to define the priority of each node in G.

Algorithm 1 gives the main steps of LIPC. In step 1, we first find the KNN neighbor of each tag based on the tag similarity defined as definition 1, and then, the local kernel density of each tag has been generated by using kernel density estimate operator. According to definition 2, a KNN directed graph G could be constructed (step 2). For each node $v \in G$, we initialize the value of *LC(v)* and *LK(v)* according to definition 3 and 4. In this way, the local information of each node has been captured and then we execute I and O operators to transit these local information on the graph G until these two operators are converged. We sort the vector LK_i to generate the priority of each tag. The node priority of tag shows the importance of the tag to its cluster. To generate the clusters: (1) we select a tag, $v \in G$, with the highest priority as the centre of a cluster, (2) we use Depth First Search (DFS) method to find the corresponding cluster members which $LK_i(v')$ value smaller than that of v, where $v' \neq v$. Steps (1) and (2) are iteratively executed until all the nodes in G are assigned to its corresponding cluster (see Box 2).

We run LIPC on Dmoz dataset (http://www.michael-noll.com/dmoz100k06/) and the priority of some tags are shown in Table 1.

From Table 1, we can find that the priority of "Favorite" is higher than Favorites, Like and Preference. This indicates that "Favorite" more likely be selected as the centre than others, that is, "Favorite" could by generally describing other tags. And the same phenomenon shows for "JAVA" and "JavaScript" tags.

Table 1. An example of tag's priority

Tag	Priority
Favorite	2.39
Favorites	2.20
Like	2.02
Preference	1.98
JAVA	2.89
JavaScript	2.84

Box 2.

```
Algorithm 1: Main steps of LIPC
Input Tagvector, number of neighbor, K
Output: cluster result, C.
1.        Generates KNN neighbors of a tag based on definition 1, and calculate the
KNN kernel density of the tag by using kernel density estimate operator.
2.        Construct a KNN directed graph G according to definition 2.
3.        Generate the local information of each node by calculate LC(v) and LK(v)
based on definition 3 and 4 respectively.
4.        Iteratively calculate LC_i and LK_i by using  I and O operators until they are
converged.
5.        Sorts LK_i.
6.        For each no visited node v∈ G.
        6.1 v=max(LK_i).
        6.2 C_v= DFS(G,v), where C_v denotes a cluster with v as centre.
        6.3 C←C∪C_v.
7.        Return C.
```

Experimental Evaluations of LIPC

To evaluate our approach, we conducted extensive experiments. We performed the experiments using Intel Core 2 Duo CPU (2.4GHz) workstation with a 4G memory, running windows XP. All the algorithms were written in Matlab 7.0. We conducted experiments on two real datasets, MedWorm (http://www.medworm.com/) and MovieLens (http://www.movielens.org/).

Datasets Description

In order to evaluate our approach, we crawled the article repository in MedWorm system during April 2010 and downloaded the contents into our local experimental environment. After stemming out the entity attributes from the data, four data files, namely users, resources, tags and quads, are obtained as the source datasets. The first three files are recorded the user, tag and document information, whereas the fourth presents the social annotation activities where for each row, it denotes a user *u* annotates a resource *r* by a tag *t*.

The second dataset is MovieLens, which is provided by GroupLens (http://www.grouplens.org/). It is a movie rating dataset. Users were selected at random for inclusion. All selected users had rated at least 20 movies. Unlike previous MovieLens datasets, no demographic information is included. Each user is represented by an id, and no other information is provided. The data are contained in three files, movies.data, rating.dat and tags.dat. Also included are scripts for generating subsets of the data to support five-old cross validation of rating predictions. The statistical results of these two datasets are listed in Table 2. These two datasets are pre-processed to filter out some noisy and extremely sparse data subjects to increase the data quality.

In this chapter, we use resources to describe tags, that is, each tag described by a set of resource

Table 2. Statistics of datasets

Property	MedWorm	MovieLens
Number of users	949	4,009
Number of resources	261,501	7,601
Number of tags	13,507	16,529
Total entries	1,571,080	95,580
Average tags per user	132	11
Average tags per resource	5	9

which assigned to it by users. In order to reduce the length of tag vector, we first omit the resources which tags are less than the average tags for each resource in Table 2.

Evaluation Measurements

Our aim of the proposed method is to find out the similar tags, which assign to different resources. Here we assume a better tag cluster composed by lots of similar tags and these tags are dissimilar to tags, which belong to other different tag clusters. In particular, we use Similarity and Dissimilarity to validate our method.

Definition 6: Given tag cluster *C*, the Similarity is defined as

$$Similarity = \frac{1}{|C|}\sum_{k=1}^{|C|} Sim(t_i, t_j),\quad t_i, t_j \in C_k$$

Definition 7: Given tag cluster *C*, the Dissimilarity is defined as

$$Dissimilarity = \frac{1}{|C|}\sum_{k=1}^{|C|} Dism(k)$$

where

$$Dism(k) = \sum_{k'=1}^{|C|} Sim(t_i, t_j),\quad t_i \in C_k, t_j \in C_{k'}, k \neq k'$$

According to the requirement of clustering, we know that higher Similarity value and smaller Dissimilarity value indicate better clustering results.

Experimental Results and Discussion

Due to the priority of each tag has a close relationship with the kernel density which depends on the KNN neighbor, there have a relationship between the number of K and the tag's priority. In order to present this relationship, we manually extract thirty tags, which form two clusters from MedWorm data set. The priority of each tag has shown in Figure 6 (the value of priority is multi by 100) with K equal to 4, 8, and 12, respectively.

From Figure 5, we can find that the change of K has no influence on the tag's priority, as well as the structure of cluster.

As we discussed in the previous sections, tagging clustering can make tags be organized into groups over clusters. That is to say, by clustering, tags can be centralized into groups. In the following, we will conduct the experiments to evaluate the effect on the tag cluster's quality. Table 3 gives the comparison result of Similarity and Dissimilar-

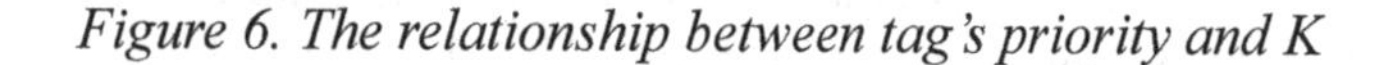
Figure 6. The relationship between tag's priority and K

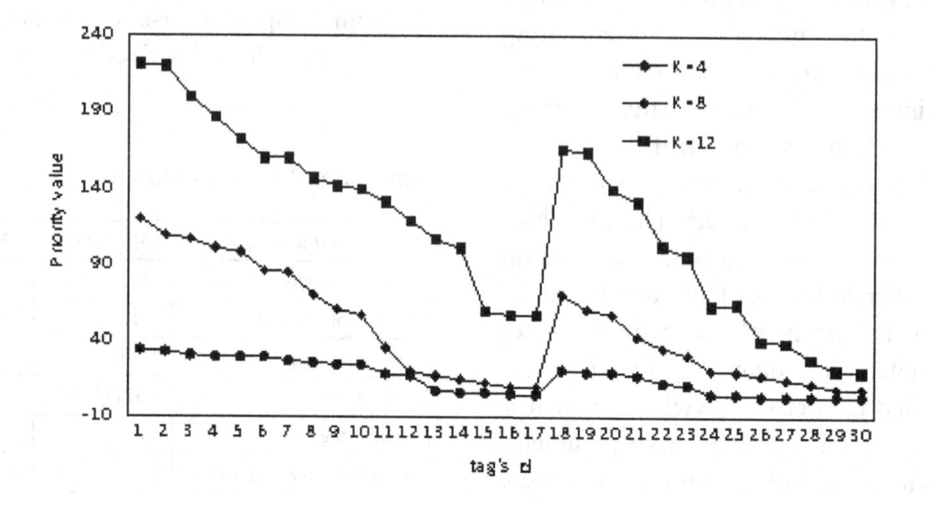

ity on Medworm and MovieLens datasets. From Table 3, we can first find that the cluster results on two datasets are nearly coincident under different K's settings. This phenomenon validates our previous experiment of the relationship between tag's priority and K's setting. As we defined in Definition 6 and 7, the higher Similarity value and the smaller Dissimilarity value indicate high quality of the tag cluster. On these two datasets, the Dissimilarity values are all small, and at the same time, the Similarity values are all high. This phenomenon shows that the quality of tag clusters obtained from Medworm and MovieLens by using our method are high. Interestingly, the clustering results derived from Medowrm look better than that of Movielens, which might be due to the tags used in Medowrm dataset is focused on a more specialized medical domain, while the domain topics span more diversely on Movielens dataset. This finding is also verified by the measures of Dissimilarity.

In order to show the effectiveness of the proposed method, we execute traditional clustering algorithm K-means and our proposed algorithm LIPC on these two real world datasets as well. The experimental results are shown in Table 4.

From Table 4, we can find that the quality of clustering results obtained by LIPC is better than that of K-means. Particularly, in one hand, the Dissimilarity values of LIPC on two datasets are smaller than that of K-means, and on the other hand, the Similarity values of LIPC on two datasets are, on the contrary, larger than that of K-means. This phenomenon indicates that LIPC algorithm has ability of finding better clustering results than that of K-means.

CONCLUSION

Tagging clustering is a useful method to find out interest tag cluster embedded in tag datasets and it has a potential in improving the effectiveness and accuracy of tag based recommender system. In this chapter, we (1) briefly discuss the background and related work of tagging clustering; (2) conclude five different tag similarity measurements which commonly used in the lectures; (3) propose a Local Information Passing Clustering algorithm (LIPC) for tags which is based on calculating the priority of tag. In LIPC, we first use the KNN neighbor and Kernel density estimate method to find out the local information of each tag, and then,

Table 3. The comparison of similarity and dissimilarity on two datasets

	Similarity		**Dissimilarity**	
	MedWorm	MovieLens	MedWorm	MovieLens
K=4	0.973	0.867	0.0245	0.450
K=8	0.974	0.870	0.0240	0.448
K=12	0.970	0.869	0.0246	0.452

Table 4. The comparison of LIPC and K-means on two datasets

	Similarity		Dissimilarity	
	MedWorm	MovieLens	MedWorm	MovieLens
K-means	0.873	0.856	0.0645	0.735
LIPC	0.966	0.885	0.0239	0.409

and define *I* and *O* operators to transit the local information over all the tags and further generate tag priority, at last, we use the tag priority to find the representative centre of various clusters. Experimental results conducted on two real world datasets have demonstrated the effectiveness and advantage of the proposed method in comparison to other traditional clustering approaches.

This chapter will benefit both academic and industry communities, who are interested in the techniques and applications of tagging clustering, recommendation, as well as Web community and social network analysis, for either in-depth academic research or industrial development in related areas.

REFERENCES

Baeza-Yates, R. (2009). *User generated content: How good is it?* Paper presented at the the 3rd Workshop on Information Credibility on the Web. Madrid, Spain.

Brooks, C. H. (2006). *Improved annotation of the blogosphere via autotagging and hierarchical clustering.* Paper presented at the the 15th International Conference on World Wide Web. Edinburgh, Scotland.

Chen, H., & Dumais, S. (2010). *Bringing order to the web: Automatically categorizing search results.* Paper presented at the the SIGCHI Conference on Human Factors in Computing Systems. Vancouver, Canada.

Durao, F., & Dolog, P. (2010). *Extending a hybrid tag-based recommender system with personalization.* Paper presented at the the 25th ACM Symposium on Applied Computing. Sierre, Switzerland.

Edwin, S. (2008). *Clustering tags in enterprise and web folksonomies.* Paper presented at the International Conference on Weblogs and Social Media. Washington, DC.

Felipe, S. M., Juan, A. P. O., & Mikel, L. F. (2006). Target-language-driven agglomerative part-of-speech tagging clustering for machine translation. *Advances in Artificial Intelligence, 4239*, 844–854.

Gemmell, J., Shepitsen, A., Mobasher, M., & Burke, R. (2008). *Personalization in folksonomies based on tagging clustering.* Paper presented at the the 6th Workshop on Intelligent Techniques for Web Personalization and Recommender Systems. Chicago, IL.

Giannakidou, E., Koutsonikola, V., Vakali, A., & Kompatsiaris, Y. (2008). *Co-clustering tags and social data sources.* Paper presented at the the 9th International Conference on Web-Age Information Management. Zhangjiajie,China.

Guan, Z., Wang, C., & Bu, J. (2010). *Document recommendation in social tagging services.* Paper presented at the the 19th International Conference on World Wide Web. Raleigh, NC.

Guan, Z. Y., Bu, J. J., Mei, Q. Z., Chen, C., & Wang, Q. (2009). *Personalized tag recommendation using graph-based ranking on multi-type interrelated objects.* Paper presented at the the 32nd International ACM SIGIR Conference on Research and Development in Information Retrieval. New York, NY.

Hayes, C., & Avesani, P. (2007). *Using tags and clustering to identify topic-relevant blogs.* Paper presented at the International Conference on Weblogs and Social Media. Boulder, CO.

Heep, M., Bachlechner, D., & Siopaes, K. (2006). *Harvesting wiki consensus - Using Wikipedia entries as ontology elements.* Paper presented at the SemWiki2006. Budva, Montenegro.

Jaschk, R., Marinho, L., Hotho, A., Schmidt-Thiem, L., & Stumme, G. (2007). *Tag recommendations in folksonomies.* Paper presented at the the 11th European Conference on Principles and Practice of Knowledge Discovery in Databases. Athens, Greece.

Jiang, Y. X., Tang, C. J., Xu, K. K., et al. (2009). *Core-tagging clustering for Web2.0 based on multi-similarity measurements.* Paper presented at the APWeb/WAIM Workshops. New York, NY.

Karydis, I., Nanopoulos, A., Gabriel, H. H., & Spiliopoulou, M. (2009). *Tag-aware spectral clustering of music items.* Paper presented at the the 10th International Society for Music Information Retrieval Conference. Kobe, Japan.

Katharina, S., & Martin, H. (2007). *Folksontology: An integrated approach for turning folksonomies into ontologies.* Paper presented at the Bridging the Gap between Semantic Web and Web 2.0 SemNet 2007. Innsbruck, Austria.

Lehwark, P., Risi, S., & Ultsch, A. (2008). *Visualization and clustering of tagged music data.* Paper presented at the the 31st Annual Conference of the German Classification Society. Freiburg, Germany.

Levenshtein, V. I. (1966). Binary codes capable of correction deletions, insertions, and reversals. *Soviet Physics, Doklady, 10*(8), 707–710.

Li, X., Guo, L., & Zhao, E. (2008). *Tag-based social interest discovery.* Paper presented at the the 17th International Conference on World Wide Web. Beijing,China.

Liu, H., Lafferty, J., & Wasserman, L. (2007). Sparse nonparametric density estimation in high dimensions using the rodeo. *Journal of Machine Learning Research.* Retrieved from http://www.cs.cmu.edu/~lafferty/pub/drodeo_aistats.pdf

Marco, L. S., & Edwin, S. (2009). *Tagging clustering with self organizing maps..* Retrieved from.http://www.hpl.hp.com/techreports/2009/HPL-2009-338.pdf

Miao, G., Tatemura, J., Hsiung, W., Sawires, A., & Moser, L. (2009). *Extracting data records from the web using tag path clustering.* Paper presented at the the 18th International Conference on World Wide Web. New York, NY.

Mika, P. (2007). Ontologies are us: A unified model of social networks and semantics. *Journal of Web Semantics, 5*(1), 5–15. doi:10.1016/j.websem.2006.11.002

Ramage, D., Heymann, P., Manning, C. D., et al. (2009). *Clustering the tagged web.* Paper presented at the the Second ACM International Conference on Web Search and Data Mining. Barcelona, Spain.

Shepitsen, A., Gemmell, J., Mobasher, B., & Burke, R. (2008). *Personalized recommendation in social tagging systems using hierarchical clustering.* Paper presented at the the 2008 ACM Conference on Recommender Systems. Bilbao, Spain.

Tso-Sutter, K. H. L., Marinho, L. B., & Schmidt-Thieme, L. (2008). *Tag-aware recommender systems by fusion of collaborative filtering algorithms.* Paper presented at the the 23rd Annual ACM Symposium on Applied Computing. Ceará, Brazil.

van Dam, J., Vandic, D., Hogenboom, F., & Frasincar, F. (2010). *Searching and browsing tag spaces using the semantic tagging clustering search framework.* Paper presented at the the 4th IEEE International Conference on Semantic Computing. Pittsburgh, PA.

Chapter 4
Grouping the Similar among the Disconnected Bloggers

Nitin Agarwal
University of Arkansas at Little Rock, USA

Debanjan Mahata
University of Arkansas at Little Rock, USA

ABSTRACT

Social interactions are an essential ingredient of our lives. People convene groups and share views, opinions, thoughts, and perspectives. Similar tendencies for social behavior are observed in the World Wide Web. This inspires us to study and understand social interactions evolving in online social media, especially in the blogosphere. In this chapter, the authors study and analyze various interaction patterns in community and individual blogs. This would lead to better understanding of the implicit ties between these blogs to foster collaboration, improve personalization, predictive modeling, and enable tracking and monitoring. Tapping interactions among bloggers via link analysis has its limitations due to the sparse nature of the links among the blogs and an exponentially large search space. The authors present two methodologies to observe interaction within the blogs via observed events addressing the challenges with link analysis-based approaches by studying the opinion and sentiments of the bloggers towards the events and the entities associated with the events. The authors present two case studies: (1) "Saddam Hussein's Verdict" and (2) "The Death of Osama Bin Laden." Through these case studies, they leverage their proposed models and report their findings and observations. Although the models offer promising opportunities, there are a few limitations. The authors discuss these challenges and envisage future directions to make the model more robust.

DOI: 10.4018/978-1-4666-2806-9.ch004

1. SOCIAL MEDIA: AN EMERGING PLATFORM FOR COMMUNICATION, INTERACTION, AND EXPRESSION

With the advent of Web 2.0 and its increasing participatory nature, the social media has become an ideal platform for analyzing the pulse of over 2 billion Web-aware people of the world (Internet World Stats, Usage, and Population Statistics, 2011), which is rapidly increasing every second. Social media includes blogs, media sharing sites, micro-blogging sites, social bookmarking sites, social networking sites, social news sites, wikis, and many other forms of media having an online presence in the Web. In recent times, social media has become more than a place to socialize. It has become a mighty platform for the common people to express his opinion and forming online communities. These virtual communities have proved to be firing grounds for various debates, protests as well as a digital instrument against tyranny and for democracy. It has led to the democratization of the Web. Social media has been integral in putting a spark to the, "Iranian Twitter Revolution" (Quirk, 2009), "Egyptian Facebook protests" (Masr, 2009) and recent movements like "The Anna Hazare Movement against corruption" (Facebook, 2011) in India. As one Cairo activist succinctly puts it, "We use Facebook to schedule the protests, Twitter to co-ordinate and You Tube to tell the world" (The Arab Spring's Cascading Effects, 2011). These Web sites are used for coordinating actions, organizing events, mobilizing crowds, disseminating news, and expressing opinions. The modern social media has revolutionized the way, the world expresses and shares opinions and views in public, making the human society, a small world to live in, and share each other's thoughts. It has become a platform to discuss a wide spectrum of topics varying from politics, economics, company products, personal experiences, science and technology to cooking recipes. Thus social media acts as an enabler to influence and propagate ideas among people who are connected to one another through these social media websites, which has further led to the realization of collective action (Tarrow, et al., 1994). A systematic methodology to study the role of social media in the contemporary forms of collective actions has been proposed in Agarwal et al. (2011) illuminating several fundamental yet theoretically obscure aspects of collective action theory.

The social nature of the Web seems to be increasing; with people getting connected to each other every single second and interacting through social media sites. Blogosphere, for instance, has been growing at a phenomenal rate of 100% every 5 months (Technorati, 2008). BlogPulse has tracked over 160 million blogs till November 2010 (BlogPulse Stats, 2011). Facebook recorded more than 800 million active users as of January 2012 (Facebook Fact Sheet, 2012); Twitter amassed nearly 200 million users in March 2011; and other social computing applications like Digg, Delicious, StumbleUpon, Flickr, YouTube, etc., are also growing at similarly terrific pace. This clearly shows the awareness and penetration of social media among individuals and their daily lives. The widespread adoption of social media certainly makes it a lucrative area for researchers converging from various disciplines such as, anthropology, sociology, political science, computer science, mathematics, economics, marketing, management, etc.

In this chapter, we focus on the blogs and try to understand the hidden ties among them in the blogosphere. Blogosphere is the virtual universe of all blogs written by people all over the world, in various languages. Blogosphere is now a huge collection of discussions, commentaries, and opinions on virtually every topic of interest. The phenomenal growth in the blogosphere is evident from the fact that more than two blog posts on an average are created every second (Blogpulse Stats, 2011). A blog is a website that allows individuals to write about different topics. A blog site typically consists of blog posts arranged in reverse chronological order, along with the comments by the readers. These sites contain a series of posts typically characterized by brief texts, which have minimal editing. A blog post may comprise only of text, or it may contain images and links to other media. Unlike websites, the contents of blogs are highly unstructured and spread across multiple topics of interest. A blog site may be owned individually, or it may be a community blog site. Blogs have become a very popular medium for expressing opinions, communicating with others, providing suggestions, sharing thoughts on different issues and also to debate over them. The large amount of valuable data contained in the blogosphere is making it an important field of research, not only for academicians but also for people from industry and other disciplines. The study of blogosphere has helped in reshaping business models (Scoble, et al., 2006), assist viral marketing (Richardson & Domingos, 2002), providing trend analysis and sales prediction (Gruhl, et al., 2005; Mishne & de Rijke, 2006), studying socio-political issues (Singh, et al., 2010), and aiding counter terrorism efforts (Thelwall, 2006). The popularity of blogs has even led the bloggers to celebrate August 31 as the World Blog Day when bloggers from different parts of the world share their blogs with one another. There have been instances when bloggers from various communities have come together and met each other at one place and discussed on various issues about their community (Bloggers Meet Up in Katmandu, 2010). The blogosphere is an ideal platform from which we can extract information, reactions on various issues, opinions, moods, and emotions on various topics, which makes it unique for conducting experiments to improve the current state-of-art in social media research.

It is in human nature to be driven by their Interests (Liu & Maes, 2005). We always look for people and things, which interest us and try to share our views with the like-minded ones. Driven by their common interests, the like-minded people get connected to one another forming clusters or groups (Watts & Strogatz, 1998). In the current era of the Internet, social media is the ideal ground for searching ones interests and people or groups having similar interests. It is likely that we will read things on the Web depending upon our taste, our mood or may be depending upon the things we need to find out, and if the Web is intelligent enough to lead us to more information on them, we are out for a really great user experience. The blogosphere is one of the best platforms where people can publicly express their views and write about their interests creating opportunities to know people and interact.

Due to the casual nature of the blogosphere, not many bloggers cite (or, link to) the original data sources they acquired the inspiration from, to create their own blog posts leading to a link sparsity problem. Unlike other social networking sites such as Facebook, Twitter, etc. where individuals explicitly mention their ties, the sparse link structure of the blogosphere pose several challenges to identify the individuals or groups discussing about similar things. Furthermore, a link-based solution to connect similar yet disconnected individuals may not be practically feasible due to the exponential search complexity. For instance, assuming an average degree 'd' of a node, the nodes that can be searched at 'k' hops are $O(d^k)$, which is computationally intractable (Agarwal, et al., 2008). Therefore, a solution is required that not only addresses the link sparsity problem but also scales to the exponential search space. By helping in identifying and constructing ties among individuals or groups in the blogosphere, our research helps to realize the vision of Sir Tim Berners-Lee, which is to ultimately connect everything on the World Wide Web (Berners Lee, 1999).

In this endeavor to connect the seemingly disconnected bloggers across the Web, we are inspired by Milgram's concept of "The Familiar Strangers" in the physical world. Milgram (1967) observed an interesting phenomenon during his experiments, which he coined as "Familiar Strangers." He defined familiar stranger as someone who we observe repeatedly for a certain period of time but no interaction occurs. These familiar strangers are indeed familiar to us, as they can be found all around us, while we commute on a bus or take our dogs for a walk. So analogous to the physical world, repeatedly observing an individual's profile on social networking websites when there is no actual connection indicates the presence of familiar strangers in virtual world. People are always curious to find those who think alike and are yet unknown to them, and they find immense pleasure to share their views and thoughts with them. These familiar strangers (Agarwal, et al., 2009) exist in the World Wide Web and the online communities. Finding such individuals is not only an interesting topic of research but extremely challenging with tremendous applications. The identified groups or individuals could be considered as recommendations to the original ones. This would foster collaboration among these otherwise disconnected yet similar bloggers. It is difficult to provide personalization to niche Long Tail community blogs due to the lack of data for catering to each and every community blog. Aggregating such niche community blogs would help in accurate personalization. Such aggregation certainly promotes predictive modeling by employing inductive algorithms to learn predictive models from the data and predict trends. Another significant application of identifying similar yet unknown community blogs and individual bloggers involves the study of interaction between them. Many researchers are interested in knowing how these groups and individuals interact within groups or between them. Such studies could lead to very interesting details on ethnic behavior to social events.

The chapter is organized as follows: Section 2 explores the background knowledge and state-of-the art research in this direction. The background is followed by the methodology of our work in Section 3. Section 4 presents two distinct case studies and discusses the findings. Finally, we conclude the chapter in Section 5 by summarizing contributions of our research delineating some future research directions in the area of social media interactions.

2. BACKGROUND AND RELATED WORK

2.1. Relationships and Interactions in Blogosphere

Although the blogosphere is an ideal platform for like-minded people to come together and express their opinions on various issues, the relationships and interactions between the bloggers are somewhat different from the other social networking websites. Unlike social networking websites, which allow users to form explicit ties, the blogosphere often does not have this functionality. However, few blog sites, such as Blogcatalog[1], provide ways for the bloggers to explicitly specify their social network. In most of the cases, ties are constructed using implicit interactions leveraging:

1. **Blogroll:** A blogroll is a list of links to blogs that the blogger likes and is usually included in the blog's sidebar. These can also be the blogs, which a blogger usually follows. Baumer and Fisher (2008) studied blogrolls in order to conclude on the topics usually read by the blogger.
2. **Blog Citations:** A blog citation is a link to other blogs, which the author cites in his blog post. The cited blog posts usually discuss the same issue the author mentioned in his post and very rarely present conflicting views. Sometimes the bloggers can also cite news articles or other websites relevant to the blog post. These citation links are good indicators of the author's perspective and sometimes represent the reactions of the author towards the topic. Adamic and Glance (2005), Adar and Adamic (2005), and Kumar et al. (2005) leveraged the blog citations in order to study

the flow of influence among the bloggers, the differences in their opinions, overlapping of the topics of discussions in bloggers belonging to different groups and formation of communities in the blogosphere.

3. **Blog Comments:** Most blogs allow visitors to leave comments on each blog post. The comments may or may not be anonymous. Blog comments are what make a blog interactive. The most popular blogs have a very interactive community who voice their opinions on posts frequently (Agarwal, et al., 2008). It is this social aspect of the blogs that helps create a sense of community and belongingness among the bloggers making blogs a powerful component of the social Web. Leaving blog comments allows readers to join in on the conversation about a topic that interests them. Herring et al. (2005) and Ali-Hasan and Adamic (2007) studied conversations and the social relationships among blogs through links and comments.

In order to understand the social interactions and relationships between the bloggers the researchers have studied all the three different ties extensively. Some of these studies are based on the implicit social networking information derived from explicit link structure in the blogosphere, which models it like the Web and thus face certain shortcomings. The most challenging problem is the sparse structure of the links connecting different blogs. The sparsity of links makes it very difficult for the researchers to model the blogosphere in the way the Web can be modeled and use the already existing state of art in Web mining to do a link analysis of the blogosphere. Moreover, doing a naive link analysis of the blogosphere gives rise to a problem involving exponential complexity (Agarwal, et al., 2008). The level of interaction in terms of comments and replies to a blog post makes the blogosphere different from the other social media sites. The existing Web models cannot simulate the highly dynamic and transient nature of the blog posts. Web models do not consider this dynamicity in the Web pages. They assume Web pages accumulate links over time. However, in a blog network, where blog posts are the nodes, it is impractical to construct a static graph like the one for the Web. These differences necessitate the need for a model more towards the characteristics of the blogosphere. A complete survey of such issues can be found in Agarwal and Liu (2008). We have taken a different approach to study these interactions and relationships in the blogosphere, which is based upon the idea of "Interaction through Observation" and takes into account the reactions, opinions and sentiments expressed by the bloggers on an event over a period of time.

2.2. Sentiment Analysis and Opinion Mining

Sentiment analysis involves classifying text based on its sentiment. While most of the approaches focus on extracting sentiments associated with polarities of positive or negative for a given document, a few focus on specific subjects from a given document. For example, the classification of a movie review (Pang, et al., 2002; Turney, 2002) assumes that all sentiment expressions in the review represent sentiments directly toward that movie, and expressions that violate this assumption (such as a negative comment about an actor even though the movie as a whole is considered to be excellent) confuse the judgment of the classification (Yi, et al., 2003). Most of the approaches to sentiment analysis involve high level of Natural Language Processing (NLP) that is very complicated and incurs a very high cost. Moreover, as mentioned earlier, blogs use a very common language that is colloquial in nature and hence NLP-based approaches are extremely challenging. Opinion mining is also related to sentiment analysis and is more or less concerned with the presence of opinionated words in the text. One can get an opinion on a movie review or on a new product that they are about to purchase. Opinion mining classifies a blog post into a positive or a negative response that enables a user to have a one-word answer from the entire post. Extracting opinions from customer reviews has been a great research focus. With the increase in the number of people writing reviews, it becomes very challenging for a potential customer to read and form a decision. Feature based opinion summarization suggested by Hu and Liu (2004) is a technique that identifies features of the products on which the customers have expressed their opinions and ranks the features according to the frequency. This technique improves the process of forming opinions on different features of a product. Such an approach can also be used in various other cases such as an event, or a topic. In our work, we have tried to capture the opinions of the bloggers on various events at event and entity level. Various sentiment extraction techniques specifically designed to tackle social media challenges are leveraged. These approaches not only identify opinions but also provide means to connect the familiar strangers. More details on the proposed approaches are presented in Section 3.

2.3. Familiar Strangers

Familiar strangers in our physical world, as coined by Stanley Milgram, do not know each other, but frequently exhibit some common patterns. It is these common patterns that determines their similarity and distinguishes them as familiar strangers. Blogosphere also exhibit such characteristics. We can find bloggers writing and expressing similar reactions about the same topics, yet they do not know each other or are disconnected from one another. The nature of the Web entails a power law distribution on the blogosphere. That is, the majority bloggers are only connected with a small number of fellow bloggers, and these bloggers are largely disconnected from each other. Familiar strangers on Blogosphere are not directly connected, but share some patterns in their blogging activities. Blogosphere contains both single authored blogs known as individual blogs and multi-authored blogs or community blogs. In individual blogs, only one author creates blog posts and readers are allowed to comment on these posts, but the readers cannot create new entries. In community blogs, several authors can create blog posts as well as comments. Readers are allowed to comment but only registered members of the community can author blog posts. In this work, we focus on individual as well as community level familiar strangers. It is highly likely that the familiar strangers occur in the Long Tail (Anderson, 2006) as depicted in Figure 1, because the bloggers in the Short Head are highly authoritative which means

Figure 1. Familiar stranger bloggers and long tail distribution

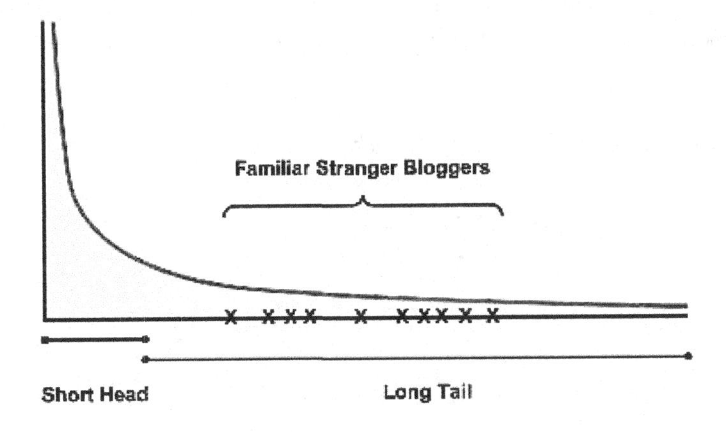

they are highly connected, hence less chances of being strangers. Moreover, existing search engines return relevant results only from the Short Head, so it is interesting and challenging to study the ones that appear in the Long Tail. Agarwal et al. (2007) presents a comprehensive discussion on the types of familiar strangers in the blogosphere and the challenges associated with finding them. In this work, we study the similarity of the bloggers' reactions towards an event in order to find the familiar strangers in the blogosphere. More details on the proposed approaches are mentioned in the next section.

3. PROPOSED MODEL AND METHODOLOGIES

3.1. Interaction Through Observation

There are an overwhelming number of blogs on the blogosphere. Bloggers in groups interact within themselves and also across various groups. These interactions could be tapped using link (more specifically, hyperlinks) analysis. The assumption is, if a blog post in a community blog or a personal blog cites another blog post published in a different community or an individual blog, then these communities and the bloggers are considered to be known to each other. Two communities or bloggers are disconnected if there is no prior interaction between them. Moreover, if two communities or bloggers interact with each other they are more likely to be similar, or talking on similar themes. A naïve link analysis based approach would enable us to identify groups or bloggers that could be similar. However, a naïve link analysis based approach to identify familiar strangers presents two challenges:

1. A naïve link analysis based approach entails an exponential search space.
2. Bloggers who connect to each other via hyperlinks may already know each other.

Thus by using link analysis, it is not very difficult to identify groups and bloggers that are similar and connected, but it is not of great help as these groups already know each other and this is not new information for them. Therefore, the dilemma of finding similar yet disconnected groups presents unique challenges. Figure 2 shows a visualization of the links flowing in and out of a blog to other blogs and websites, which can be analyzed for finding these explicit links between the blog posts. The link analysis approach brings in exponential complexity in tackling the problem (Agarwal, et al., 2008). So we propose to approach the problem in a different way, which is an interaction based approach referred as, "Interaction through Observation." The proposed approach takes into account the opinion, sentiments, views, and moods that form the reactions, expressed in the blog posts by the bloggers on a particular issue or an event.

This kind of interaction is different from the interaction that the bloggers have within them or between them. This is also not tapped through

Figure 2. Flow of links between various blogs and the cited resources

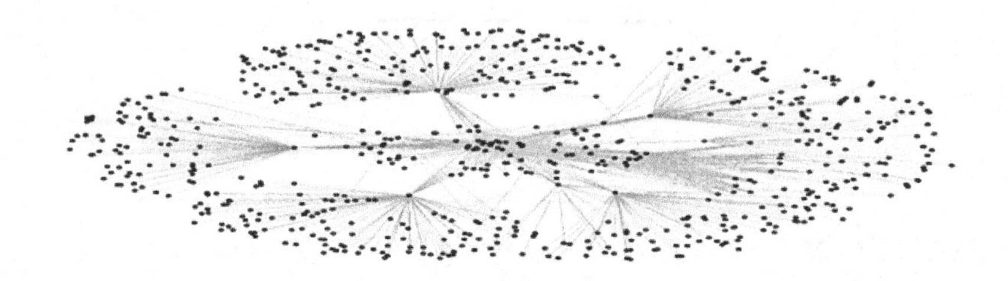

link analysis. First significant advantage of interaction through observation is the non-utilization of link analysis, which is known to be exponential in computation. Second, since blogs have very sparse link structure as compared to the websites, exploiting link analysis to identify similar community blogs gives poor results (Kritikopoulos, et al., 2006). These types of interactions are illustrated in Figure 3. Based on the reactions that individual bloggers and communities have on an event/issue one can identify whether two communities or bloggers are similar or not. Intuitively, if communities and bloggers consistently express similar feelings on an issue or an event then they tend to be similar. This concept forms the bottom-line of our proposed approach for identifying similar yet disconnected community and individual blogs via interaction through observation. Next we explain the two methodologies based on the idea of 'interaction through observation,' that can be used to analyze the possible interactions among bloggers and discovering familiar strangers from the reactions and opinions expressed by them on an event and the entities associated with the event.

3.2. Event-Opinion-Based Methodology

Our approach initially requires identifying an event that incites reactions of sufficient posts to analyze. From the blog posts, we summarize the text using the tool Subject Search Summarizer. SSSummarizer[2] (Coombs, et al., 2008) generates and displays summaries as a list of key sentences the product extracts from documents. By presenting and translating sentences, it reflects the subject of a given document thus providing the key information. The SSSummarizer allows you to choose the number of sentences to display in the summary. We tried 10, 20, and 30 sentences for different posts and compared the performance of the tool by manual analysis. We were convinced that when the number of sentences is set to 20, the tool provides relevant information that is good enough for further analysis.

The stop words are eliminated from the summarized text and it is fed into a tag cloud generator that spits out the representative words from the summarized text. The tag cloud generator[3] is an online tool from artviper that generates tag clouds

Figure 3. Different types of interactions among blogs

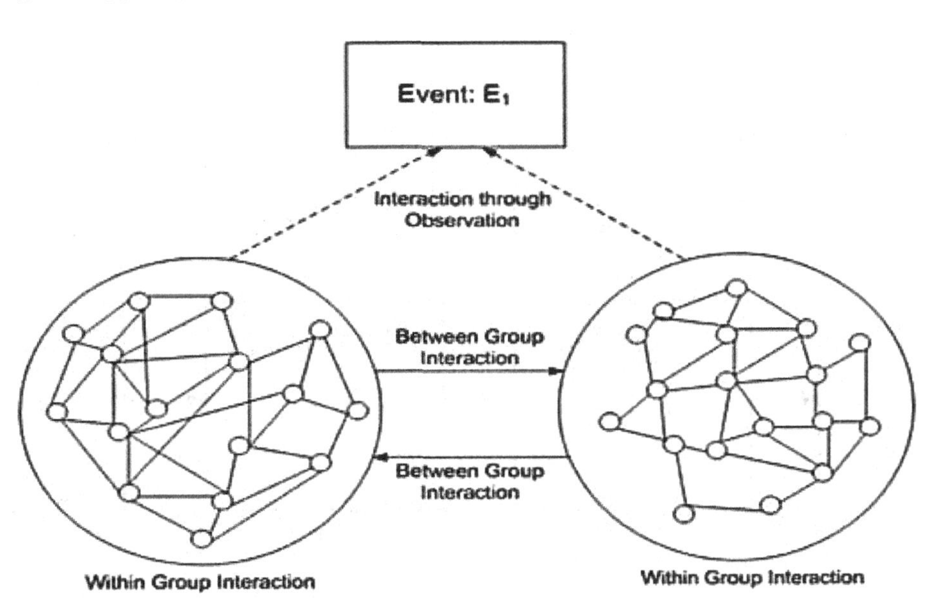

for the given text. The tag cloud generator works well when it is used after the SSSummarizer. If used without the tool and raw posts are given directly as input, the identified tags are extremely noisy and irrelevant and appear in larger font increasing its significance. Moreover, the number of tags generated is huge and if in case we restrict the number then the quality of results is compromised.

We then compare the identified tags with the list of sentiment key words given by WeFeelFine[4], augmented by Thesaurus[5], and tag the matching words as sentimental words. WeFeelFine provides an API that contains a list of sentiments that have been identified from the blogosphere. Each word has a number associated with it that indicates the number of times it has been identified as sentiment in the blogs. We consider a word to be a "Sentiment word" only if this number is greater than 10. Thesaurus.com has an online searchable collection of words and is grouped together with antonyms. These groups of words are used to augment the sentiment words obtained from WeFeelFine.

From this point by manually analyzing each of these words we tag them as either a positive sentiment or a negative sentiment. From this collection of positive and negative sentiment words, we will be able to decide the type of reaction of the communities for that event. If the reactions of these communities remain consistently similar then these communities are more likely to be similar. Thus the approach based on interaction through observation, enables us to address the challenges due to link analysis and thus find two similar communities that are disconnected. A summarized flow chart of the whole approach is illustrated in Figure 4. We present a case study in the section Case Study-1 on real world blogs using this methodology and report the interesting observations.

This methodology gives us the overall sentiment of the blog for the event. However, this approach may miss out the views expressed by the blogger for various entities associated with the event and discussed by the blogger in the blog post. Thus this approach provides us a coarse analysis of the sentiments expressed in a blog post and misses out the finer details, which may lead us to incorrect results. For a more granular analysis, we propose another methodology as discussed in the next section.

Figure 4. Flowchart of different components of event-opinion methodology

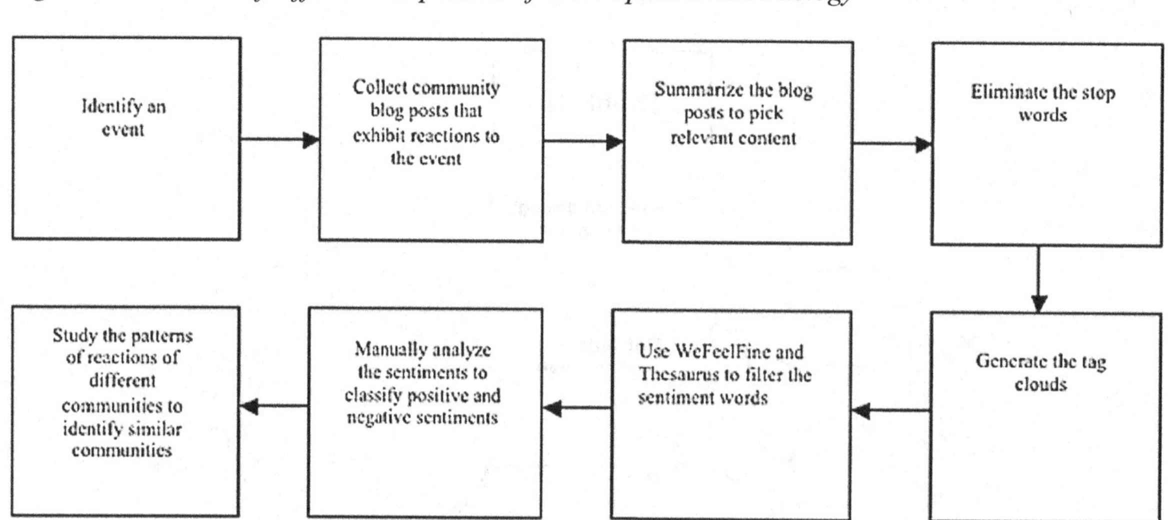

3.3. Event-Entity-Opinion-Based Methodology

Similar to the previous methodology, in this approach, we identify an event and the reactions expressed by blog posts. More specifically, we collect top 100 blog posts related to the event using Google blog search[6]. We use Alchemy API[7] to extract the text from the blogs and identify the various entities mentioned in the blog posts by the bloggers. These entities can be name of a person, place, organization, etc. as returned by the Alchemy API. We take the k most frequently occurring entities for our analysis. These entities are considered to be related with the event and the bloggers have expressed opinions on them. Therefore, sentiments or opinions are attached with these entities mentioned in the blog post. The Alchemy API also returns the sentiment score categorizing it into positive, negative or neutral. While examining the sentiment score we study the context of the post to fine-tune the sentiments such as, ensuring that the sentiments correspond to the extracted entities, consider sarcasm and negation, etc[8]. We study the patterns of the reactions expressed by different bloggers for the top k entities in their blog posts and use this pattern to identify the blog posts that provide similar reactions to the entities related to the event. A summarized flow chart of the whole approach is described in Figure 5. We present a case study in the section Case Study-2 on real world blogs using this methodology and report the interesting observations.

This approach helps us to do a more granular and an in-depth analysis of the bog posts and helps us to find the sentiment expressed by the blogger at the entity level. From this, we can also find the entities associated with the event and the reactions of the bloggers towards these entities. In this way, we can group the bloggers who express similar opinions for a set of entities. The views of the bloggers on the entities also give us an idea of the orientation of the bloggers towards the event and the various entities associated with it.

4. ANALYSIS OF THE PROPOSED MODEL AND METHODOLOGIES

Both the above methodologies help us to find the familiar strangers in the blogosphere. The bloggers who are not directly connected with one another yet express similar reactions and views for an

Figure 5. Flowchart of the various components of event-entity-opinion-based methodology

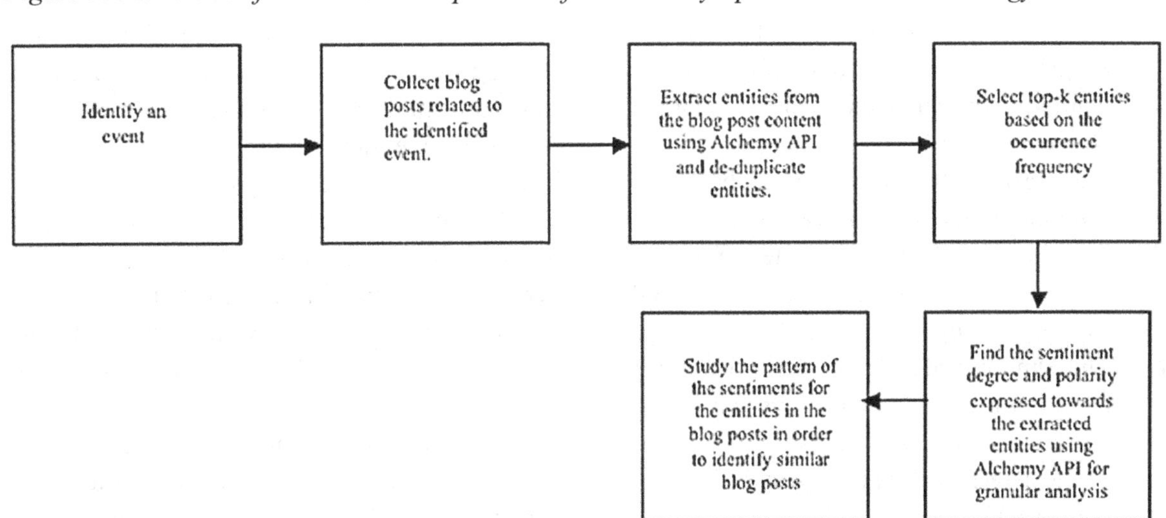

event and the different entities are the familiar strangers. It is highly likely for these bloggers to be connected and form pools of different views and opinions towards the event and the entities that play an important role in it. We present two case studies, which show how these methodologies can be realized in the real world for the analysis of various events.

4.1. Case Study 1: "Saddam Hussein's Verdict"

We take the event of Saddam Hussein's verdict as our first case study. The Sunnis[9] opposed the event stating it to be ridiculous. At the same time, the Shiites[10] felt it was a good decision and they were supporting the event. Such interactions could be found in the blogs but there is no direct way to identify how each group reacted to such events without reading the full post. Though there are several tools available to summarize, identify concepts and themes, there is no such tool to directly observe ideological differences from the blogosphere. By identifying sentiments from these blog posts, we can observe their feelings and reactions.

Identifying similar groups that are disconnected through observing an event involved the following steps.

1. We obtain the posts from the three sites – Iraq The Model, Baghdad Burning, and East Kurd in the month of the event i.e., Saddam Hussein's Verdict.
2. We use the SSSummarizer to obtain the summary of the posts for each site.
3. The Stop words are eliminated from the summarized text of the blog posts obtained from these blogs.
4. The summarized text after stop word elimination is given as input to the Tag Cloud Generator to identify the tags.
5. The generated tags are then checked with the API provided by WeFeelFine (augmented by the Thesaurus) and the matching words are tagged as Sentiment words.
6. The words identified as sentiment words are then tagged manually as either positive or negative. From these words that have been identified for the three blogs, we are able to observe that they have different feelings towards the event.

The sentiment words clearly revealed that one website, i.e. Baghdad Burning opposed the event while the other two i.e., Iraq The Model and East Kurd, were in favor of the event. We also aligned our findings with the ground truth obtained from the news site and came to a conclusion that Iraq The Model aligns well with the Shia, Baghdad Burning aligns well with the Sunnis and East Kurd aligns well with the Kurds. Figure 6 shows the reactions of Iraq The Model and Baghdad Burning to Saddam Hussein's verdict. Iraq The Model had a very positive reaction to the event as evident by the tags like 'accept,' 'agree,' 'building,' and 'patriotic' whereas Baghdad Burning had a very negative reaction to the event as evident by the tags like 'bad,' 'dead,' 'demonstration,' 'shut,' and 'stupidity.'

We also considered another event that was not very famous as the Saddam's verdict; the series of suicide bombings in Iraq during the month of April 2006. This event did not have as many blog post reactions as compared to that of the Saddam Hussein's verdict. We considered posts from the same three blog sites to identify their reaction to this event. We identified that all the three sites had posts that indicated a negative sentiment. This clearly indicates that the three blog sites strongly opposed the event. This was further manually verified by reading the blog posts. The blogs revealed their grief in the events and people expressed how much they were affected by these bombings. Based on these findings we could observe that East Kurd and Iraq The Model are very similar in terms of their reaction to these events. This analysis could also be useful in an-

Figure 6. Blog reactions to Saddam Hussein's verdict

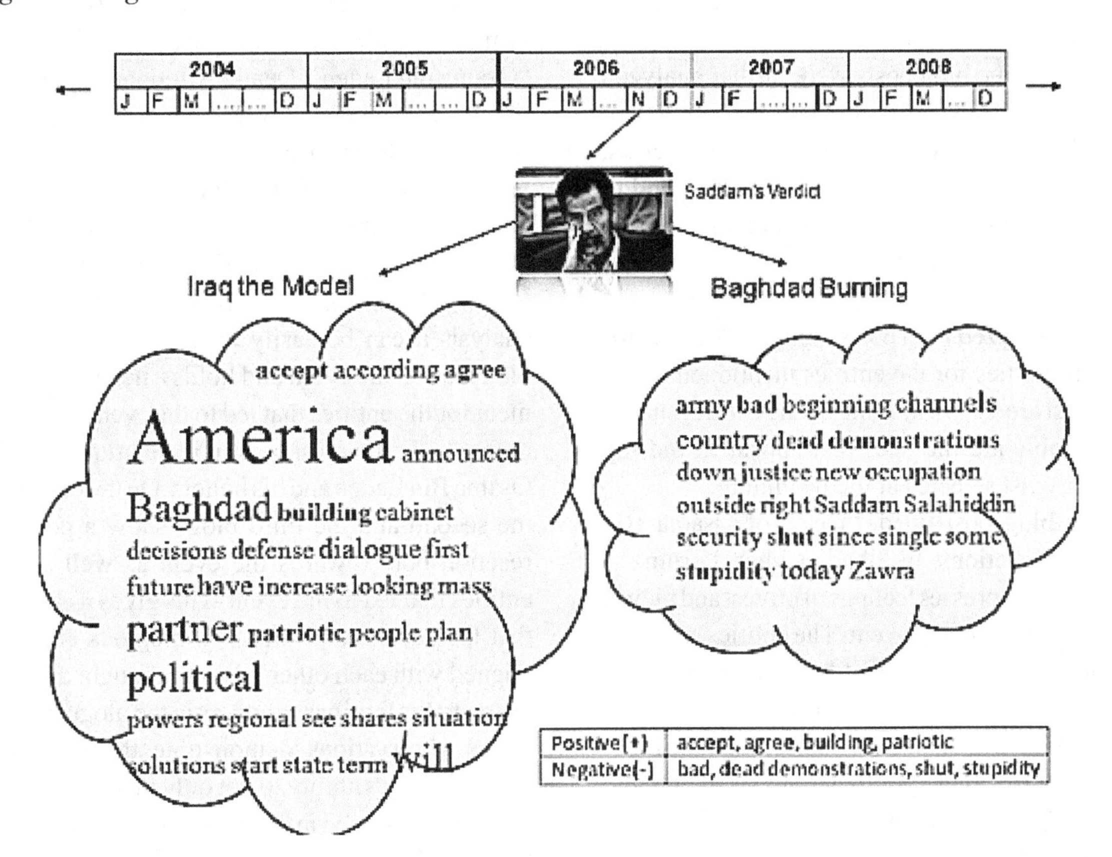

Positive(+)	accept, agree, building, patriotic
Negative(-)	bad, dead demonstrations, shut, stupidity

ticipating reactions towards future events. The results for both the events are summarized in Table 1. The table represents a summary of the results obtained. (+) sign indicates a positive sentiment and (-) sign indicates a negative sentiment towards the event.

4.2. Case Study 2: "The Death of Osama Bin Laden"

For our second case study, we took the event, "The Death of Osama Bin Laden," which resulted in tremendous online reactions in various forms. The bloggers from all over the world expressed their varied reactions for the different entities (people, organizations, and countries) associated with the event. We extract the entities mentioned in the blog posts and analyze the reactions expressed on them, which not only gives us an overall picture of the collective reactions expressed for a particular entity but also gives the idea of the reactions expressed by the individual bloggers for the various entities. In this way we find the like-minded bloggers who have similar views associated with the entities involved in the event. We take the following steps for achieving our purpose:

1. We collect the top 100 blog posts returned by Google Blog Search for the event.
2. We use Alchemy API to find the top 20 entities mentioned in the blog posts returned in the previous step.
3. We take three blog posts from three different sources expressing views on the event and find the entities and the sentiments expressed towards the entities using Alchemy API. The API returns a sentiment score and the

polarity as positive, negative, and neutral for each entity.

4. The three blog posts were further analyzed manually in order to study the context of the post to fine-tune the sentiments such as, ensuring that the sentiments correspond to the extracted entities, consider sarcasm and negation of sentiments, etc.

The analyzed blog posts along with the sentiment polarities for the entities mentioned in the blog post are shown in Table 2. The entities shown in the table are the ones that appear in the top 20-entity list selected at the beginning.

The blog post titled, "Death of Osama Bin Laden: Reactions In Jihadi Cyber Forums – Analysis," expresses feelings of protest and views, which are against the event. The entities are more aligned towards the line of belief of Osama Bin Laden, whereas the blog post titled, "Bin Laden's Death Met with Fanfare, Sobriety at U of C," expresses the views and reactions of the people of a university in United States of America. In this blog, "President Barack Obama," "The White House," "United States of America" are treated with positive reactions and "Osama Bin Laden" is treated with negative ones. The third blog, "Osama bin Laden's Death: Aftermath and Reaction, as it Happened" writes about the reactions of the people from all over the world and expresses positive reactions for "President Barack Obama," "United States of America," "The White House" and negative reactions for "Osama Bin Laden," "Pakistan," "Abbottabad." From this analysis it can be easily observed that the first blog resents the event and holds a negative sentiment for the entities that led to the event. Furthermore, the first blog shows an inclination towards Osama Bin Laden and his beliefs. On the contrary, the second and the third blogs show a positive reaction both towards the event as well as the entities that led to the event. This gives us an idea that the second and the third bloggers are well aligned with each other's line of thought and can be treated as familiar strangers in the blogosphere. These observations demonstrate the proposed methodology's ability to not only identify overall opinions expressed in the blog posts over the event but also the sentiments towards the entities associated with the event. These sentiments can be

Table 1. Summary of results obtained from case study 1

Event	Iraq The Model	Baghdad Burning	East Kurd
Nov 2006 – Saddam Verdict – death sentence	Accept and support the verdict. (+)	Oppose the verdict. Feels its lynching. (-)	Accepts and supports the verdict. (+)
August 2006 – Series of Suicide bomb explosions	Feel bad for it and oppose mildly. (-)	Feels bad and oppose the event strongly. (-)	Feels bad and opposes. (-)

Table 2. Summary of results obtained from case study 2

Death of Osama Bin Laden: Reactions In Jihadi Cyber Forums- Analysis	Bin Laden's death met with fanfare, sobriety at U of C	Osama bin Laden's death: aftermath and reaction. As it happened
Osama Bin laden (+)	Osama Bin Laden (-)	Osama Bin Laden (-)
Taliban (+)	President Barack Obama (+)	United States of America (+)
Pakistan (+)	Facebook (+)	White House (+)
United States of America (-)	United States of America (+)	Pakistan (-)
Al-Qaeda (+)	White House (+)	Abbottabad (-)

further used to identify like-minded bloggers at event as well as entity level.

4.3. Opinion-Entity Graph

We also create an opinion-entity graph as shown in Figure 7, for the event: "Death of Osama Bin Laden." The graph is created from the main entities mentioned in the three blogs analyzed in Case Study- 2, where the nodes of the graph are represented by the entities. The labels of the nodes also contain the sentiment polarities and scores for the entities. For instance, the entity 'President Barack Obama' has a positive sentiment with score of 0.151 and a negative sentiment with score of 0. An edge is drawn between two entities if both of them occur in the same blog post. The weights of the edges are the Pointwise Mutual Information (PMI) scores between the two entities. The edges having weights below a certain threshold value are discarded. The edge weights indicate the strength of the connection between two entities. The graph not only shows the polarities of the opinions towards the entities for the event, but also shows how the entities are connected to one another. One can also get an idea of the influence of the entities on one another from the edge strengths. In this way, it is possible to study the influence between the entities with respect to the sentiments associated with the entities. Various clusters can also be found from the graph comprising of a set of entities that are tightly bound to one another and have the possibility to have similar opinion polarities for the event. Given the entities and their polarities, it is possible to predict the opinions of a particular blog post towards a future event by extracting the most frequently occurring entities in the post. Entity opinion graphs can be designed further to create event-entity-opinion-blog hypergraph, which can be analyzed in order to address the following research questions:

1. To study the collective sentiment of the bloggers towards the event.
2. To observe the influence of bloggers on one another and how it shapes collective sentiment.

Figure 7. Opinion entity graph for the event: "Death of Osama Bin Laden," obtained from the analyzed blogs

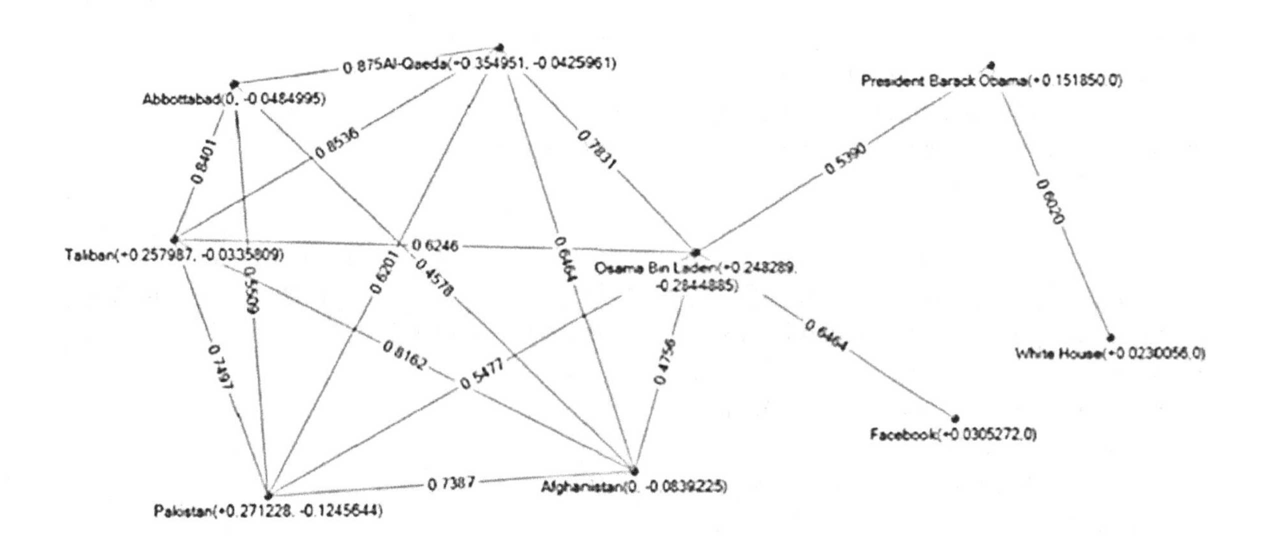

3. To find the set of entities that have similar reactions for an event and to predict the reactions for those entities in a future event.
4. To find similar bloggers based on their reactions on the entities associated with an event.
5. To predict the reactions of the bloggers towards the entities for a similar future event.

5. CONCLUSION AND FUTURE WORK

In this chapter, we present a novel problem of identifying similar yet disconnected communities and individuals in the social media who are also known as familiar strangers. We formulate the problem and identify the challenges that are commonly associated with tapping the interaction via link analysis, especially in the blogosphere. The motivation for studying the blogosphere is three-folds: (1) it is one of the fastest growing platforms in the social media, (2) Blogosphere is one of the most widely used platforms for individuals and communities to share, express, and discuss view, opinions, and reactions towards various events, and (3) the interactions on the blogosphere are publicly available as open data sources. The proposed approach observes the interaction between blogs via their expressed reactions towards an event. The underlying assumption of the proposed approach is that the blogs that have same or very similar opinions on a varied set of events are more likely to be similar, hence they would benefit from connecting with each other, if they are not already connected. We proposed two methodologies that help realize these latent interactions. Similarity between the communities and individuals is evaluated using the identified sentiments and opinions. We present two case studies on real world blogs and report our findings and observations with the models.

As part of the future work, we plan to identify and investigate more community and individual blog sites to perform a more rigorous analysis of the proposed model and the methodologies. We also plan to leverage advanced techniques for identifying positive and negative sentiments. Moreover, we consider augmenting the sentiment words provided by the various APIs used in the work with more words that are colloquially used to express feelings with respect to a specific domain. We are also exploring ways to categorize sentiment words into groups, such as religion, Law/Order, etc. to build a taxonomy of sentiments for varying level of analysis. The current study opens up several opportunities such as, study the diffusion of sentiments between the bloggers; predict group behavior towards a future event based on their reaction to similar events in the past; among others.

ACKNOWLEDGMENT

This research was funded in part by the National Science Foundation's Social-Computational Systems (SoCS) Program and Human Centered Computing (HCC) Program within the Directorate for Computer and Information Science and Engineering's Division of Information and Intelligent Systems (Award numbers: IIS - 1110868 and IIS - 1110649) and by the U.S. Office of Naval Research (Grant number: N000141010091). We gratefully acknowledge this support.

REFERENCES

Adamic, L. A., & Glance, N. (2005). The political blogosphere and the 2004 US election: Divided they blog. In *Proceedings of the 3rd International Workshop on Link Discovery,* (pp. 36–43). IEEE Press.

Adar, E., & Adamic, L. A. (2005). Tracking information epidemics in blogspace. In *Proceedings of the 2005 IEEE/WIC/ACM International Conference on Web Intelligence*, (pp. 207–214). IEEE Press.

Agarwal, N., Lim, M., & Wigand, R. T. (2011). Collective action theory meets the blogosphere: A new methodology. *Networked Digital Technologies*, *136*(3), 224–239. doi:10.1007/978-3-642-22185-9_20

Agarwal, N., & Liu, H. (2008). Blogosphere: Research issues, tools, and applications. *SIGKDD Explorations*, *10*(1), 18–31. doi:10.1145/1412734.1412737

Agarwal, N., Liu, H., Murthy, S., Sen, A., & Wang, X. (2009). A social identity approach to identify familiar strangers in a social network. In *Proceedings of the 3rd International AAAI Conference of Weblogs and Social Media*. AAAI.

Agarwal, N., Liu, H., Salerno, J., & Yu, P. S. (2007). Searching for familiar strangers on blogosphere: Problems and challenges. In *Proceedings of the NSF Symposium on Next-Generation Data Mining and Cyber-enabled Discovery and Innovation (NGDM)*. NGDM.

Agarwal, N., Liu, H., Salerno, J. J., & Sundarajan, S. (2008). Understanding group interaction in blogosphere: A case study. In *Proceedings of the 2nd International Conference on Computational Cultural Dynamics (ICCCD 2008)*. Washington, DC: ICCCD.

Ali-Hasan, N., & Adamic, L. A. (2007). Expressing social relationships on the blog through links and comments. In *Proceedings of International Conference on Weblogs and Social Media*. IEEE.

Anderson, C. (2006). *The long tail: Why the future of business is selling more for less*. New York, NY: Hyperion.

Baumer, E., & Fisher, D. (2008). Smarter blogroll: An exploration of social topic extraction for manageable blogrolls. In *Proceedings of the Hawaii International Conference on System Sciences*, (pp. 155–155). IEEE.

Berners-Lee, T. (1999). *Weaving the web: The original design and ultimate destiny of the world wide web by its inventor*. San Francisco, CA: Harper San Francisco.

Facebook. (2011). *Anna Hazare - Movement against corruption*. Retrieved 17 September, 2011 from http://www.facebook.com/AnnaHazareMovement

Granovetter, M. S. (1973). The strength of weak ties. *American Journal of Sociology*, *78*(6), 1360–1380. doi:10.1086/225469

Gruhl, D., Guha, R., Kumar, R., Novak, J., & Tomkins, A. (2005). The predictive power of online chatter. In *Proceedings of the Eleventh ACM SIGKDD International Conference on Knowledge Discovery in Data Mining*, (pp. 78–87). ACM Press.

Herring, S. C., Kouper, I., Paolillo, J. C., Scheidt, L. A., Tyworth, M., Welsch, P., et al. (2005). Conversations in the blogosphere: An analysis. In *Proceedings of the Thirty-Eighth Hawai'i International Conference on System Sciences (HICSS)*. IEEE.

Hu, M., & Liu, B. (2004). Mining and summarizing customer reviews. In *Proceedings of the Tenth ACM SIGKDD International Conference on Knowledge Discovery and Data Mining*, (pp. 168–177). ACM Press.

Internet World Stats, U., & Statistics, P. (2011). *Website*. Retrieved 17 September, 2011 from http://www.internetworldstats.com/stats.htm

Kritikopoulos, A., Sideri, M., & Varlamis, I. (2006). BlogRank: Ranking weblogs based on connectivity and similarity features. In *Proceedings of the 2nd International Workshop on Advanced Architectures and Algorithms for Internet Delivery and Applications,* (p. 8). IEEE.

Kumar, R., Novak, J., Raghavan, P., & Tomkins, A. (2005). On the bursty evolution of blogspace. *World Wide Web (Bussum)*, *8*(2), 159–178. doi:10.1007/s11280-004-4872-4

Liu, H., & Maes, P. (2005). *Interestmap: Harvesting social network profiles for recommendations*. Paper presented at Beyond Personalization. Los Angeles, CA.

Masr, B. (2009). *Stop, look, what's that sound – The death of Egyptian activism*. Retrieved February 8, 2009 from http://bikyamasr.wordpress.com/2009/08/02/bm-opinion-stop-look-whats-that-sound-the-death-of-egyptian-activism/

Milgram, S. (1967). The small world problem. *Psychology Today*, *2*(1), 60–67.

Mishne, G., & de Rijke, M. (2006). Deriving wishlists from blogs show us your blog, and we'll tell you what books to buy. In *Proceedings of the 15th International Conference on World Wide Web,* (pp. 925–926). IEEE.

Nepali Blogger. (2011). *Bloggers meet up in Kathmandu*. Retrieved 18 September, 2011 from http://nepaliblogger.com/bloggers/bloggers-meet-up-in-kathmandu/1174/

Pang, B., & Lee, L. (2008). Opinion mining and sentiment analysis. *Foundations and Trends in Information Retrieval*, *2*(1-2), 1–135. doi:10.1561/1500000011

Pulse, B. (2011). *Stats 2011*. Retrieved 17 September, 2011 from http://www.blogpulse.com/

Quirk, P. W. (2009). *Iran's Twitter revolution: Foreign policy in focus*. Retrieved June 17, 2009 from http://www.fpif.org/articles/irans_twitter_revolution

Richardson, M., & Domingos, P. (2002). Mining knowledge-sharing sites for viral marketing. In *Proceedings of the Eighth ACM SIGKDD International Conference on Knowledge Discovery and Data Mining,* (pp. 61–70). ACM Press.

Scoble, R., Israel, S., & Corporation, E. (2006). *Naked conversations: How blogs are changing the way businesses talk with customers*. New York, NY: John Wiley.

Singh, V., Mahata, D., & Adhikari, R. (2010). Mining the blogosphere from a socio-political perspective. In *Proceedings of the 6th International Conference on Next Generation Web Services Practices*, (pp. 365-370). Gwalior, India: IEEE Press.

Statistics, F. (2012). *Webpage.* Retrieved 27 January, 2012 from http://www.facebook.com/press/info.php?statistics

Tarrow, S., & Tollefson. (1994). *Power in movement: Social movements, collective action and politics*. Cambridge, UK: Cambridge University Press.

Technorati's State of the Blogosphere. (2011). *Website.* Retrieved 17 September, 2011 from http://technorati.com/state-of-the-blogosphere/

Thelwall, M. (2006). Bloggers under the London attacks: Top information sources and topics. In *Proceedings of the 3rd Annual Workshop on Webloging Ecosystem: Aggregation, Analysis and Dynamics*. IEEE.

Turney, P. D. (2002). Thumbs up or thumbs down? Semantic orientation applied to unsupervised classification of reviews. In *Proceedings of the 40th Annual Meeting on Association for Computational Linguistics*, (pp. 417–424). ACL.

Watts, D. J., & Strogatz, S. H. (1998). Collective dynamics of "small-world" networks. *Nature*, *393*(6684), 440–442. doi:10.1038/30918

Yi, J., Nasukawa, T., Bunescu, R., & Niblack, W. (2003). Sentiment analyzer: Extracting sentiments about a given topic using natural language processing techniques. In *Proceedings of the third IEEE International Conference on Data Mining, 2003*, (pp. 427–434). IEEE Press.

ENDNOTES

1. Blogcatalog: www.blogcatalog.com
2. http://www.kryltech.com/summarizer.htm
3. http://www.artviper.net/texttagcloud/
4. http://www.wefeelfine.org/
5. http://www.thesaurus.com/
6. http://www.google.com/blogsearch
7. www.alchemyapi.com
8. Since this fine-tuning is done manually, we leave this line of research as a future direction, which is out of the scope of the book chapter and also independent of the main contributions of this work.
9. http://en.wikipedia.org/wiki/Sunni_Islam
10. http://en.wikipedia.org/wiki/Shia_Islam

Chapter 5
Topic Modeling for Web Community Discovery

Kulwadee Somboonviwat
King Mongkut's Institute of Technology Ladkrabang (KMITL), Thailand

ABSTRACT

The proliferation of the Web has led to the simultaneous explosive growth of both textual and link information. Many techniques have been developed to cope with this information explosion phenomenon. Early efforts include the development of non-Bayesian Web community discovery methods that exploit only link information to identify groups of topical coherent Web pages. Most non-Bayesian methods produce hard clustering results and cannot provide semantic interpretation. Recently, there has been growing interest in applying Bayesian-based approaches to discovering Web community. The Bayesian approaches for Web community discovery possess many good characteristics such as soft clustering results and ability to provide semantic interpretation of the extracted communities. This chapter presents a systematic survey and discussions of non-Bayesian and Bayesian-based approaches to the Web community discovery problem.

INTRODUCTION

In recent years, the World Wide Web has become a popular platform for disseminating and searching for information. Due to the explosive growth of the Web, the low precision of Web search engine, and the lack of a data model for the Web data, it is increasingly difficult for the users to search for and access the desired information. Motivated by this problem, a lot of research has been done to discover the implicit communities of topically related Web pages or *Web communities* (e.g. Gibson, et al., 1998; Kumar, et al., 1999; Flake, et al., 2000). The Web communities provide invaluable, reliable, timely, and up-to-date topic specific information resources for users interested in them. Furthermore, a set of extracted Web communities can be used as a key building block in Web applications and value added services such as focused crawling, Web portals, Web search ranking, Web spamming detection, Web recommendation, and Web personalization (such as Flake, et al., 2000;

DOI: 10.4018/978-1-4666-2806-9.ch005

Pierrakos, et al., 2003; Otsuka, et al., 2004; Li, et al., 2010).

Conceptually, a Web community is defined as a set of Web pages on a specific topic created by people sharing the same interests. The Web community usually manifests itself as a subgraph with dense connections and coherent content. Most work on Web community discovery focused on the efficient detection of community structure based purely on link information between Web pages using non-Bayesian approaches such as spectral methods, graph partitioning, and clustering (e.g. Kumar, et al., 1999; Flake, et al., 2000; Toyoda & Kitsuregawa, 2001). These non-Bayesian link based methods suffer from the lack of semantic interpretation (most implementation uses a simple top-k most frequent keyword to summarize a topic of a Web community). Furthermore, most non-Bayesian approaches for Web community discovery do not allow a Web page to be assigned to more than one community.

On the other hand, probabilistic topic models (e.g. Blei, Ng, & Jordan, 2003; Griffiths & Steyvers, 2002, 2003, 2004; Hofmann, 1999, 2001) have recently gained much popularity as a suite of algorithmic tools to help organizing, searching, and understanding large collections of text documents. The key idea underlying these models is that a document is generated from a probabilistic process, and consists of multiple topics, where a topic is a probability distribution over words taken from a fixed set of vocabularies. This representation naturally captures the hidden topical structure in text, and can then be used in text mining tasks to discover topics from a large text collection.

With the explosive growth of the Web and linked data sets, some recent work on topic modeling has extended the basic topic models by taking into consideration the link structure information. The work in this area can be classified into five directions. The first line of work (e.g. PHITS-PLSA by Cohn & Hofmann, 2001; LDA-Link-Word by Erosheva, et al., 2004; Link-PLSA-LDA by Nallapati, et al., 2008) incorporates the notion of link information into the document generative model. The second line of work (relational or supervised topic models) models textual content and link separately by representing the link between documents as a binary random variable conditioned on their content (e.g. Chang & Blei, 2009). The third line of work regularizes topic models with a discrete regularizer defined based on the link structure of the data set (e.g. NetPLSI by Mei, et al., 2008). The fourth line of work (e.g. iTopicModel by Sun, et al., 2009) model the relationship between documents using a multivariate Markov Random Field (MRF). Lastly, Yang et al. (2009) proposed the PCL model, which is a discriminative model for combining link and content information for community detection.

Probabilistic topic modeling possesses many characteristics that are desirable for the Web community discovery problem. Firstly, they produce soft clustering results therefore it is possible for a Web page to belong to multiple communities (i.e. they can produce overlapping community structure). Secondly, by leveraging the textual information, they are able to provide semantic interpretation of the extracted communities.

This chapter presents a systematic review and discussions of research efforts on Web community discovery problem. The plan of the chapter is as follows. First, we present background concepts related to Web community discovery and probabilistic topic models. Next, we formulate the Web community discovery problem, and present some representative non-Bayesian community discovery algorithms. Then, we describe key concepts underlying probabilistic topic models in details, and discuss some recent work on Bayesian based methods for Web community discovery. Finally, we highlight some interesting future research directions and conclude the chapter.

BACKGROUND

Graph Terminologies

Definition 1 (Directed graph): A *directed graph* $G(V,E)$ consists of a set V of nodes (or vertices) and a set E of ordered pairs of nodes, called edges. Each edge is an ordered pair of nodes (u,v) representing a directed connection from node u to node v.

Definition 2 (In-degree and Out-degree): The *in-degree* of a node u is the number of distinct incoming edges incident to u (i.e. the number of distinct edges (v_1,u), (v_2,u), ..., (v_k, u). The *out-degree* of a node u is the number of distinct outgoing edges incident to u (i.e. the number of distinct edges (u,v_1), (u,v_2), ..., (u,v_k).

Definition 3 (Path and Distance): A *path* from node u to node v is a sequence of edges (u,u_1), (u_1,u_2), ... (u_k,v). The *length of the path* is the number of edges along the path. The *distance* from node u to node v is equal to the length of the smallest path from u to v for which such a path exists. If no path exists, the distance from u to v is defined to be infinity.

Definition 4 (Diameter): A *diameter* of a graph $G(V,E)$ is the greatest *distance* between any pair of nodes (u,v) in G.

Definition 5 (Webgraph): A *webgraph* is a directed graph $G(V,E)$ where V is a set of nodes corresponding to Web pages, and E is a set of ordered pairs (u,v) corresponding to hyperlinks from page u to page v.

Definition 6 (Directed bipartite graph): A *directed bipartite graph* is a directed graph $G(V,E)$ whose node set can be partitioned as $V = F \cup C$, with the property that every edge $e \in E$ has one end in F and the other end in C.

Definition 7 (Complete directed bipartite graph): A *complete directed bipartite graph* is a directed bipartite graph $G_C(V,E)$ whose node set can be partitioned as $V = F \cup C$, such that every node in F has an edge to every node in C. A complete bipartite graph is denoted as $G_{Cf,c}$ where $f = |F|$ and $c = |C|$.

A complete directed bipartite graph, $G_{C4,4}$ is shown in Figure 1. The node set of the graph $G_{C4,4}$ can be divided into two disjoint sets $F = \{f_1, f_2, f_3, f_4\}$ *and* $C = \{c_1, c_2, c_3, c_4\}$, such that each node f_i (i=1,2,3,4) in F has edges to every node c_j (j=1,2,3,4) in C.

Definition 8 (Dense directed bipartite graph) A *dense directed bipartite graph* is a directed bipartite graph $G_D(V,E)$ whose node set can be partitioned as $V = F \cup C$, such that a node in F must has edges to at least γ_C $(1 \leq \gamma_C \leq |C|)$ nodes in C, and at least γ_F $(1 \leq \gamma_F \leq |F|)$ nodes in F link to every node in C. A dense directed bipartite graph is denoted as $G_D\gamma_C\gamma_F$

A dense directed bipartite graph, $G_{D3,2}$ is shown in Figure 2. The node set of the graph $G_{D3,2}$ can be divided into two disjoint sets $F = \{f_1, f_2, f_3, f_4\}$ *and* $C = \{c_1, c_2, c_3, c_4\}$, such that a node f_i in F has edges to at least γ_C=3 in C, and at least γ_F=2 in F link to every node in C.

HITS Algorithm

HITS (Hypertext Induced Topic Search) (Kleinberg, 1998) is a search query dependent ranking algorithm that produces two rankings of Web

Figure 1. A complete directed bipartite graph $G_{C4,4}$

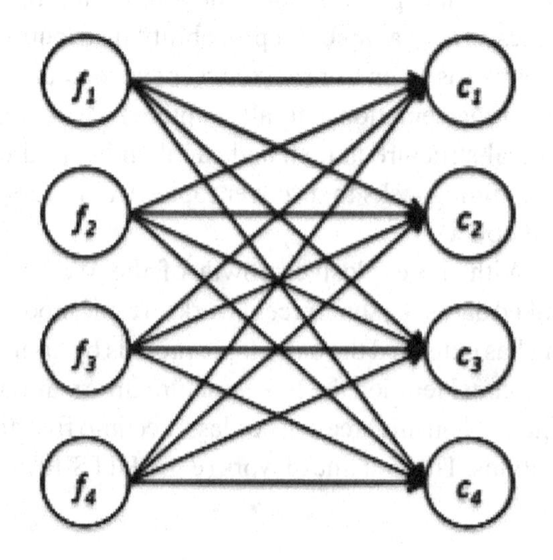

Figure 2. A dense directed bipartite graph $G_{D3,2}$.

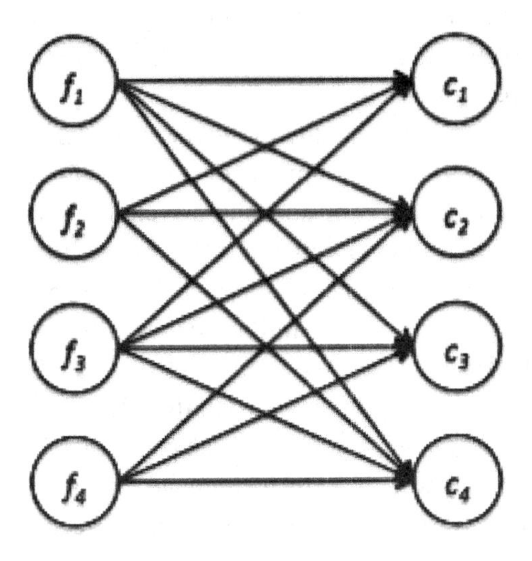

pages in response to a user query: an *authority ranking* and a *hub ranking*. An *authority* is a page with authoritative content on a topic and is linked to by many pages. A *hub* is a page with many links to good authority pages on a topic. The key idea underlying HITS is the concept of *mutual reinforcement relationship* between authority and hub pages (i.e. a good hub points to many good authorities, and a good authority is pointed to by many good hubs). Topologically, the mutual reinforcement relationship usually manifests itself as a set of *dense bipartite subgraph* $G(V,E)$; $V=F\cup C$ where F *is a set of hub pages*, and C *is a set of authority pages*.

The HITS algorithm consists of two phases. Given an input query from a user, the first phase constructs a base set S, which is a focused Web subgraph containing many good authorities using Web search engines. The second phase computes an authority score and a hub score for every page in the base set S. Let n be the number of pages in S; $G(V,E)$ denotes the hyperlink graph induced from S, and L is an adjacency matrix of the hyperlink graph $G(V,E)$. Further, let the authority score of page i be $a(i)$ and the hub score of page i be $h(i)$. We define a column vector of hub scores $h=(h(1),...,h(n))^T$ and a column vector of authority scores $\boldsymbol{a}=(a(1),...,a(n))^T$. Based on the mutual reinforcement relationship, the authority and the hub scores can be defined by

$$a = L^T h \quad (1)$$

$$h = La \quad (2)$$

The authority and hub scores in Equation (1) and (2) can be computed using an iterative algorithm (see Algorithm 1). Note that, the following equations in Algorithm 1

$$\begin{aligned} a_k &= L^T h_{k-1} \\ h_k &= La_k \end{aligned} \quad (3)$$

can be simplified to

$$\begin{aligned} a_k &= L^T L a_{k-1} \\ h_k &= LL^T h_{k-1} \end{aligned} \quad (4)$$

The two equations in (4) define the *power iteration method* for computing the *principal eigenvector* for the matrices L^TL and LL^T. Therefore, the computation of the authority and hub scores boils down to finding the principal right-hand eigenvectors of L^TL (the authority matrix) and LL^T (the hub matrix), respectively. Readers interested in detailed discussions of the HITS algorithm are directed to Kleinberg, 1998; Liu, 2007; Langville & Meyer, 2006 (see Algorithm 1).

The Small-World Phenomenon

The *small-world phenomenon* of the Webgraph is described by the fact that the diameter of the Webgraph is on average small relative to the size of the overall graph (i.e. the diameter is bounded by a polynomial in log n where n is the number of nodes in the graph). The small-world phenomenon has been reported in many subgraphs of the Web spanning from a university, a country, and a global

Algorithm 1. HTS algorithm

Let a_k, h_k denote authority and hub scores at the kth iteration

Initialize: $h_0 = (1, 1, ..., 1)^T$

Until convergence, do

$a_k = L^T h_{k-1}$

$h_k = L a_k$

$k = k + 1$

Normalize a_k and h_k (e.g. by using the 1-norm):

$a_k = a_k / \| a_k \|_1$

$h_k = h_k / \| h_k \|_1$

Web (see, for example Albert, et al., 1999; Barabasi & Albert, 1999; Broder, et al., 2000; Boldi, et al., 2002; Tamura, et al., 2007; Somboonviwat, 2008).

Due to the small diameter of the Webgraph, it is straightforward that the distance between any two nodes in a Webgraph is rather short. An important implication of the existence of relatively short distances between nodes in the Webgraph is that there exist potentially many small densely connected clusters of Web pages in the Webgraph. As we shall see later, a densely connected cluster of Web pages is the identifying characteristics of a Web community.

To demonstrate the existence of rather small distances between same types (i.e. same topics or same languages) of nodes in a Webgraph, let us consider Figure 3 which is a plot of the distance between nearest *Thai Web pages* in a Webgraph determined from a 4 million pages crawl of the Thai Web in 2004 (see Tamura, et al., 2007; Somboonviwat, 2008 for more details). Note that, the *Thai Web pages* refer to the Web pages that, based on the classification by a language classifier, are likely to be written in the Thai language. According to Figure 3, most Thai pages are located near to the other Thai pages in the Webgraph. The maximum distance between any two Thai pages in the data sets is only 10, which is relatively small compared to the size of the Thai Webgraph induced from our Thai Web dataset (the graph consists of 9,953,318 nodes and 123,836,342 non-duplicated directed

Figure 3. Distribution of distances between nearest Thai pages

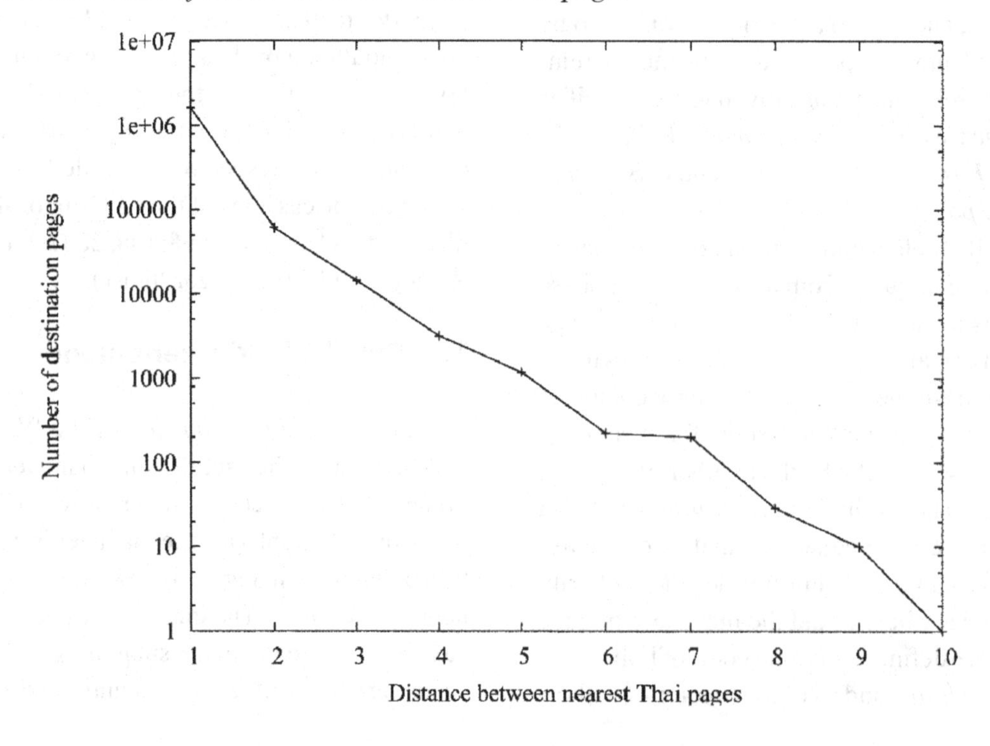

edges). This result suggests the occurrence of the small-world phenomenon in the Thai Webgraph. Therefore, we can anticipate the existence of many small dense clusters of Thai Web pages (i.e. the communities of Thai pages) in the Webgraph associated with the Thai dataset. In fact, according to a Web community extraction experiment conducted by Somboonviwat (2008), more than 67,000 communities were identified on this Thai Webgraph. The distribution of community sizes as reported in Somboonviwat (2008) is shown in Figure 4.

Probabilistic Graphical Models

Probabilistic graphical models are graphical representations of probability distributions. They provide a simple way to visualize the structure of a probabilistic model. In a probabilistic graphical model, directed graphs are used to express causal relationships between random variables. Each node in a graph represents a random variable (or a group of random variables); a shaded node represents an observed random variable. Each edge in a graph represents probabilistic relationships between random variables. A directed edge from node u to node v corresponds to a conditional distribution $p(v|u)$. A *plate notation* can be used to more compactly represent a replicated structure in the model. Figure 5(a) shows a graphical model corresponding to a joint distribution in Equation (5). The equivalent plate notation for this graph is shown in Figure 5(b). Note that, there are N observed random variables in this graph i.e. the shaded nodes $t_1, t_2, \ldots, t_{N-1}, t_N$.

Figure 4. Distribution of community sizes extracted from a Thai Web crawl

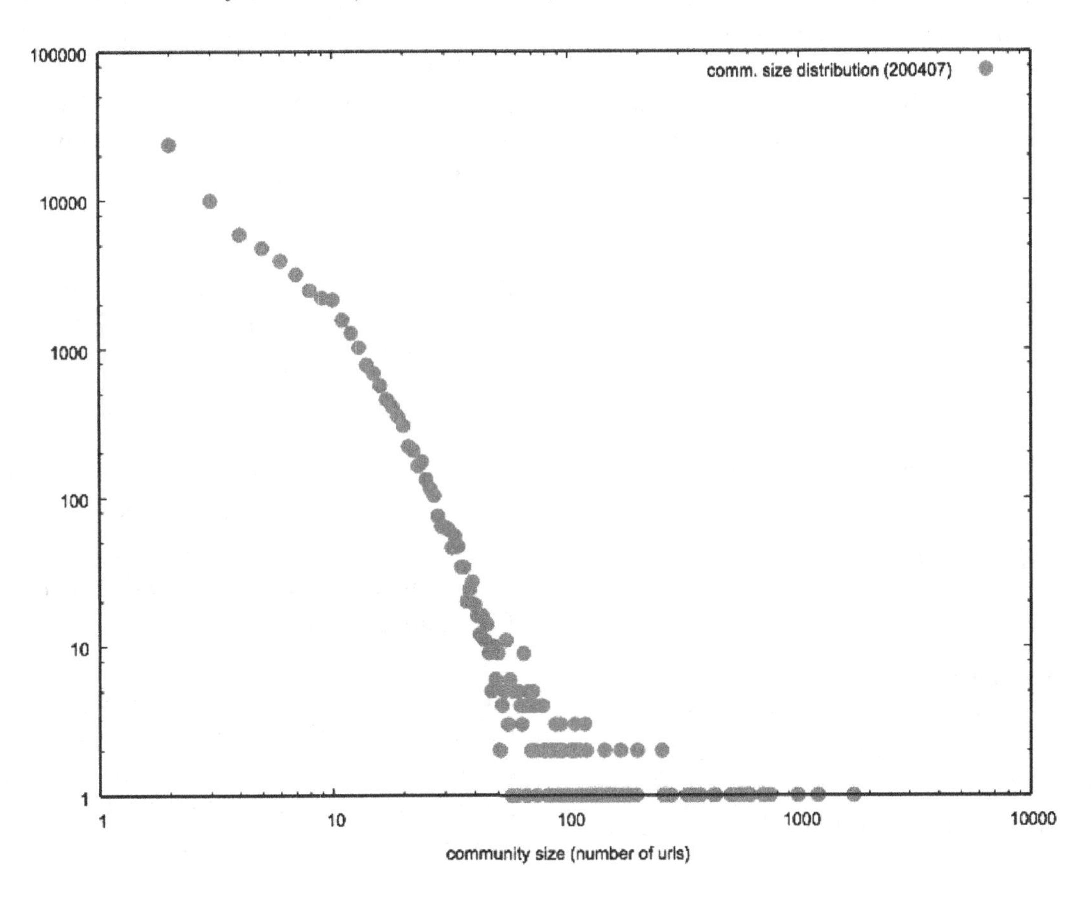

Figure 5. Graphical models: (a) directed graphical model representing the joint distribution in (5); (b) the same graphical model depicted using a plate notation

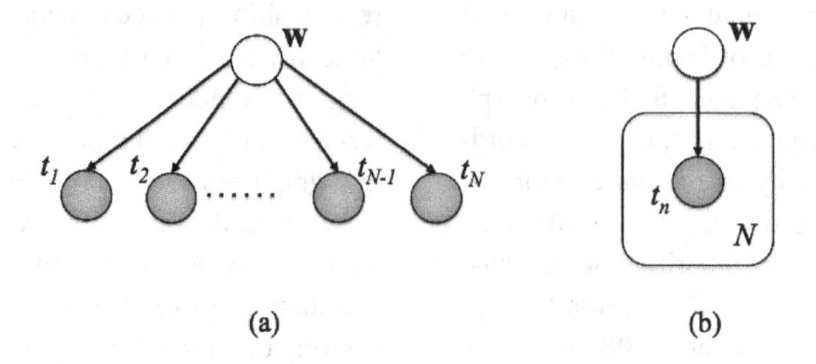

$$p(w,t_1,t_2,\ldots,t_{N-1},t_N) = p(w)\prod_{n=1}^{N} p(t_n|w) \qquad (5)$$

The Dirichlet Distribution

The *Dirichlet distribution* is an exponential family distribution that is a conjugate prior of the multinomial distribution. Given a multinomial distribution $\mu=(\mu_1, \ldots, \mu_K)$, such that $0 \le \mu_k \le 1$, $\sum_k \mu_k = 1$; and a parameter $\alpha=(\alpha_1, \ldots, \alpha_K)^T$. A K dimensional Dirichlet distribution for the multinomial μ is defined over a simplex of dimensionality $K-1$, and its normalized form is defined by

$$Dir(\mu|\alpha) = \frac{\Gamma(\sum_{k=1}^{K} a_k)}{\prod_{k=1}^{K}\Gamma(\alpha_k)} \prod_{k=1}^{K} \mu_k^{a_k - 1} \qquad (6)$$

Here $\Gamma(x)$ is a gamma function defined by

$$\Gamma(x) = \int_0^\infty u^{x-1} e^{-u} du \qquad (7)$$

The parameter $\alpha=(\alpha_1, \ldots, \alpha_K)^T$ of the Dirichlet distribution controls the mean shape and the sparsity of the Dirichlet distribution as illustrated in Figure 6. For $\alpha < 1$ (see Figure 6[a]), the modes of the distribution are located at the corners of the simplex, leading to more sparse distribution of the density. For $\alpha > 1$ (see Figure 6[c]), the mode of the distribution is located away from the corners of the simplex, leading to more smooth distribution of the density.

Finally, we note that because the Dirichlet distribution is a conjugate prior of the multinomial distribution, therefore it follows that given a multinomial distribution μ, the posterior distribution of μ has the same functional form as the prior, i.e. the Dirichlet distribution. The conjugacy of the Dirichlet priors lead to greatly simplified Bayesian analysis.

NON-BAYESIAN APPROACHES TO WEB COMMUNITY DISCOVERY

Problem Formulation

Conceptually, a *community* is a set of entities (e.g. Web pages, people, organizations) sharing a common interest. Based on a conceptual definition of a community given in Liu (2007), we define a *Web community* as follows.

Definition 9 (Web Community): Given a finite set of $P = \{p_1, p_2, \ldots, p_n\}$ of Web pages, a *community* is a pair $C=(T, M)$, where T is the *community topic* and $M \subseteq P$ is the set of all pages in P that shares the topic T. If $p_i \in M$, then p_i is said to be a *member of the community* C.

Figure 6. The three dimensional Dirichlet distribution. The two horizontal axes are coordinates in the plane of the simplex, and the vertical axis is the density. (a) $\alpha = (0.1,0.1,0.1)^T$, (b) $\alpha = (1,1,1)^T$, (c) $\alpha = (10,10,10)^T$.

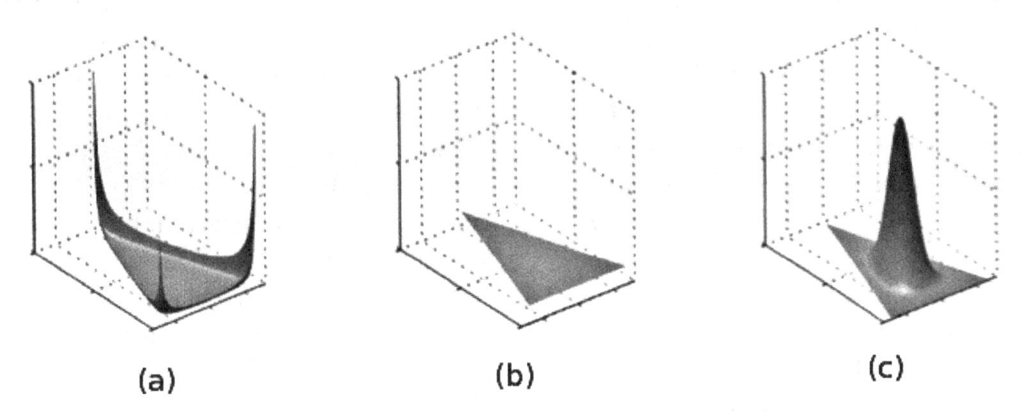

Based on the concepts of Web communities defined above, we can formalize the task of *link based Web community discovery* as follows.

Definition 10 (Link Based Web Community Discovery): Given a collection of Web pages P and a Web graph G induced from P, the *Link Based Web Community Discovery task* is to extract k Web communities $\{C_1, ..., C_k\}$, where each C_i is a subgraph in G with dense connections and coherent content.

Definition 9 and 10 are conceptual definitions of a Web community and link based Web community discovery task. Different Web community discovery algorithms define their own operational definition of a Web community based on an assumption made regarding the specific forms of manifestation of Web communities in the dataset.

Bipartite Cores-Based Approach

As shown by Gibson et al. (1998), HITS can detect some Web communities based on broad topic queries using eigenvector computation. This result suggests that the footprint of a Web community can be recognized as a *dense bipartite subgraph* arises from the co-citation process, where related pages do not reference one another but are frequently referenced together. This could be caused by the fact that the Web sites are competitors or they do not share a common viewpoint, or simply because they are not aware of each other. Based on this observation, Kumar et al. (1999) argue that co-citation is a characteristic of well-developed explicitly known communities and is also an early indicator of emerging Web communities. Nevertheless, eigenvector computation as used by Gibson et al. (1998) is inefficient for iterating all bipartite Web communities. Kumar et al. (1999) proposed a heuristic based technique for automatically and efficiently enumerating all bipartite core communities from a large Web crawl. They used the term *trawling the Web* to denote this process. The notion of dense bipartite subgraphs representing Web communities can be formulated mathematically as follows.

Let $K_{ij} = G_C(V,E)$ be a *complete bipartite graph* whose nodes can be divided into two sets denoted F and C (where $V = F \cup C$; $|F| = i$, $|C| = j$), and every directed edge in E is directed from a node $f \in F$ to a node $c \in C$. Define an *(i, j)* core as a complete bipartite graph K_{ij} with optionally additional edges other than edges from F to C. According to this definition, an *(i, j) core* community contains a set of i pages all of which point to another set of j pages. Intuitively, the i pages are pages created by members of the community who wants to link their pages to the most valuable resources (i.e. the *j* pages) for that community. Due to this reason, the *i* pages are called *fans*, and the *j* pages are called *centers*.

The trawling algorithm is based on two assumptions. The first assumption states that *any sufficiently strong Web community will be highly likely to contain an (i, j) core.* The second assumption states that *almost all occurrences of (i, j) cores are due to the existence of a community rather than random.* Based on these two assumptions, the main idea of the trawling algorithm is to identify a community by finding its core, and then to expand the core to the rest of the community. The trawling algorithm is presented in Algorithm 2. Note that dense subgraph extraction arises in many real-world graph analysis problems. Gibson et al. (2006) provides a detailed discussion on this problem, especially in the context of Web community discovery.

Maximum Flow-Based Approach

Flake et al. (2000) define a Web community based on density of links between Web pages. Given a Webgraph $G=(V, E)$, a community is a subset C of V such that each $v \in C$ has at least as many neighbors in C as in $V - C$. According to this definition, identifying a community is intractable because it maps into a family of NP-complete graph partitioning problems (Garey & Johnson, 1979). Flake et al. (2000) show that by assuming the existence of one or more seed websites and utilizing regularities of the Web graph (Bernado, et al., 1998; Barabasi, et al., 1999), a community can be identified by calculating the $s-t$ maximum flow of G and identifying nodes that are reachable from s to be the Web community.

The $s - t$ maximum flow problem is defined as follows. Given a directed graph $G=(V, E)$, with a source node $s \in V$, a target node $t \in V$, and edge capacities $c(u,v) > 0$, find the maximum flow that can be directed from the source node s to the target node t, without exceeding the capacity constraints on any edge. The Max Flow-Min Cut theorem of Ford and Fulkerson (1956) proves that the maximum flow of the network is identical to the capacity of the minimum cut that separates s and t. Many implementations exist for solving the maximum flow problem in polynomial time (Ravindra, et al., 1997). The Max Flow Web Community Discovery algorithm (Flake, et al., 2000, 2002) is illustrated in Algorithm 3. The procedure Max-Flow-Community (G, S, k) augments the

Algorithm 2. Trawling algorithm

STEP 1: Identifying the community cores *(i, j)*
STEP 1-1: Iterative pruning.
Until no node is qualified for deletion, do
Delete potential fans with out-degree less than *j*
Delete potential centers with in-degree less than *i*
STEP 1-2: Inclusion-exclusion pruning // Inclusion-exclusion based on fan pages
Until no further inclusion/exclusion, do
Choose a fan *u* with out-degree exactly *j*
Let $\Gamma(u)$ be a set of centers to which *u* points to
If there are *i-1* other fans all pointing to each center in $\Gamma(u)$
Include a new *(i, j)* core to the output
Otherwise
Exclude *u* from further contention (as a fan)
STEP 1-3: Exhaustive enumeration
Fix *j*, start with the *(1, j)* cores, one can construct all *(2, j)* cores by checking every fan that also points to any center in a *(1, j)* core. Similarly, all *(3, j)* can be constructed by checking every fan that points to any center in a *(2, j)* core. The procedure proceeds in a similar manner until the desired number of communities is extracted from the graph.
STEP 2: Core Expansion
Use algorithms, such as HITS (Kleinberg, 19998) or Clever (Chakrabarti et al., 1998), to expand the core.

input Webgraph G with an artificial source s and an artificial sink t. After augmenting the graph, a residual flow is generated by calling a maximum flow procedure Max-Flow (G, s, t). Then, all nodes reachable from s through non-zero positive edges form the output community. In comparison to the bipartite cores based approach, the max-flow based community discovery can extract larger, more complete communities. Like the bipartite-based approach, however, it cannot find the topic, and the relationships of Web communities.

Other Non-Bayesian Approaches

A great deal of work has been devoted to discovering communities in networks based on non-Bayesian methods. Gibson et al. (1998) show that the HITS algorithm can be used to extract some communities by computing the non-principal eigenvectors of the authority and hub matrices. Their results suggest that a Web community can manifest itself as a dense bipartite subgraph. Based on this intuition, Kumar et al. (1999) propose a heuristic based technique for finding all bipartite core communities efficiently from a large Web crawl dataset. Flake et al. (2000) cast the Web community discovery problem into the framework of the maximum flow model, and presented a maximum flow community discovery algorithm (see Algorithm 3). Toyoda and Kitsuregawa (2001, 2003) introduce a Web community chart that can identify relationships among Web communities, and present an algorithm to extracting relationship between Web communities and their evolution from a series of Japanese Web archives. Other recent Web community detection methods include e.g. Ino et al. (2005) and Anderson and Lang (2006).

Communities in networks may also be viewed as clusters, there are numerous methods from graph partitioning, clustering, and matrix reordering that have been applied to community detection. These include, for example, the METIS method (Karypis, et al., 1999), spectral analysis methods (Chung, 1997), multi-level clustering (Dhillon, et al., 2007), co-clustering (Dhillon, et al., 2007), matrix factorization (Zhu, et al., 2007), and Kernel fusion (Yu, et al., 2009). Readers interested in detailed discussions of Web communities analysis and construction are directed to Zhang et al. (2006).

Although link based community discovery has been very successful and applied in many real-world applications, a limitation exist when we not only want to know in an automated fashion the community memberships of each vertex in the network but also the semantics of the extracted

Algorithm 3. Max flow community discovery (Flake, et al., 2002)

Algorithm Find-Community
input: a set S of seed Web pages
while number of iteration is less than desired
do
Set $G = (V,E)$ to a fixed depth crawl from S.
Set $k = |S|$.
call: C =Max-Flow-Community(G, S, k)
Rank all $v \varepsilon C$ by number of edges in C.
Add the highest ranked non-seed nodes to S.
end while
output: all $v \varepsilon V$ still connected to the source s.

procedure Max-Flow-Community
input: graph $G=(V,E)$; a set $S \subset V$; integer k.
Create artificial nodes s and t, and add to V.
for all $v \varepsilon S$ **do**
Add (s,v) to E with $c(s,v)=\infty$.
end for
for all $(u,v) \varepsilon E, u \neq s$ **do**
Set $c(u, v) = k$.
if $(v, u) \notin E$ **then**
add (v, u) to E with $c(v, u) = k$.
end if
end for
for all $v \varepsilon V, v \notin S \cup \{s, t\}$ **do**
Add (v,t) to E with $c(v,t)=1$.
end for
call: Max-Flow(G, s, t).
output: all $v \varepsilon V$ still connected to s.

communities. One promising approach to providing the semantics in Web community extraction is the probabilistic topic modeling approach (e.g. Nallapati, et al., 2008; Chang & Blei, 2009; Mei, et al., 2008; Sun, et al., 2009; Yang, et al., 2009). The remaining of this chapter describes the basic concepts of topic modeling and recent efforts on the application of topic modeling to Web community discovery problem.

BAYESIAN APPROACHES TO WEB COMMUNITY DISCOVERY

Topic modeling has been successfully applied to discovering the topical structure of large text corpora. In this section, we first introduce two well-known traditional topic modeling methods, i.e. the Probabilistic Latent Semantic Indexing (PLSI) by Hoffman, 1999, and the Latent Dirichlet Allocation (LDA) by Blei et al. (2003). Then, we discuss several enhancements to the traditional topic modeling methods for Web community discovery.

Probabilistic Topic Modeling

Topic modeling (Hofmann, 1999, 2001; Blei, et al., 2003; Griffiths & Steyvers, 2002, 2003, 2004) is a suite of algorithms that extract a set of latent topics hidden inside a text corpus. The key idea of topic models is to represent a document in a latent space with a finite mixture model of *k* topics (i.e. a document can be viewed as consisting of multiple topics), where each topic is a probabilistic distribution over a set of words. The topic model defines a *generative model* for generating documents. The parameters in the generative model can be estimated by fitting the data with the model using posterior computation algorithms such as Gibbs sampling, and variational inference algorithms. Two well-known topic models are the Probabilistic Latent Semantic Analysis (PLSA) proposed by Hoffman (1999) and the Latent Dirichlet Allocation (LDA) proposed by Blei et al. (2003).

The PLSA model (Hofmann, 1999) is based on a latent variable model for co-occurrence data, called the aspect model (Hofmann, et al., 1999). The aspect model associates an unobserved latent class variable $z \in Z = \{z_1, ..., z_K\}$ with each occurrence of a word $w \in W = \{w_1, ..., w_M\}$ in a document $d \in D = \{d_1, ..., d_N\}$. A generative model of PLSA is defined as follows.

- Pick a document *d* with probability *P(d)*.
- Pick a latent topic z with probability *P(z|d)*.
- Generate a word *w* with probability *P(w|z)*.

The graphical model for PLSA is as shown in Figure 7(a). The generative process of the PLSA can be translated into a joint probability model as follows.

$$P(d, w) = P(d)P(w|d), where \tag{8}$$

$$P(w|d) = \sum_{z \in Z} P(w|z)P(z|d) \tag{9}$$

Let *n(d,w)* denotes the number of occurrences of the term *w* in a document *d*. The log likelihood of a document collection *D* to be generated is given by

$$L(D) = \sum_{d \in D} \sum_{w \in W} n(d, w) \log \sum_{j=1}^{K} p(w \mid z_j) p(z_j \mid d) \tag{10}$$

The posterior computation of PLSA can be done using the standard Expectation Maximization (EM) (Dempster, et al., 1977) algorithm. The EM algorithm alternates between an *expectation step* (E-step) and a maximization step (M-step). In the E-step, the posterior probabilities are calculated for the latent variables using the current estimates of the parameters. In the M-step, the model parameters are updated based on a local maximum

Figure 7. The graphical models for (a) PLSA, (b) LDA

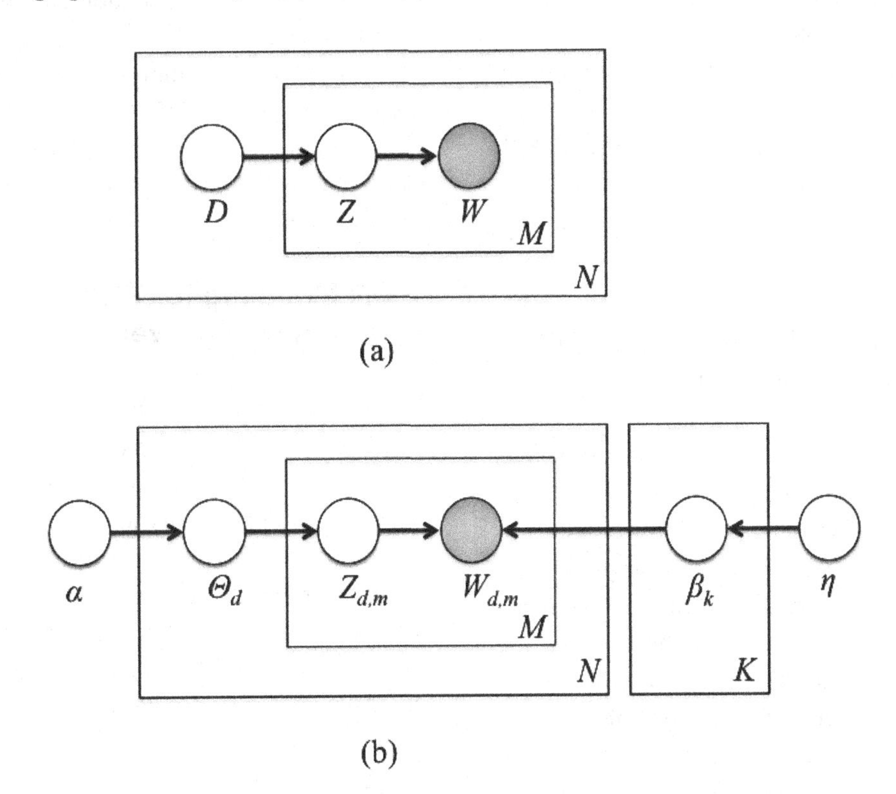

of the log-likelihood in (10), which depends on the posterior calculated in the latest E-step. The posterior probabilities and the re-estimation equations for the E-step and the M-step for PLSA are given in Equation (11) - (14) below.

Expectation Step (E-step):

$$P(z \mid d, w) = \frac{P(z)P(d|z)P(w|z)}{\Sigma_{z'} P(z')P(d|z')P(w|z')} \tag{11}$$

Maximization Step (M-step):

$$P(w \mid z) = \frac{\Sigma_d n(d,w)P(z|d,w)}{\Sigma_{d,w'} n(d,w')P(z|d,w')} \tag{12}$$

$$P(d \mid z) = \frac{\sum_w n(d,w)P(z \mid d,w)}{\sum_{d',w} n(d',w)P(z \mid d',w)} \tag{13}$$

$$P(z) = \frac{1}{R}\sum_{d,w} n(d,w)P(z \mid d,w), R \equiv \sum_{d,w} n(d,w) \tag{14}$$

According to Equation (10), It can be seen that PLSA estimates the probability distribution of each document on the latent topics independently. There are $NK+MK$ parameters $\{P(z_j|d), P(w|z_j)\}$ in PLSA model, consequently the number of parameters grow linearly with the number of training documents N. This behavior suggests that PLSI is prone to overfitting (Cai, et al., 2008).

This problem has been addressed in the LDA model. LDA model treats the probability distribution of each document on the latent topics as a *K-parameter* hidden latent variables generated from the same Dirichlet distribution. The generative process of LDA is given below.

- Select a document d, with probability $P(d)$.
- Pick a latent topic $z_{d,m}$:
  - Generate a topic proportion $\theta_d \sim Dir(\alpha)$,
  - Pick a latent topic $z_{d,m}$ with probability $P(z_{d,m} \mid \theta_d)$.
- Generate a word w with probability $P(w_{d,m}|z_{d,m})$.

where, Dir(α) is a K-dimensional Dirichlet distribution. The graphical model for LDA is as shown in Figure 7(b).

Formally, let the topics be $\beta_{1:K}$, where each β_K is a distribution over a set of words ($\beta_k \sim Dir(\eta)$). The topic proportions for the dth document are θ_d, where $\theta_{d,k}$ is the topic proportion for topic k in document d. The topic assignments for the dth document are z_d, where $z_{d,m}$ is the topic assignment for the mth word in document d. And, the observed words for document dth are w_d, where $w_{d,m}$ is the m*th* word in document d. According to this notation, the LDA generative process corresponds to the following joint distribution of the hidden and observed variables (Blei, et al., 2003).

$$p(\beta_{1:K}, \theta_{1:N}, z_{1:N}, w_{1:N}) = \prod_{j=1}^{K} p(\beta_j) \prod_{d=1}^{D} p(\theta_d) (\prod_{m=1}^{M} p(z_{d,m} \mid \theta_d) p(w_{d,m} \mid \beta_{1:K}, z_{d,m})) \quad (15)$$

The posterior of LDA is

$$p(\beta_{1:K}, \theta_{1:N}, z_{1:N} \mid w_{1:N}) = \frac{p(\beta_{1:K}, \theta_{1:N}, z_{1:N}, w_{1:N})}{p(w_{1:N})} \quad (16)$$

The exact value of this posterior is intractable to compute due to the exponentially large number of possible latent topic structures. Generally, the posterior of LDA can be computed using either sampling based algorithms (e.g. Gibbs sampling, see Heinrich, 2009) or variational algorithms (e.g. coordinate ascent variational inference, see Hoffman, et al., 2010).

Notice that the number of parameters in a *K-topic* LDA is $K+MK$, which does not grow with the number of documents N in the corpus. Consequently, LDA does not have the overfitting issue as PLSA.

Topic Modeling for Web Community Discovery

Recall that, previously in the context of link based community discovery, a Web community is defined as a set of Web pages on a *specific topic*. Now, we turn our discussion to the discovery of Web community based on the Bayesian approach. Let us first re-define some notations and formulate the Web community discovery problem in our new context. This formulation mainly follows the work of Mei et al. (2008).

Problem Formulation

Definition 11 (Document): A *document d* in a text collection D is a bag of words $\{w_1 w_2 \ldots w_{|d|}\}$, where w_i is a word from a fixed vocabulary. We use $n(d,w)$ to denote the occurrences of word w in d.

Definition 12 (Network): A *network* associated with a text collection D is a graph $G(V, E)$, where V is a set of nodes and E is a set of edges. A node $u \in V$ corresponds to a document $d_u \in D$. An edge $<u,v>$ corresponds to a binary relation between nodes u and v, and $w(u,v)$ is used to denote the weight of $<u,v>$. Note that, an edge can be either directed or undirected.

Definition 13 (Topic): A semantically coherent *topic* in a text collection D is represented by a *topic model* θ, which is a probability distribution of words.

Definition 14 (Topical Web Community Discovery): Given a collection D *and* its associated network structure G, the task of *topical Web community discovery* is to extract k topical communities $\{V_1, \ldots, V_k\}$, where each $V_i = \{v_{i1}, \ldots, v_{il}\}$ has a coherent semantic summary θ_i, which is one of the k major topics in D.

A Survey of Topic Modeling-Based Methods for Web Community Discovery

Traditional topic models, e.g. PLSA (Hofmann, 1999) and LDA (Blei, et al., 2003), analyze content of a given corpus to uncover hidden topics within the corpus. These topics can be naturally interpreted as a community. However, the content analysis alone cannot accurately identify Web communities because the content information usually contains words that are irrelevant to the target topics.

Many link based probabilistic models have been developed for Web community discovery. Cohn et al. (2000) proposed PHITS, a PLSA-like topic model for identification of communities of hubs and authorities in a document network. The PHITS model defines generative processes for both text and hyperlinks. The PHITS model assigns high probability of hyperlinking to a document d with respect to a topic k if the document d is pointed to by several documents that are relevant to the topic k (which are those documents sharing the same word distribution as d). Based on this same assumption, Erosheva et al. (2004) proposed the Link-LDA model that uses LDA as the basic generative building block instead of the PLSA. Recently, Yang et al. (2010) has proposed a probabilistic model for *directed* network community detection, called PPL that captures both incoming links and outgoing links differentially.

Next, let us describe the work that combines link and content analysis into the probabilistic topic modeling framework. The work in this area can be classified into five directions.

The first line of work (e.g. PHITS-PLSA by Cohn & Hofmann, 2001; LDA-Link-Word by Erosheva, et al., 2004; Link-PLSA-LDA by Nallapati, et al., 2008; Dietz, et al., 2007; Gruber, et al., 2008) incorporates the notion of link information into the document generative model. Such methods need expert knowledge in order to translate the semantic of links between documents and embed it into the model, and thus are not generalize to different types of datasets.

The second line of work (relational or supervised topic models) models textual content and link separately by representing the link between documents as a binary random variable conditioned on their content. Prior work in this direction is the Relational Topic Model or RTM by Chang and Blei (2009). Note that, the RTM model does not support weighted graph.

The third line of work regularizes topic models with a discrete regularizer defined based on the link structure of the data set (NetPLSA by Mei, et al., 2008). NetPLSA combines link and text information into a unified framework via the combination of two objective functions, one based on textual data and another one based on the network structure.

The fourth line of work (iTopicModel by Sun, et al., 2009) models the relationship between documents using a multivariate Markov Random Field (MRF). The iTopicModel constructs a two-layer graphical model structure. The top layer is a multivariate MRF that capture the dependency relationship among documents in the network. The bottom layer is a traditional topic model. The assumption underpinning the iTopicModel is the *topical locality* of documents in the network i.e. that the documents in the same neighborhood in the network should be topically similar (Davidson, 2000).

Lastly, the fifth direction, Yang et al. (2009) proposed the PCL model, which is a discriminative model for combining link and content information for community detection.

FUTURE RESEARCH DIRECTIONS

A central problem in the study of semantically coherent implicit Web community that we have discussed so far is how to efficiently extract those communities given a snapshot of a Web dataset. Because the Web is dynamic and constantly changing, it is crucial to understand the dynamic nature of the Web. Therefore, the *temporal* aspect of Web community identification problem is one of important research directions to pursue. Another key research problem is the evaluation of the community discovery algorithms and the *quality* of extracted communities. Mei et al. (2008) compared the performance of NetPLSA with the traditional PLSA using the cut edge weights and ratio cut metrics. However, these metrics only consider the quality based on the link density of the community. Leskovec et al. (2010) proposed an idea to consider the quality of community as a function of its size and conducted large-scale empirical comparison of algorithms for network community detection. However, they only considered non-Bayesian methods.

CONCLUSION

In this chapter, we have explored two complementary approaches to Web community identification: non-Bayesian and Bayesian. The non-Bayesian approaches exploit merely the network structure and extract communities based on some identifying signatures of Web communities. Although they have proved successful, the major limitation of the non-Bayesian based approaches is the lack of semantic interpretation for the extracted communities. The Bayesian approaches on the other hand can be enhanced to combine both textual and link information, and have become popular as algorithmic tools for extracting Web communities from large document network corpora. There is still a lot of opening research issues in this area, and we have identified two interesting future research directions: community dynamics and evaluation of Web community extraction methods.

REFERENCES

Albert, R., Jeong, H., & Barabasi, A. (1999). The diameter of the world wide web. *Nature*, *401*(130).

Anderson, R., & Lang, K. (2006). Communities from seed sets. In *Proceedings of the 15th International Conference on World Wide Web (WWW 2006)*. IEEE.

Barabasi, A., & Albert, R. (1999). Emergence of scaling in random networks. *Science*, 286.

Bernardo, A. H., Peter, P., James, E. P., & Rajan, M. L. (1998). Strong regularities in world wide web surfing. *Science*, *280*(5360), 95–97. doi:10.1126/science.280.5360.95

Blei, D. M., Ng, A. Y., & Jordan, M. I. (2003). Latent dirichlet allocation. *Journal of Machine Learning Research*, *3*, 993–1022.

Boldi, P., Codenotti, B., Santini, M., & Vigna, S. (2002). Structural properties of the African web. In *Proceedings of the 11th International Conference on World Wide Web (WWW 2002)*. IEEE.

Broder, A., Kumar, R., Maghoul, F., Raghavan, P., Rajagopalan, S., & Stata, R. … Wiener, J. (2000). Graph structure in the web. In *Proceedings of the 9th International Conference on World Wide Web (WWW 2000)*. IEEE.

Cai, D., Mei, Q., Han, J., & Zhai, C. (2008). Modeling hidden topics on document manifold. In *Proceedings of ACM 17th Conference on Information and Knowledge Management (CIKM 2008)*. ACM Press.

Chakrabarti, S., Dom, B., Gibson, D., Kumar, R., Raghavan, P., Rajagopalan, S., & Tomkins, A. (1998). Experiments in topic distillation. In *Proceedings of the SIGIR Workshop on Hypertext Information Retrieval on the Web*. ACM.

Chang, J., & Blei, D. (2009). Relational topic models for document networks. In *Proceedings of the 12th International Conference on Artificial Intelligence and Statistics (AISTATS 2009)*. IEEE.

Chung, F. R. K. (1997). *Spectral graph theory*. New York, NY: AMS Bookstore.

Cohn, D., & Chang, H. (2000). Learning to probabilistically identify authoritative documents. In *Proceedings of the International Conference on Machine Learning (ICML 2000)*. ICML.

Cohn, D., & Hofmann, T. (2001). The missing link - A probabilistic model of document content and hypertext connectivity. In *Proceedings of the International Conference on Neural Information Processing Systems (NIPS 2001)*. NIPS.

Davidson, B. D. (2000). Topical locality in the web. In *Proceedings of the 23rd Annual International ACM SIGIR Conference on Research and Development in Information Retrieval (SIGIR 2000)*. ACM Press.

Dempster, A., Laird, N., & Rubin, D. (1977). Maximum likelihood from incomplete data via the EM algorithm. *Journal of the Royal Statistical Society. Series B. Methodological*, *39*, 1–38.

Dhillon, I., Guan, Y., & Kullis, B. (2007). Weighted graph cuts without Eigen vectors: A multilevel approach. *IEEE Transactions on Pattern Analysis and Machine Intelligence*, *29*(11), 1944–1957. doi:10.1109/TPAMI.2007.1115

Dhillon, I. S., Mallela, S., & Modha, D. S. (2003). Information-theoretic co-clustering. In *Proceedings of the 9th ACM SIGKDD International Conference on Knowledge Discovery and Data Mining (KDD 2003)*. ACM Press.

Dietz, L., Bickel, S., & Scheffer, T. (2007). Unsupervised prediction of citation influences. In *Proceedings of the International Conference on Machine Learning (ICML 2007)*. ICML.

Erosheva, E., Fienberg, S., & Lafferty, J. (2004). Mixed membership models of scientific publications. *Proceedings of the National Academy of Sciences of the United States of America*, *101*, 5220–5227. doi:10.1073/pnas.0307760101

Flake, G. W., Lawrence, S., & Gile, C. L. (2000). Efficient identification of web communities. In *Proceedings of the 6th ACM SIGKDD International Conference on Knowledge Discovery and Data Mining (KDD 2000)*. ACM Press.

Flake, G. W., Lawrence, S., Gile, C. L., & Coetzee, F. (2002). Self-organization of the web and identification of communities. *IEEE Computer*, *35*(3), 66–71. doi:10.1109/2.989932

Ford, L., & Fulkerson, D. (1956). Maximal flow through a network. *Canadian Journal of Mathematics*, *8*(3), 399–404. doi:10.4153/CJM-1956-045-5

Garey, M. R., & Johnson, D. S. (1979). *Computers and intractability: A guide to the theory of NP-completeness*. New York, NY: W. H. Freeman.

Gibson, D., Kleinberg, J., & Raghavan, P. (1998). Inferring web communities from link topology. In *Proceedings of 9th ACM Conference on Hypertext and Hypermedia*. ACM Press.

Gibson, D., Kumar, R., McCurley, K. S., & Tomkins, A. (2006). Dense subgraph extraction. In *Mining Graph Data* (pp. 411–441). New York, NY: Wiley. doi:10.1002/9780470073049.ch16

Griffiths, T. L., & Steyvers, M. (2002). A probabilistic approach to semantic representation. In *Proceedings of the 24th Annual Conference of the Cognitive Science Society*. IEEE.

Griffiths, T. L., & Steyvers, M. (2003). Prediction and semantic association. In *Neural Information Processing Systems*. Cambridge, MA: MIT Press.

Griffiths, T. L., & Steyvers, M. (2004). Finding scientific topics. *Proceedings of the National Academy of Sciences of the United States of America*, *101*, 5228–5235. doi:10.1073/pnas.0307752101

Gruber, A., Rosen-Zvi, M., & Weiss, Y. (2008). Latent topic models for hypertext. In *Proceedings of the 24th Conference on Uncertainty in Artificial Intelligence (UAI 2008)*. UAI.

Heinrich, G. (2009). *Parameter estimation for text analysis. Technical Report*. Darmstadt, Germany: Fraunhofer IGD.

Hoffman, M., Blei, D., & Bach, F. (2010). Online learning for latent dirichlet allocation. *Neural Information Processing Systems*. Retrieved from http://www.cs.princeton.edu/~blei/papers/HoffmanBleiBach2010b.pdf

Hofmann, T. (1999). Probabilistic latent semantic analysis. In *Proceedings of the Fifteenth Conference on Uncertainty in Artificial Intelligence*. IEEE.

Hofmann, T. (2001). Unsupervised learning by probabilistic latent semantic analysis. *Machine Learning Journal*, *42*(1), 177–196. doi:10.1023/A:1007617005950

Hofmann, T., Puzicha, J., & Jordan, M. I. (1999). Advances in Neural Information Processing Systems: *Vol. 11. Unsupervised learning from dyadic data*. Cambridge, MA: MIT Press.

Ino, H., Kudo, M., & Nakamura, A. (2005). Partitioning of web graphs by community topology. In *Proceedings of the 14th International Conference on World Wide Web (WWW 2005)*. IEEE.

Karypis, G., & Kumar, V. (1999). Parallel multilevel k-way partitioning for irregular graphs. *SIAM Review*, *41*(2), 278–300. doi:10.1137/S0036144598334138

Kleinberg, J. (1998). Authoritative sources in a hyperlinked environment. In *Proceedings of 9th ACM-SIAM Symposium on Discrete Algorithms*. ACM Press.

Kumar, R., Raghavan, P., Rajagopalan, S., & Tomkins, A. (1999). Trawling the web for emerging cyber-communities. In *Proceedings of the 8th International Conference on World Wide Web*. IEEE.

Langville, A. N., & Meyer, C. D. (2006). *Google's PageRank and beyond: The science of search engine rankings*. Princeton, NJ: Princeton University Press.

Leskovec, J., Lang, K. J., & Mahoney, M. W. (2010). Empirical comparison of algorithms for network community detection. In *Proceedings of the 19th International Conference on World Wide Web (WWW 2010)*. IEEE.

Li, L., Otsuka, S., & Kitsuregawa, M. (2010). Finding related search engine queries by web community based query enrichment. *World Wide Web (Bussum)*, *13*(1-2), 121–142. doi:10.1007/s11280-009-0077-1

Liu, B. (2007). *Web data mining: Exploring hyperlinks, contents, and usage data*. Berlin, Germany: Springer.

Mei, Q., Deng, C., Zhang, D., & Zhai, C. (2008). Topic modeling with network regularization. In *Proceedings of the 17th International Conference on World Wide Web (WWW 2008)*. IEEE.

Nallapati, R. M., Ahmed, A., Xing, E. P., & Cohen, W. W. (2008). Joint latent topic models for text and citations. In *Proceedings of The 14th ACM SIGKDD International Conference on Knowledge Discovery and Data Mining (KDD 2008)*. ACM Press.

Otsuka, S., Toyoda, M., Hirai, J., & Kitsuregawa, M. (2004). Extracting user behavior by web communities technology on global web logs. In *Proceedings of the 15th International Conference on Database and Expert Systems Applications (DEXA 2004)*. DEXA.

Pierrakos, D., Paliouras, G., Papatheodorou, C., Karkaletsis, V., & Dikaiakos, M. D. (2003). Web community directories: A new approach to web personalization. In *Proceedings of the 1st European Web Mining Forum (EWMF 2003)*. EWMF.

Somboonviwat, K. (2008). *Research on language specific crawling and building of Thai web archive*. (Unpublished Doctoral Dissertation). University of Tokyo. Tokyo, Japan.

Sun, Y., Han, J., Gao, J., & Yu, Y. (2009). iTopicModel: Information network-integrated topic modeling. In *Proceedings of 2009 International Conference on Data Mining (ICDM2009)*. ICDM.

Tamura, T., Somboonviwat, K., & Kitsuregawa, M. (2007). A method for language-specific web crawling and its evaluation. *Systems and Computers in Japan*, *38*(2), 10–20. doi:10.1002/scj.20693

Toyoda, M., & Kitsuregawa, M. (2001). Creating a web community chart for navigating related communities. In *Proceedings of the 12th ACM Conference on Hypertext and Hypermedia (HT 2001)*. ACM Press.

Toyoda, M., & Kitsuregawa, M. (2003). Extracting evolution of web communities from a series of web archives. In *Proceedings of the 14th ACM Conference on Hypertext and Hypermedia (HT 2003)*. ACM Press.

Yang, T., Chi, Y., Zhu, S., & Jin, R. (2010). Directed network community detection: A popularity and productivity link model. In *Proceedings of the 2010 SIAM International Conference on Data Mining (SDM 2010)*. SIAM.

Yang, T., Jin, R., Chi, Y., & Zhu, S. (2009). Combining link and content for community detection: A discriminative approach. In *Proceedings of The 15th ACM SIGKDD International Conference on Knowledge Discovery and Data Mining (KDD 2009)*. ACM Press.

Yu, S., Moor, B. D., & Moreau, Y. (2009). Clustering by heterogeneous data fusion: Framework and applications. In *Proceedings of the NIPS Workshop*. NIPS.

Zhang, Y., Xu Yu, J., & Hou, J. (2006). *Web communities: Analysis and construction*. Berlin, Germany: Springer.

Zhu, S., Yu, K., Chi, Y., & Gong, Y. (2007). Combining content and link for classification using matrix factorization. In *Proceedings of the 30th Annual International ACM SIGIR Conference on Research and Development in Information Retrieval (SIGIR 2007)*. ACM Press.

Chapter 6
Analyzing Social Networks to Mine Important Friends

Carson K.-S. Leung
University of Manitoba, Canada

Irish J. M. Medina
University of Manitoba, Canada

Syed K. Tanbeer
University of Manitoba, Canada

ABSTRACT

The emergence of Web-based communities and social networking sites has led to a vast volume of social media data, embedded in which are rich sets of meaningful knowledge about the social networks. Social media mining and social network analysis help to find a systematic method or process for examining social networks and for identifying, extracting, representing, and exploiting meaningful knowledge—such as interdependency relationships among social entities in the networks—from the social media. This chapter presents a system for analyzing the social networks to mine important groups of friends in the networks. Such a system uses a tree-based mining approach to discover important friend groups of each social entity and to discover friend groups that are important to social entities in the entire social network.

INTRODUCTION

Due to advances in technology, Web-based communities and social networking sites have emerged. They have facilitated collaboration and information sharing between users, and have led to a vast volume of social media data. Intuitively, *social media* can be considered as forms of electronic communication (e.g., Web sites such as Twitter for social networking and micro-blogging) through which users create online Web-based communities to share information, ideas, personal messages, and other contents such as images and videos. More formally, *social media* (Kaplan & Haenlein, 2010) refer to groups of electronic communications (e.g., for Web-based or mobile-based applications [Matera, 2009]) that (a) build on the ideological and technological foundations of Web 2.0 and (b) allow the creation and exchange of user-generated content. Common social media

DOI: 10.4018/978-1-4666-2806-9.ch006

include (a) blogs or tweets (e.g., Blogger, Twitter [Kelsey, 2010]), (b) collaborative projects (e.g., Wikipedia), (c) content communities (e.g., Flickr, SlideShare, YouTube [Lacy, 2008]), (d) social networking sites (e.g., Facebook, Google+, LinkedIn, MySpace), (e) virtual game worlds (e.g., multiplayer online role-playing games like EverQuest and its successors, as well as World of Warcraft), and (f) virtual social worlds (e.g., Second Life [Rymaszewski, et al., 2007]). Embedded in these social media data are rich sets of meaningful knowledge. This leads to *social media mining*, which aims to extract, represent, and exploit the rich sets of meaningful knowledge from vast volumes of social media data ranging from those in digital textual forms to those in rich multimedia formats.

Social media mining can also be considered as an interaction between data mining and social computing. *Data mining* (Frawley, et al., 1991) refers to non-trivial extraction of implicit, previously unknown, and potentially useful information from data (e.g., social media data); *social computing* intersects social behaviour and computing systems in the sense that it computationally facilitates social studies and human-social dynamics in social networks, creates social conventions through the use of computer software, and designs information and communication technologies to deal with social context. An important social media-mining task is to discover meaningful knowledge about the social networks residing in the social media data. A *social network* (Schwagereit & Staab, 2009) is a structure made of social entities (e.g., individuals, corporations, collective social units, or organizations) that are linked by some specific kinds of interdependency (e.g., friendship, kinship, common interest, beliefs, or financial exchange). A social entity is connected to another entity as his friend, next-of-kind, collaborator, co-author, classmate, co-worker, team member, or business partner. As such, *social network analysis* (Wasserman & Faust, 1994)—which often refers to a systematic method or process for identifying useful social information like interdependency among social entities (i.e., social relationships and connections) from the social networks—is in demand.

Nowadays, various social computing applications such as blogs, email, instant messaging, social bookmarking, social networking, and wikis have been widely popularized so that people could interact socially via computing space. For instance, a Facebook user can create a personal profile, add other users as friends (who can be further categorized into different customized lists such as close friends, acquaintances, or family), exchange private messages, post messages on friends' walls, and join common-interest user groups. Similarly, a LinkedIn user can create a professional profile, establish connections to other users (who can be further annotated with tags corresponding to overlapping categories such as colleagues, classmates, business partners, and friends), exchange messages, recommend other users, and join common-interest user groups. A Google+ user can create a profile and add other users in one or more of his circles (e.g., circles for his friends, family, acquaintances, and followers). He can also share some posts, photos, or videos among users in the same circles.

Regardless which of the above social networking sites are used by users, there are some commonalities among the social experience of these users. Specifically, a user may have many friends in his social network. The number of friends may vary from one user to another. It is not uncommon for a user to have hundreds or thousands of friends. From the *ego-centric prospective* of a single user, some of his friends are more important than some others. It is desirable to discover his important groups of friends. To a further extent, from the *socio-centric prospective* of all users in the social networks, some of their groups of friends are important to some others. It is also desirable to discover groups of friends that are important in the networks. Hence, an objective of this chapter is to propose a tree-based system that (a) discovers important friend groups for a social entity and (b) discovers friend groups that are important to the entities of interest in the social networks.

The remainder of this chapter is organized as follows. We first provide background. Then, we focus on our proposed social network analysis and social media mining system. Afterwards, we outline our plan for future extensions of our system, and summarize the key features of our system in the conclusion.

BACKGROUND

Social media mining and social network analysis—which can be considered as a fusion of data mining and social computing—have become emerging research topics in the field of computer science. This is evidenced by numerous works (Cameron, et al., 2011; Bhagat, et al., 2012; Leung & Tanbeer, 2012) presented in (a) well-established conferences (e.g., ACM SIGKDD International Conference on Knowledge Discovery and Data Mining, IEEE International Conference on Data Mining [ICDM], Pacific-Asia Conference on Knowledge Discovery and Data Mining [PAKDD], European Conference on Principles of Data Mining and Knowledge Discovery [PKDD], and SIAM International Conference on Data Mining [SDM]), (b) new conferences and workshops (e.g., International Conference on Advances in Social Networks Analysis and Mining, [ASONAM], International Conference on Computational Aspects of Social Networks [CASoN], International Conference on Social Computing and its Applications [SCA], IEEE International Conference on Social Computing [SocialCom], International Conference on Social Informatics [SocInfo], and International Workshop on Social Networks and Social Web Mining [SNSM]), as well as (c) journals over the past few years. Examples of these research works include clustering and classification of tweets (Pennacchiotti & Popescu, 2011; Dalvi, et al., 2012; Hong, et al., 2012), mining and analysis of co-authorship networks (Leung, et al., 2011; Barbosa, et al., 2012), and visualization of social networks (Leung & Carmichael, 2010; Górecki, et al., 2011). In this chapter, we focus on a different aspect of social media mining and social network analysis—namely, *frequent pattern mining* on social networks.

Frequent pattern mining usually searches traditional transactional databases (e.g., sales transactions capturing customers' purchases) for implicit, previously unknown, and potentially useful patterns consisting of frequently co-occurring merchandise items. The classical algorithm—namely, Apriori (Agrawal & Srikant, 1994)—uses the candidate-and-test paradigm to mine frequent patterns using the *support* measure (which counts the presence of merchandise items) as the criterion for determining the interestingness of the patterns returned to the users. Numerous algorithms have been proposed over the past two decades. For example, *tree-based algorithms* (Han, et al., 2000) speed up the mining process by avoiding the candidate-and-test paradigm of Apriori-based algorithms. Several alternative interestingness measures have also been proposed. For example, the use of *share* (Carter, et al., 1997; Barber & Hamilton, 2003) as the interestingness criterion not only captures the presence of merchandise items but also their quantities. However, many of the existing frequent pattern-mining algorithms—whether they are Apriori-based or tree-based, using support or share as the interestingness criterion—mine traditional transactional databases but *not* social networks or social media data.

OUR SOCIAL NETWORK MINING SYSTEM

In this section, we present our tree-based mining system that mines social networks embedded in social media data. Specifically, the algorithm analyzes social networks to discover important friend groups. In particular, the system provides users with two functions: (a) Finding important groups of friends in an ego-centric prospective (i.e., friend groups that are important to a specific user) and (b) finding important groups of friends in a socio-centric prospective (i.e., friend groups that are important to the entire social networks).

Key Concepts for Finding Ego-Centric Friend Groups

Let us consider a social network for a social networking site (e.g., Facebook). Then, interdependency relationships among social entities can be captured in a collection of *friend lists*. From the ego-centric prospective, a friend list of a particular social entity contains his friends in the social network. Recall that friends can be categorized into close friends, acquaintances, family, or other categories. Friends within each category are not necessarily of the same importance. Some of his friends are more important than others. Hence, for each social entity E, we capture in his friend list not only his *friends* but also the *importance of his friends with respect to E*. For simplicity, the importance of a friend F with respect to E is measured by the number of messages F posted on E's wall. Note that, we are not confined to this importance measure, other measures or criteria are possible.

Table 1 shows an illustrative example of seven related friend lists. The first row shows the friend list for Don, who has three friends: Amy, Ed, and Gary. The list also captures the number of postings made by Don's friends on his wall. Specifically, Amy, Ed, and Gary, respectively, made 30, 40, and 10 postings on Don's wall. With respect to Don, the *importance* of friend groups {Amy}, {Ed}, and {Gary} are 30, 40, and 10, respectively. To a further extent, with respect to Don, the importance of friend groups {Amy, Ed}, {Amy, Ed, Gary}, {Amy, Gary}, and {Ed, Gary} are 30+40=70, 30+40+10=80, 30+10=40, and 40+10=50, respectively. We use imp(FG, E) to denote the *importance of a friend group FG with respect to a social entity E*. For example, imp({Amy}, Don) = 30 and imp({Ed, Gary}, Don) = 50.

In an ego-centric social network of a social entity E, a group of friends of E is considered to be *important* to E if members of this group have posted enough messages on E's wall (i.e., the total number of messages posted by members of this group on E's wall is equal to or higher than a user-specified threshold *minImp*). For example, let us consider Don's friend list shown in Table 1. If the user-specified threshold *minImp* = 25%, then {Amy} is an *important friend group to Don* because she made 30 postings ≥ 25% of a total of 30+40+10 = 80 postings on Don's wall. Similarly, {Amy, Ed} is also an *important friend group to Don* because both Amy and Ed together made 30+40 = 70 postings ≥ 25% of a total of 80 postings on Don's wall. For completeness, {Amy},

Table 1. An example of friend lists

Social entity E	Friend list of E
Don	{Amy (30), Ed (40), Gary (10)}
Ed	{Amy (20), Bob (40), Cara (10), Don (50)}
Fred	{Amy (20), Bob (10), Cara (10), Don (20), Holly (10)}
Gary	{Amy (30), Bob (50), Cara (30)}
Amy	{Bob (20), Don (20), Ed (10), Holly (10)}
Cara	{Bob (10), Gary (20)}
Bob	{Amy (20), Ed (30), Fred (40)}

{Amy, Ed}, {Amy, Ed, Gary}, {Amy, Gary}, {Ed}, and {Ed, Gary} are all the important friend groups to Don. Moreover, it is interesting to observe the following:

1. If a friend group FG is important to a social entity E, then any friend group formed by adding new members into FG is guaranteed to be important to E too.
2. If a friend group FG is important to a social entity E, then it is *not* necessary that all subsets of FG are important to E (i.e., the "importance" measure does *not* satisfy the Apriori property).

For instance, knowing {Amy} is an important friend group to Don implies that all supersets of {Amy} including {Amy, Ed}, {Amy, Ed, Gary}, {Amy, Gary} are also important friend groups to Don. However, the converse is *not* true. For instance, the friend group {Ed, Gary} is important to Don, but its subset {Gary} is unimportant to Don. In other words, based on Observation 2, knowing that {Gary} is unimportant to Don, we still need to examine whether or not its superset (e.g., {Ed, Gary}) is important to Don. Fortunately, finding ego-centric friend groups is manageable as we are dealing with only a single friend list. Moreover, due to Observation 1, once we found an important friend group FG to a social entity E, we do not need to examine its superset FG′ (because $imp(FG', E) \geq imp(FG, E) \geq minImp$ for $FG' \supseteq FG$).

Key Concepts for Finding Socio-Centric Friend Groups

Besides finding ego-centric friend groups, our system also provides users with the functionality of finding socio-centric friend groups. Here, the system finds the friend groups that are important to the entire social networks consisting of multiple friend lists. In a socio-centric social network, a group of friends is considered to be *important* to the network if members of this group have put together enough wall postings on the walls of one or more social entities (i.e., the total number of messages posted by members of this group is equal to or higher than a user-specified threshold *minImp*). For example, let us consider all friend lists shown in Table 1. If the user-specified threshold *minImp* = 25%, then although {Amy} is an important friend group to Don, she is *not* an important friend group in the entire network because she made only 30+20+20+30+20 = 120 postings < 25% of a total of 560 wall postings in the entire network. As another example, {Amy, Bob} is an important friend group in the entire network because they together made 20+40=60 postings on Ed's wall, 20+10=30 postings on Fred's wall and 30+50=80 postings on Gary's wall, for a total of 60+30+80 = 170 postings in the entire network ≥ 25% of a total of 560 wall postings in the entire network. As the third example, {Amy, Bob, Holly} is *not* an important friend group in the entire network because they together made 20+10+10 = 40 postings only on Fred's wall for a total of 40 postings in the entire network < 25% of a total of 560 wall postings in the entire network. We use imp(FG) to denote the *importance of a friend group FG* with respect to the entire social network. For example, imp({Amy}) = 120, imp({Amy, Bob}) = 170, and imp({Amy, Bob, Holly}) = 40. It is interesting to observe the following:

3. If a friend group FG is important to the entire social network, then it is *not* necessary that all subsets of FG are also important to the entire social network (i.e., the "importance" measure does *not* satisfy the Apriori property).
4. If a friend group FG is important to the entire social network, then it is *not* necessary that all supersets of FG are also important to the entire social network.

For instance, the friend group {Amy, Bob} is an important socio-centric friend group in the entire

Table 2. An example of the importance of friend lists

Social entity E	Friend list of E	Importance of friend list
Don	{Amy (30), Ed (40), Gary (10)}	80
Ed	{Amy (20), Bob (40), Cara (10), Don (50)}	120
Fred	{Amy (20), Bob (10), Cara (10), Don (20), Holly (10)}	70
Gary	{Amy (30), Bob (50), Cara (30)}	110
Amy	{Bob (20), Don (20), Ed (10), Holly (10)}	60
Cara	{Bob (10), Gary (20)}	30
Bob	{Amy (20), Ed (30), Fred (40)}	90

network, but its subset {Amy} is unimportant. In other words, based on Observation 3, knowing that {Amy} is unimportant, we still need to examine whether or not its superset (e.g., {Amy, Bob}) is important. What makes the mining of *socio-centric* friend groups more challenging than the mining of *ego-centric* friend groups is due to Observation 4. Knowing that {Amy, Bob} is important, we still need to examine whether or not its superset (e.g., {Amy, Bob, Holly}) is important. When dealing with a single friend list, adding more members to the friend group would increase the value for the "importance" measure. However, an increase in the number of members of the group would lead to a decrease in the chance of all members of the group contribute to the posting on the same wall, which in turn would decrease the value for the "importance" measure. For instance, adding Holly to the friend group {Amy, Bob} increases the "importance" value from 20+10=30 to 20+10+10=40 with respect to postings on Fred's wall, but Holly does not post on Ed's or Gary's wall. Hence, pruning is unavailable; exhaustive enumeration of all friend groups is impractical.

As the mining of important socio-centric friend groups is more computationally intensive and challenging, we need a better and practical solution. To facilitate effective pruning, our system finds candidates for important socio-centric friend groups by using an upper bound of the "importance" measure. Specifically, the system computes the upper bound by summing the importance *of relevant friend lists* (instead of the importance of friend group). If an upper bound for a friend group FG (i.e., sum of the importance of relevant friend lists containing FG) is lower than *minImp*, then the upper bound for any superset of FG (i.e., sum of the importance of relevant friend lists containing the superset of FG) is guaranteed to be lower than *minImp*. In other words, such an upper bound satisfies the Apriori property so that pruning is possible. The following summarizes our observation:

5. If an upper bound of the importance of friend group FG meets or exceeds the user-specified *minImp* threshold, then the upper bound of the importance of any subset of FG is guaranteed to meet or exceed the user-specified *minImp* threshold.

Once all candidates are found, we perform a post-processing step to filter out the false positives and return to the user the true positives (i.e., friend groups that are important to the entire network). For instance, given the seven friend lists shown in Table 1, we compute the importance of these friend lists (as shown in Table 2) and use them as an upper bound for the importance of socio-centric friend groups. As Amy appears on Don's, Ed's, Fred's, Gary's and Bob's friend lists, the upper bound of the importance of the friend group {Amy} is the sum of importance of these five friend lists, i.e., 80+120+70+110+30 = 410. Similarly, as both Amy and Bob appears on Ed's, Fred's, and Gary's friend lists, the upper bound of the importance of the friend group {Amy, Bob} is the sum of importance of these three friend lists, i.e., 120+70+110=300. As Amy, Bob and Holly all appear only on Fred's friend list, the upper bound of the importance of the friend group {Amy, Bob, Holly} is the importance of Fred's friend list, i.e., 70. Based on Observation 5, our system prunes those friend groups with low values for the upper bounds because their supersets would have even lower values for the upper bounds.

To tighten the upper bound, we could remove those friends that are unimportant to the entire network because they would not contribute to the importance of any of their supersets. Table 3 shows the tightened upper bound for the importance of friend lists in Table 2. For instance, as Holly does not contribute too much to the postings in the entire network, she is not considered important to the network. By pruning her, the upper bound of the importance of the friend group {Amy} is tightened from 80+120+70+110+30 = 410 to 70+120+60+110+10=370. Let us use UB(imp(FG)) to denote the *tightened upper bound of the importance of a friend group FG* with respect to the entire social network. For example, UB(imp({Amy})) = 70+120+60+110+10 = 370.

Capturing Social Networks by an Important-Friend Tree (IF-Tree)

In the previous sections, we discussed the key concepts for finding both ego-centric and socio-centric friend groups. An observant reader may notice that, between the two mining tasks performed by our proposed system (i.e., mining of ego-centric friend groups, mining of socio-centric friend groups), mining of socio-centric friend group is more computationally intensive. In the remainder of this section, we focus on the algorithmic details of how we use a tree-based approach to mine friend groups that are important to the entire network. In particular, we discuss how we build an important-friend tree (IF-tree) to capture the relevant information about the social network and how we use the IF-tree for the mining of important friend groups.

Each node of an IF-tree consists of (1) a friend F, (2) the tightened upper bound of the importance of F, (3) a parent pointer pointing to its parent node, (4) a list of child pointers pointing to its child nodes, and (5) a node pointer linking to the next node of the same social entity E in the IF-tree. In addition, the name of the social entity E is also maintained at the last node of the friend list of E.

The IF-tree is constructed with two scans. First, the system scans the friend list of each social entity E once to compute the upper bound of the importance of social entity E in the network. Then, the system removes unimportant social

Table 3. An example of the tightened importance of friend lists

Social entity E	Friend list of E	Tightened importance of friend list
Don	{Amy (30), Ed (40), ~~Gary (10)~~}	~~80~~ 70
Ed	{Amy (20), Bob (40), Cara (10), Don (50)}	120
Fred	{Amy (20), Bob (10), Cara (10), Don (20), ~~Holly (10)~~}	~~70~~ 60
Gary	{Amy (30), Bob (50), Cara (30)}	110
Amy	{Bob (20), Don (20), Ed (10), ~~Holly (10)~~}	~~60~~ 50
Cara	{Bob (10), ~~Gary (20)~~}	~~30~~ 10
Bob	{Amy (20), Ed (30), ~~Fred (40)~~}	~~90~~ 50

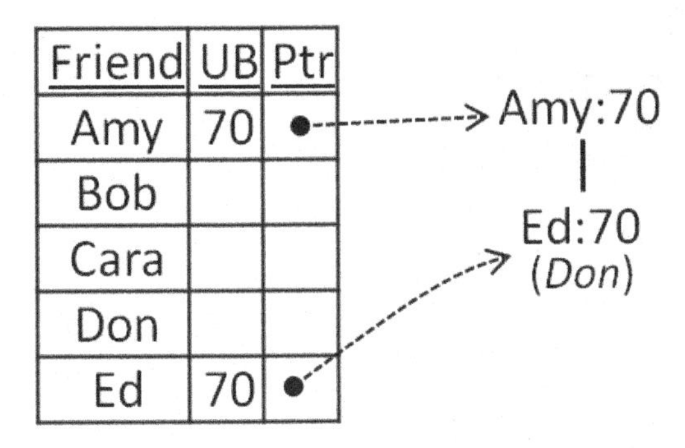

Figure 1. The content of the IF-tree after inserting Don's friend list

entities (i.e., entities with importance measure < user-specified *minImp* threshold) and sort all remaining important social entities according to their importance values. For instance, to construct an IF-tree for the friend lists shown in Table 1 with *minImp*=25%, our system scans these lists and obtains upper bound of the importance of social entities. As the upper bound of the importance of Fred, Gary and Holly are 90, 80+30 = 110 and 70+60 = 130 (which are all lower than *minImp* = 25% of a total of 560 wall postings in the entire network), their importance values are lower than the user-specified *minImp* threshold. So, they are pruned. Note that these pruned unimportant friends have no influence on the computation of important friend groups. The removal helps to tighten the upper bound and to reduce the number of false positives (before the final post-processing step to keep only those true positives). The remaining social entities are sorted according to their upper bounds: UB(imp({Amy}))=70+120+60+110+50 = 410, UB(imp({Bob})) = 120+60+110+50+10 = 350, UB(imp({Cara})) = 120+60+110 = 290, UB(imp({Don})) = 120+60+50 = 230, and UB(imp({Ed})) = 70+50+50 = 170.

Our system then scans the friend lists the second time to construct the IF-tree. With this scan, all the friend lists are inserted into the IF-tree one-by-one (with except of those unimportant friends identified in the first scan). For example, after we discarded those unimportant friends (namely, Fred, Gary, and Holly), our system scans the friend lists the second time. The first list to be scanned is Don's list. The system ignores Gary (an unimportant social entity) from the list, and inserts ⟨Amy:70, Ed:70 (Don)⟩—which indicates that Amy and Ed are friends of Don with a tightened upper bound of the importance of any entity in Don's list—into the IF-tree. See Figure 1.

Next, the system scans Ed's list and tries to insert ⟨Amy:120, Bob:120, Cara:120, Don:120 (Ed)⟩ into the IF-tree. We note that the previously inserted list (i.e., Don's friend list) and this list (i.e., Ed's friend list) share the common prefix (namely, Amy). To make the IF-tree more compact (and thus save memory space), the system increments the upper bound value of Amy by 120 (which increases the value from 70 to 70+120=190) and inserts ⟨Bob:120, Cara:120, Don:120 (Ed)⟩ as a child of Amy:120. The header table is also updated accordingly, as shown in Figure 2.

The other five friend lists are scanned and inserted in a similar fashion. Figure 3 shows the contents of our IF-trees after the insertion of each of the remaining five friend lists. It is interesting to observe the following:

6. The system only needs to perform the projection for each social entity in the header table once.

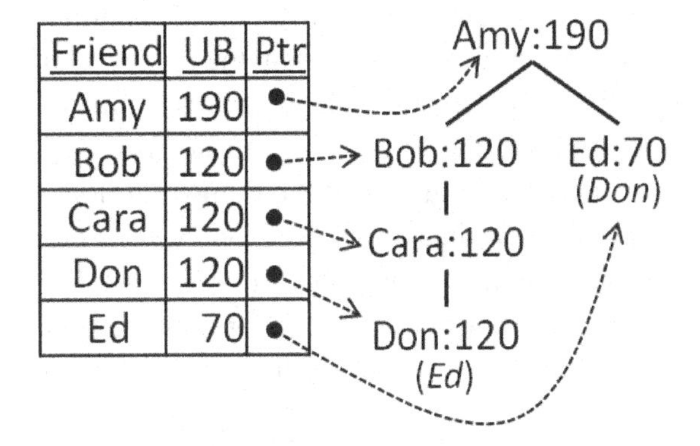

Figure 2. The content of the IF-tree after inserting Ed's friend list

Figure 3. The contents of IF-trees after subsequent insertions of friend lists

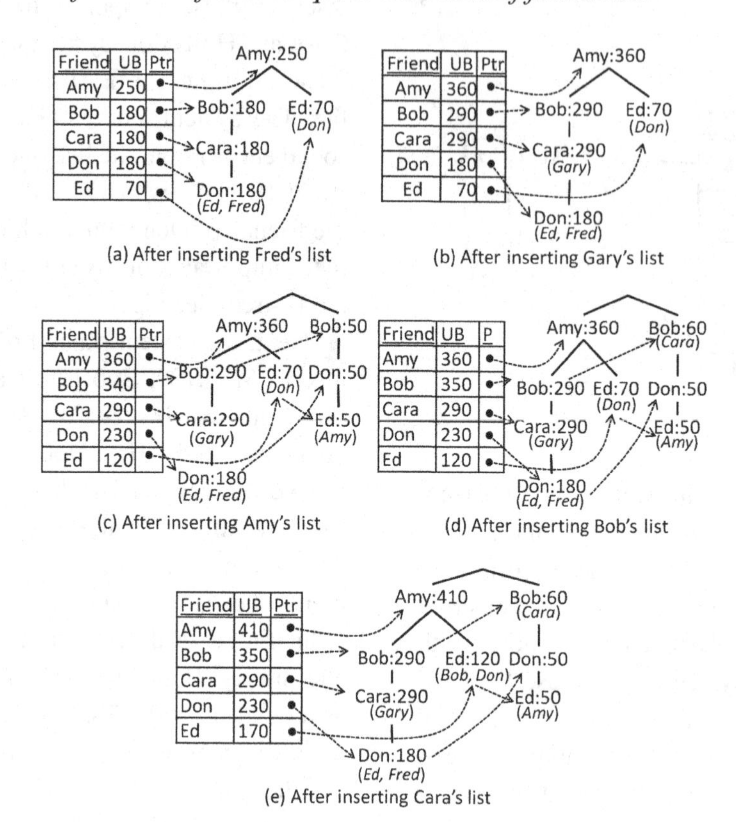

7. The upper bound value stored in any node in an IF-tree is greater than or equal to the sum of the upper bound values of its children.

Mining of Important Socio-Centric Friend Groups from the IF-Tree

Observed from the above tree construction step, the IF-tree keeps the friend information about the social network in a highly compact tree structure. This enables a fast subsequent mining of important socio-centric friend groups from the IF-tree. Our system finds important friend groups by recursively mining the IF-tree to generate candidate important friend groups without any additional scans of the original friend lists. Hence, the basic operations in IF-tree mining are (1) the construction of the projected network for a potentially important friend group and (2) the recursive mining of the further potentially important friend extensions of that group. Our algorithm performs these operations by creating conditional pattern-base (or pattern-base in short) for each potentially important friend group and examining all such pattern-bases for conditional IF-trees consisting of the set of potentially important friend groups occurring with a suffix group.

Let us continue with our running example. Our system starts the mining process by constructing the projected network for the last social entity (namely, Ed) in the header table of the IF-tree. The projected network for Ed is constructed by taking all the branches from the IF-tree with suffix Ed. See Figure 4(a). After creating the projected network, the IF-tree is adjusted by pushing up the information at all nodes for Ed to their respective parent. For example, the information of node "Ed:120" (i.e., Bob, Don) and node "Ed:50"

Figure 4. The contents of conditional IF-trees

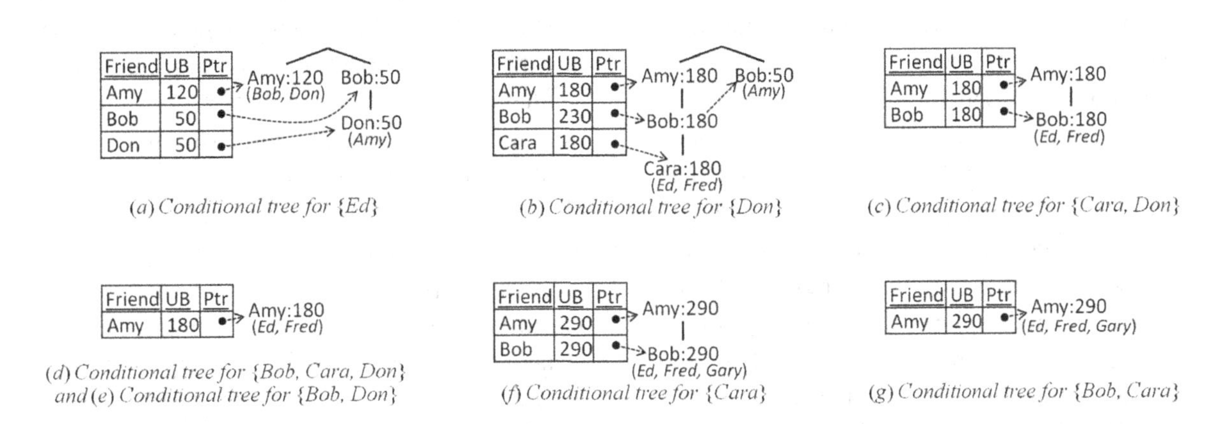

(i.e., Amy) are pushed to their respective parent nodes, as shown in Figure 4(a). Such an operation ensures that the system obtains the correct information for any node in the IF-tree during the whole mining phase.

The conditional IF-trees are constructed by removing all unimportant friends from the projected network. Because none of the friends in the header table of Ed's projected network is important (i.e., upper bound values are below *minImp*), we remove all friends from Ed's pattern-base. Hence, no further candidate important friend group will be generated from Ed's conditional database, which terminates mining for Ed.

Mining for the next friend in the header table (namely, Don) is then performed by constructing Don's pattern-base and conditional trees (as shown in Figure 4[b]) in the same fashion from the IF-tree. After constructing the pattern-base for Don, the original IF-tree is again adjusted by pushing the information at each node for Don to its parent. The candidate friend group {Cara,

Table 4. Important groups of friends

Candidate friend groups FG	UB(imp(FG))	imp(FG)	Important friend group?
{Amy, Bob}	290	**170**	**Yes**
{Amy, Bob, Cara}	290	**220**	**Yes**
{Amy, Bob, Cara, Don}	180	**180**	**Yes**
{Amy, Bob, Don}	180	**160**	**Yes**
{Amy, Cara}	290	120	*No*
{Amy, Cara, Don}	180	130	*No*
{Amy, Don}	180	110	*No*
{Bob, Cara}	290	**150**	**Yes**
{Bob, Cara, Don}	180	**140**	**Yes**
{Bob, Don}	230	**160**	**Yes**
{Cara, Don}	180	90	*No*

Don} is generated first from the conditional tree for Don in Figure 4(b) with an upper bound value of 180, and a similar recursive mining process (of constructing the pattern-bases and conditional trees) is followed for {Cara, Don} and others. The pattern-base for {Cara, Don} is shown in Figure 4(c). Note that, to ensure the recursive mining, the tree adjustment phase of pushing the information towards the parent for a deleted node is also performed, while constructing conditional trees from a pattern-base. The pattern-base and conditional tree for the next extension of {Cara, Don}, i.e., {Bob, Cara, Don} is shown in Figure 4(d). Consequently, the set of all candidate important socio-centric friend groups generated by mining the IF-tree is shown in Table 4. Our system applies an additional scan of friend lists as a post-processing step, and eliminates all unimportant friend groups (i.e., false positives) from the set of candidate important friend groups.

FUTURE RESEARCH DIRECTIONS

So far, we have presented a system for analyzing social networks for mining important friend groups. The system discovers the important friend groups based on the frequency and importance of friends within the social networks. In many real-life situations, users may want to focus the mining on certain friend groups by specifying some constraints expressing their interest. A naïve approach of handling these user-specified constraints is to first find all important friend groups and then apply a post-processing step to filter out those that do not satisfy the user-specified constraints. As future work, we plan to adapt the *constraint-based association rule mining* techniques (Leung, 2009) to social media mining and social network analysis. Specifically, we plan to extend our system by pushing constraints into the mining process so that the users could freely express their interest by specifying constraints and the resulting system could mine from social media for all and only those *important friend groups that satisfy the user-specified constraints* directly—without having a post-processing step to filter out those important friend groups that do not satisfy the user-specified constraints.

Currently, our presented system returns to users all important friend groups in a textual form, i.e., a *textual list* of all friend groups. As suggested by the adage "a picture is worth a thousand words," visual representation of the mining results is usually more comprehensive to users. Thus, it is desirable to show the mining results interactively by applying the concepts of visual analytics. As such, to enhance user experience in exploring the social networks, our second future research direction is to extend our system by incorporating *visual analytics and interactive technologies* (Leung & Carmichael, 2011) so that the resulting interactive visual analytics system could mine social networks, from which the mined important friend groups are returned to the users in visual forms.

CONCLUSION

The emergence of social networking sites such as Facebook has led to a vast volume of social media data, embedded in which are rich sets of meaningful knowledge such as important friends of social entities in a social network. In this chapter, we applied social media mining and social network analysis techniques, which are non-trivial integration of data mining and social computing. Moreover, we also presented a system for analyzing social networks to mine important groups of friends. Such a tree-based system systematically examines social networks to identify, extract, represent, and exploit friendship relationships among social entities. Specifically, the system first computes the importance of friend groups for each social entity within the social network of interest. The system then captures the relevant importance of each friend group within the network in the IF-tree, from which important friend groups are mined. By recursively extracting appropriate paths from the IF-tree, the system finds meaningful knowledge such as friend groups that are important to the social network.

REFERENCES

Agrawal, R., & Srikant, R. (1994). Fast algorithms for mining association rules in large databases. In J. B. Bocca, M. Jarke, & C. Zaniolo (Eds.), *Proceedings of the 20th International Conference on Very Large Data Bases* (*VLDB*), (pp. 487-499). San Francisco, CA: Morgan Kaufmann.

Barber, B., & Hamilton, H. J. (2003). Extracting share frequent itemsets with infrequent subsets. *Data Mining and Knowledge Discovery*, *7*(2), 153–185. doi:10.1023/A:1022419032620

Barbosa, E. M., Moro, M. M., Lopes, G. R., & de Oliveira, J. P. M. (2012). VRRC: Web based tool for visualization and recommendation on co-authorship network. In K. S. Candan, Y. Chen, R. Snodgrass, L. Gravano, & A. Fuxman (Eds.), *Proceedings of the 2012 ACM SIGMOD International Conference on Management of Data,* (p. 865). New York, NY: ACM Press.

Bhagat, S., Goyal, A., & Lakshmanan, L. V. S. (2012). Maximizing product adoption in social networks. In E. Adar, J. Teevan, E. Agichtein, & Y. Maarek (Eds.), *Proceedings of the Fifth ACM International Conference on Web Search and Data Mining* (*WSDM*), (pp. 603-612). New York, NY: ACM Press.

Cameron, J. J., Leung, C. K.-S., & Tanbeer, S. K. (2011). Finding strong groups of friends among friends in social networks. In J. Chen, S. Lawson, & N. Agarwal (Eds.), *Proceedings of the 2011 International Conference on Social Computing and Its Applications* (pp. 824-831). Los Alamitos, CA: IEEE Computer Society.

Carter, C. L., Hamilton, H. J., & Cercone, N. (1997). Share based measures for itemsets. In J. Komorowski & J. Zytkow (Eds.), *Proceedings of the First European Symposium on Principles of Data Mining and Knowledge Discovery* (*PKDD*) (pp. 14-24). Berlin, Germany: Springer.

Dalvi, N., Kumar, R., & Pang, B. (2012). Object matching in tweets with spatial models. In E. Adar, J. Teevan, E. Agichtein, & Y. Maarek (Eds.), *Proceedings of the Fifth ACM International Conference on Web Search and Data Mining* (*WSDM*), (pp. 43-52). New York, NY: ACM Press.

Frawley, W. J., Piatetsky-Shapiro, G., & Matheus, C. J. (1991). Knowledge discovery in databases: An overview. In Piatetsky-Shapiro, G., & Frawley, W. J. (Eds.), *Knowledge Discovery in Databases* (pp. 1–30). Cambridge, MA: The MIT Press.

Górecki, J., Slaninová, K., & Snášel, V. (2011). Visual investigation of similarities in global terrorism database by means of synthetic social networks. In A. Abraham, E. Corchado, R. Alhaj, & V. Snášel (Eds.), *Proceedings of the 2011 International Conference on Computational Aspects of Social Networks* (*CASoN*), (pp. 255-260). Los Alamitos, CA: IEEE Computer Society.

Han, J., Pei, J., & Yin, Y. (2000). Mining frequent patterns without candidate generation. In W. Chen, J. Naughton, & P. A. Bernstein (Eds.), *Proceedings of the 2000 ACM SIGMOD International Conference on Management of Data,* (pp. 1-12). New York, NY: ACM Press.

Hong, L., Ahmed, A., Gurumurthy, S., Smola, A. J., & Tsioutsiouliklis, K. (2012). Discovering geographical topics in the twitter stream. In A. Mille, F. Gandon, J. Misselis, M. Rabinovich, & S. Staab (Eds.), *Proceedings of the 21st International World Wide Web Conference* (*WWW*), (pp. 769-778). New York, NY: ACM Press.

Kaplan, A. M., & Haenlein, M. (2010). Users of the world, unite! The challenges and opportunities of social media. *Business Horizons*, *53*(1), 59–68. doi:10.1016/j.bushor.2009.09.003

Kelsey, T. (2010). *Social networking spaces: From Facebook to Twitter and everything in between*. New York, NY: Apress.

Lacy, S. (2008). *The stories of Facebook, YouTube & MySpace: The people, the hype and the deals behind the giants of web 2.0*. Richmond, UK: Crimson Publishing.

Leung, C. K.-S. (2009). Constraint-based association rule mining . In Wang, J. (Ed.), *Encyclopedia of Data Warehousing and Mining* (2nd ed., pp. 307–312). Hershey, PA: IGI Global.

Leung, C. K.-S., & Carmichael, C. L. (2010). Exploring social networks: A frequent pattern visualization approach. In J. Zhan (Ed.), *Proceedings of the Second IEEE International Conference on Social Computing* (*SocialCom*), (pp. 419-424). Los Alamitos, CA: IEEE Computer Society.

Leung, C. K.-S., & Carmichael, C. L. (2011). iVAS: An interactive visual analytic system for frequent set mining. In Q. Zhang, R. S. Segall, & M. Cao (Eds.), *Visual Analytics and Interactive Technologies: Data, Text and Web Mining Applications,* (pp. 213-231). Hershey, PA: IGI Global.

Leung, C. K.-S., Carmichael, C. L., & Teh, E. W. (2011). Visual analytics of social networks: Mining and visualizing co-authorship networks. In D. Schmorrow & C. M. Fidopiastis (Eds.), *Proceedings of the Sixth International Conference of Foundations of Augmented Cognition* (pp. 335-345). Berlin, Germany: Springer.

Leung, C. K.-S., & Tanbeer, S. K. (2012). Mining social networks for significant friend groups. In P.-N. Tan, S. Chawla, C. K. Ho, & J. Bailey (Eds.), *Proceedings of the Third International Workshop on Social Networks and Social Web Mining* (pp. 180-192). Berlin, Germany: Springer.

Li, L., Xiao, H., & Xu, G. (2012). Finding related micro-blogs based on WordNet. In P.-N. Tan, S. Chawla, C. K. Ho, & J. Bailey (Eds.), *Proceedings of the Third International Workshop on Social Networks and Social Web Mining* (pp. 115-122). Berlin, Germany: Springer.

Matera, M. (2009). Social applications . In Liu, L., & Özsu, M. T. (Eds.), *Encyclopedia of Database Systems* (p. 2667). New York, NY: Springer.

Pennacchiotti, M., & Popescu, A.-M. (2011). Democrats, republicans and Starbucks afficionados: User classification in twitter. In C. Apte, J. Ghosh, & P. Smyth (Eds.), *Proceedings of the 17th ACM SIGKDD International Conference on Knowledge Discovery and Data Mining* (pp. 430-438). New York, NY: ACM Press.

Rymaszewski, M., Au, W. J., Wallace, M., Winters, C., Ondrejka, C., & Batstone-Cunningham, B. (2007). *Second life: The official guide*. Indianapolis, IN: Wiley Publishing.

Schwagereit, F., & Staab, S. (2009). Social networks . In Liu, L., & Özsu, M. T. (Eds.), *Encyclopedia of Database Systems* (pp. 2667–2672). New York, NY: Springer.

Wasserman, S., & Faust, K. (1994). *Social network analysis: Methods and applications*. Cambridge, UK: Cambridge University Press. doi:10.1017/CBO9780511815478

ADDITIONAL READING

Abraham, A., Corchado, E., Alhaj, R., & Snasel, V. (Eds.). (2011). *Proceedings of the 2011 international conference on computational aspects of social networks* (*CASoN*). Los Alamitos, CA: IEEE Computer Society.

Abraham, A., Snášel, V., & Węgrzyn-Wolska, K. (Eds.). (2009). *Proceedings of the 2009 international conference on computational aspects of social networks (CASoN)*. Los Alamitos, CA: IEEE Computer Society.

Alhajj, R., Memon, N., & Ting, I.-H. (Eds.). (2011). *Proceedings of the 2011 international conference on advances in social networks analysis and mining (ASONAM)*. Los Alamitos, CA: IEEE Computer Society.

Bolc, L., Makowski, M., & Wierzbicki, A. (Eds.). (2010). *Proceedings of the second international conference on social informatics (SocInfo)*. Berlin, Germany: Springer.

Carrington, P. J., Scott, J., & Wasserman, S. (Eds.). (2005). *Models and methods in social network analysis*. Cambridge, UK: Cambridge University Press. doi:10.1017/CBO9780511811395

Datta, A., Shulman, S., Zheng, B., Lin, S.-D., Sun, A., & Lim, E.-P. (Eds.). (2011). *Proceedings of the third international conference on social informatics (SocInfo)*. Berlin, Germany: Springer.

Furht, B. (Ed.). (2010). *Handbook of social network technologies and applications*. New York, NY: Springer. doi:10.1007/978-1-4419-7142-5

Han, J., Kamber, M., & Pei, J. (2011). *Data mining: Concepts and techniques* (3rd ed.). San Francisco, CA: Morgan-Kaufmann.

Karray, F., & Polat, F. (Eds.). (2012). *Proceedings of the 2012 IEEE/ACM international conference on advances in social networks analysis and mining (ASONAM)*. Los Alamitos, CA: IEEE Computer Society.

Knoke, D., & Yang, S. (2007). *Social network analysis* (2nd ed.). Thousand Oaks, CA: SAGE.

Liu, H., Yu, P. S., Agarwal, N., & Suel, T. (2010). Guest editors' introduction: Social computing in the blogosphere. *IEEE Internet Computing*, *14*(2), 12–14. doi:10.1109/MIC.2010.39

Memon, N., & Alhajj, R. (Eds.). (2009). *Proceedings of the 2009 international conference on advances in social networks analysis and mining (ASONAM)*. Los Alamitos, CA: IEEE Computer Society.

Memon, N., & Alhajj, R. (Eds.). (2010). *Proceedings of the 2010 international conference on advances in social networks analysis and mining (ASONAM)*. Los Alamitos, CA: IEEE Computer Society.

Nohuddin, P. N. E., Coenen, F., Christley, R., Setzkorn, C., Patel, Y., & Williams, S. (2012). Finding "interesting" trends in social networks using frequent pattern mining and self organizing maps. *Knowledge-Based Systems*, *29*, 104–113. doi:10.1016/j.knosys.2011.07.003

Oliveira, M., & Gama, J. (2012). An overview of social network analysis. *Wiley Interdisciplinary Reviews: Data Mining and Knowledge Discovery*, *2*(2), 99–115. doi:10.1002/widm.1048

Schuller, B., & Smith, M. (Eds.). (2012). *Proceedings of the fourth ASE/IEEE international conference on social computing (SocialCom)*. Los Alamitos, CA: IEEE Computer Society.

Schwagereit, F., & Staab, S. (2009). Social networks . In Liu, L., & Özsu, M. T. (Eds.), *Encyclopedia of Database Systems* (pp. 2667–2672). New York, NY: Springer.

Scott, J., & Carrington, P. J. (Eds.). (2011). *The SAGE handbook of social network analysis*. London, UK: SAGE.

Tan, P.-N., Chawla, S., Ho, C. K., & Bailey, J. (Eds.). (2012). *Proceedings of the third international workshop on social networks and social web mining*. Berlin, Germany: Springer.

Tang, L., & Liu, H. (2010). *Community detection and mining in social media*. San Rafael, CA: Morgan & Claypool. doi:10.2200/S00298ED1V01Y201009DMK003

Ting, I.-H., Hong, T.-P., & Wang, L. S.-L. (Eds.). (2012). *Social network mining, analysis and research trends: Techniques and applications*. Hershey, PA: IGI Global.

Vinciarelli, A., Pantic, M., Bertino, E., & Zhan, J. (Eds.). (2011). *Proceedings of the third IEEE international conference on social computing (SocialCom)*. Los Alamitos, CA: IEEE Computer Society.

Xu, G., Zhang, Y., & Li, L. (2011). *Web mining and social networking: Techniques and applications*. New York, NY: Springer. doi:10.1007/978-1-4419-7735-9

Xu, J., Yu, G., Zhou, S., & Unland, R. (Eds.). (2011). *Proceedings of the second international workshop on social networks and social web mining on the web*. Berlin, Germany: Springer.

Yoshikawa, M., Meng, X., Yumoto, T., Ma, Q., Sun, L., & Watanabe, C. (Eds.). (2010). *Proceedings of the first international workshop on social networks and social web mining on the web*. Berlin, Germany: Springer.

Zeng, J., Pan, J.-S., & Abraham, A. (Eds.). (2010). *Proceedings of the 2010 international conference on computational aspects of social networks (CASoN)*. Los Alamitos, CA: IEEE Computer Society.

Zhan, J. (Ed.). (2010). *Proceedings of the second IEEE international conference on social computing (SocialCom)*. Los Alamitos, CA: IEEE Computer Society.

KEY TERMS AND DEFINITIONS

Data Mining: Data mining refers to non-trivial extraction of implicit, previously unknown, and potentially useful information from data.

Frequent Pattern Mining: Frequent pattern mining searches from vast volumes of data for implicit, previously unknown, and potentially useful patterns consisting of frequently co-occurring events or objects such as merchandise items.

Social Computing: Social computing aims to computationally facilitate social studies and human-social dynamics in social networks as well as to design and use information and communication technologies for dealing with social context.

Social Media: Social media refer to groups of Internet-based applications that build on the ideological and technological foundations of Web 2.0 and that allow the creation and exchange of user-generated content. Common social media include blogs, collaborative projects, content communities, social networking sites, virtual game worlds, and virtual social worlds.

Social Media Mining: Social media mining aims to extract, represent, and exploit the rich sets of meaningful knowledge from vast volumes of social media data ranging from those in digital textual forms to those in rich multimedia formats.

Social Network: A social network is a structure made of social entities that are linked by some specific kinds of interdependency such as friendship.

Social Network Analysis: Social network analysis refers to the method or process for identifying meaningful social information—such as interesting social relationships and connections, as well as influential social entities—from social networks.

Chapter 7
News Document Summarization Driven by User-Generated Content

Luca Cagliero
Politecnico di Torino, Italy

Alessandro Fiori
IRC@C: Institute for Cancer Research and Treatment at Candiolo, Italy

ABSTRACT

The outstanding growth of the Internet has made available to analysts a huge and increasing amount of Web documents (e.g., news articles) and user-generated content (e.g., social network posts) coming from social networks and online communities that are worth considering together. On one hand, the need of novel and more effective approaches to summarize Web document collections makes the application of data mining techniques established in different research contexts more and more appealing. On the other hand, to generate appealing summaries the data mining and knowledge discovery process cannot disregard the major Web users' interests.

This chapter presents a novel news document summarization system, namely NeDocS, that focuses on generating succinct, not redundant, yet appealing summaries by means of a data mining and knowledge discovery process driven by messages posted on social networks. NeDocS retrieves from the Web and summarizes news document collections by exploiting (1) frequent itemsets, i.e., recurrences that frequently occur in the analyzed data, to capture most significant correlations among terms and (2) a sentence relevance evaluator that takes into account term significance in a collection of social network posts ranging over the same news topics. This approach allows not disregarding sentences whose terms rarely occur in the news collection but are deemed relevant by Web users. To the best of our knowledge, the combined usage of frequent itemsets and user-generated content in news document summarization is an appealing research direction that has never been investigated so far.

Experiments performed on real collections of news articles and driven by on-topic Twitter posts show the effectiveness of the proposed approach.

DOI: 10.4018/978-1-4666-2806-9.ch007

INTRODUCTION

The increasing availability of Web documents (e.g., news articles, scientific papers, books, and magazines) and the popularity of the social network communities, such as Twitter and Facebook, have relevantly changed the life style of Web users. Nowadays, social network sites help users to find people with similar interests and goals, provide means of news exchange, and facilitate multimedia content sharing. Furthermore, the huge amount of news document collections available on the Web represents a powerful source of knowledge for both industrial and academic purposes.

An interesting research direction focuses on conveying the huge mass of electronic document content into concise representations, i.e., the summaries. Multi-document summarization addresses the selection of the most relevant and not redundant sentences belonging to a collection of textual documents. Previous approaches commonly rely on either information retrieval (Carenini, Ng, & Zhou, 2007; Radev, 2004) or data mining approaches (Thakkar, Dhareskar, & Chandak, 2010; Wang & Li, 2010; Wang, Zhu, Li, Chi, & Gong, 2011). In fact, most of them are based on (1) clustering algorithms (e.g., Thakkar, et al., 2010; Wang & Li, 2010; Wang, et al., 2011), (2) graph-based methods (e.g., Radev, 2004), or (3) linear programming algorithms (e.g., Takamura & Okumura, 2009b). While clustering is exploited to group sentences belonging to a document collection, graph-based methods try to represent correlations among sentences by means of a graph-based model. To select most representative sentences according to the generated model, well-established graph-based indexing strategies are usually exploited (Radev, 2004). Differently, linear programming approaches formalize the summarization problem as a min-max optimization problem (Takamura & Okumura, 2009b). However, all the aforementioned approaches consist in general-purpose summarizers applicable to document collections coming from any source and, thus, do not consider the real social interest to effectively accomplish the summarization task.

The outstanding growth of online communities and social networks has made available to analysts a powerful and huge amount of User-Generated Content (UGC). Some preliminary attempts to convey the information provided by user-generated content into the document summarization process have been performed. Previous works entail (1) the exploitation of the Wikipedia content to identify key concepts in document collections (Gong, Qu, & Tian, 2010; Miao & Li, 2010), (2) the use of social annotations (tags) in graph-based text summarization (Zhu, et al., 2009), and (3) the learning of classification models driven by the main social network data features to evaluate document sentences (Yang, et al., 2011). However, they still show a limited effectiveness due to (1) the hardness in capturing most significant correlations among multiple terms at the same time and (2) the low quality of the training data used for sentence selection in supervised methods.

The chapter proposes a novel multi-document system, namely NeDocS (News Document Summarizer), that addresses the summarization of news document collections, coming from the Web, driven by on-topic user-generated messages posted on social network sites. To this aim, it exploits a well-established unsupervised data mining technique, i.e., frequent itemset mining, whose application to transactional data summarization is well established. Frequent itemsets represent correlations among data whose frequency of occurrence (i.e., the support) exceeds a given threshold. The discovery of the most informative and not redundant subset of frequent itemsets allows highlighting the most relevant correlations among multiple terms at the same time. NeDocS adopts an efficient and effective approach, recently proposed in (Mampaey, et al., 2011), that allows evaluating on the fly the significance of the itemsets during the extraction process, without the need of a post-pruning step. Moreover, the selection of the most representative sentences to include in the summary is driven by both the top-k selected frequent itemsets and the user-generated content. A relevance term evaluator, based on tf-idf statistics, is exploited to evaluate the importance of terms occurring in the news document collection by also considering their significance in a collection of social network posts, ranging over the same news topics. This approach allows not disregarding sentences whose terms rarely occur in the news collection but are deemed relevant by Web users. The problem of selecting the minimum set of sentence that best covers the set of the selected itemsets is modeled as a set covering optimization problem and addressed by means of a branch-and-bound algorithm. To the best of our knowledge, the combined usage of frequent itemsets and user-generated content in news document summarization is an appealing research direction that has never been investigated so far.

The effectiveness of NeDocS in generating concise yet representative summaries has been validated on real news document collections retrieved from the Web. To drive the summarization process, on-topic collections of Twitter posts (i.e., tweets) are exploited. Experimental results, reported in Section "Experimental evaluation" show that the combined usage of frequent itemsets and a sentence relevance evaluator based on UGC allow generating succinct, not redundant, yet appealing news document summaries. NeDocS significantly outperforms mostly used previous summarizers in terms of precision, recall, and F-measure, computed by means of the ROUGE toolkit (Lin & Hovy, 2003). Furthermore, they showed that the usage of the user-generated content significantly improves the performance of the summarization process.

This Chapter is organized as follows. Section "Related Works" overviews most relevant related works and compares our approach with the most recent related approaches. Section "The NeDocS System" presents the architecture of the proposed summarization system and describes its main blocks. Section "Experimental Evaluation" assesses the effectiveness of NeDocS in summarizing real news document collections, while Sections "Future Research Directions" and "Conclusions" respectively present future developments of this work and draw conclusions.

RELATED WORKS

The past few years have witnessed the rapid proliferation of social network sites. This has resulted in User-Generated Content (UGC), produced by online community users, becoming a popular and everyday part of the Internet culture. Thus, many research efforts have been devoted to knowledge discovery from UGC by means of data mining techniques.

A notable research issue is the problem of data summarization. The huge amount of textual document collections (e.g., news, articles) available on the Web prompts the need of generating succinct summaries to ease the analysis task of domain experts. A parallel effort has been devoted to summarizing large data collections coming from different research fields (e.g., biological, context, or network traffic data) by means of frequent itemsets, which are patterns that frequently occur in the source data.

This section overviews previous works concerning the application of data mining and knowledge discovery approaches to the UGC coming from online communities and social networks. Furthermore, with the aim at providing a strong foundation to the data mining approaches adopted in this chapter, we also discuss most relevant literature concerning both document summarization and data summarization by means of frequent itemsets.

Data Mining and Knowledge Discovery from User-Generated Content

In the last years, many data mining algorithms have been proposed to address knowledge discovery from the User-Generated Content (UGC). Significant research efforts have been devoted to (1) developing recommender systems to enhance the quality of targeted suggestions to final users, (2) improving the understanding and the categorization of online resources, and (3) building query engines based on semantic models, e.g., folkosonomies and ontologies. Recommender systems are focused on identifying the objects (e.g., products, news) that are mostly correlated with the user behavior and preferences. Based on the kind of the analyzed data and the items to be suggested, different approaches have been proposed. For instance, Li, Wang, Chen, and Lin (2010) propose a news recommendation system that exploits a graph-based model to analyze the relationships held among user comments. Similarly, in Wang, Li, and Chen (2010) relationships among user comments are exploited to recommend relevant pieces of information in a forum-based social media. Differently, Phelan, McCarthy, and Smyth (2009) recommend RSS news by identifying emerging topics and breaking events in Twitter posts. RSS stories are ranked based on a weighted score that combines the Lucene tf-idf score of each article term with the knowledge hidden in Twitter posts.

The analysis of the single user interests and/or the social community trends may enhance the quality of tag-based recommended systems. For instance, Basile, Gendarmi, Lanubile, and Semeraro (2007) propose a classification system to identify the documents and the associated tags that are more likely to be of user's interest. Differently, Shepitsen, Zhang, and Giles (2008) identify correlations among groups of tags by means of a hierarchical clustering algorithm. To improve the quality of the suggested tags in different tagging systems (i.e., Del.icio.us, CiteULike, and BibSonomy), Song et al. (2011) propose two document-centered approaches. The first one identifies the document topics by partitioning a graph-based model, which represents the co-occurrence of documents and tags. The second approach aims at discovering the most representative documents within the data collections by exploiting a sparse multi-class Gaussian classifier. An overview of the most popular recommender systems is given in Herlocker, Konstan, Terveen, and Riedl (2004). Moreover, in Adomavicius and Tuzhilin (2011), the relevance and the exploitation of the contextual information to enhance the quality of recommender systems is thoroughly discussed.

The user-generated content can be also exploited to improve the understanding of online resources (e.g., Web pages, photos, or videos). The analysis of the co-occurrence of the user tags may be useful for capturing the topics of major interest. For instance, Mathioudakis and Koudas (2010) perform trend detection from Twitter streams. The "bursty" keywords, i.e., keywords that appear in tweets at an unusually high rate, are identified by means of co-occurrence measures. The usage of user tags to enhance Web resource categorization is investigated in Yin, Li, Mei, and Han (2009). To address this issue, it models, by means of a graph-based representation, the relationships among Web objects coming from del.icio.us and their social tags. Furthermore, the propagation of training sample categories from one domain to another is investigated as well.

Heymann, Ramage, and Gacia-Molina (2008) first suggest to exploit semantics-based model, i.e., folksonomies and ontologies, generated from data coming from social network sites to improve the performance of query engines. Similarly, Bender and others (2008) exploit the information provided by social network posts to enhance the quality of the online news searches. Keyword searches are expanded based on the correlations among tags associated with media resources. Differently, Abrol and Khan (2010) analyze the geographical contextual information associated with Twitter posts to both determine most relevant ones, according to the user query, and retrieve a set of semantically related keywords. Finally, to further improve search queries on long document sequences, Lappas, Arai, Platakis, Kotsakos, and Cunopulos (2009) propose to apply burstiness detection methods. Unlike previously discussed approaches, this chapter focuses on exploiting the UGC coming from social networks to drive the news document summarization process.

Document Summarization

Document summarization aims at generating a short and concise summary that describes the content of a single document or a collection of textual documents. Depending on the type of generated summaries, two main approaches to text summarization have been proposed. The sentence-based approach consists in partitioning the document(s) into sentences and selecting the most informative ones to include in the summary (e.g., Carenini, et al., 2007; Goldstein, Mittal, Carbonell, & Kantrowitz, 2000; Wang & Li, 2010; Wang, et al., 2011). Differently, the keyword-based approach focuses on detecting salient keywords that effectively summarize the document content by adopting, for instance, co-occurrence measures (Lin & Hovy, 2003) or latent semantic analysis (Dredze, Wallach, Puller, & Pereira, 2008). The summarization system proposed in this chapter is focused on a sentence-based approach.

A number of sentence-based approaches exploit clustering algorithms (e.g., Radev, Jing, Sty, & Tam, 2004; Wang & Li, 2010; Wang, et al., 2011). Clustering is a well-established data mining technique that may be effectively exploited to group sentences belonging to a document collection and select the most authoritative representatives among each group (Thakkar, et al., 2010; Wang & Li, 2010). For instance, Radev et al. (2004) propose the MEAD text summarizer that clusters documents instead of single sentences and evaluates the corresponding cluster centroids. A centroid, in this context, is a pseudo-document that consists of words having a significant tf-idf value (Tan, Steinbach, & Kumar, 2006), i.e., their tf-idf value exceeds a predefined threshold within the cluster documents. A score is assigned to each sentence according to (1) its similarity to the centroids, (2) its position into the documents, and (3) its length. Differently, Wang and Li (2010) face the issue of updating summaries over time by using an incremental hierarchical clustering algorithm similar to that proposed in Guha, Mishra, Motwani, and O'Callagham (2000), while Wang et al. (2011) propose to use both document-term and sentence-term matrices to simultaneously cluster and summarize documents. However, clustering algorithms suffer from the high computational complexity and the sub-optimality of the devised models. Thus, in some cases, they may be less effective in selecting a small and accurate subset of highly informative sentences.

The selection of the document sentences to include in the summaries could be based on graph-based models (Radev, 2004; Wan & Yang, 2006; Carenini, et al., 2007; Thakkar, et al., 2010), in which graph nodes represent document sentences while edges between nodes are weighted by a similarity measure computed on each node pair. To reduce the computational complexity, early edge pruning is commonly performed by enforcing a minimum similarity threshold. For instance, Radev (2004) ranks sentences according to the eigenvector centrality computed on the sentence linkage matrix by means of the well-known PageRank algorithm (Brin & Page, 1998). Differently, Wan and Yang (2006) consider both sentence novelty and information richness based on the sentence graph. The effectiveness of graph-based approaches strictly depends on both the goodness of the similarity measure and the choice of the minimum similarity threshold value. Lower similarity thresholds averagely provide better results but make the computational effort more severe. Thus, a trade-off between efficiency and effectiveness is needed.

Document summarization has been also addressed by means of linear programming algorithms to identify most relevant sentences by formalizing the summarization task as a min-max optimization problem (Takamura & Okumura, 2009a, 2009b). For instance, Filatova and Hatzivassiloglou (2004) represent sentences as set of words and formalizes the summarization task as a maximum coverage problem with Knapsack constraints. Similarly, Takamura and Okumura, 2009a, 2009b) search for a linear programming solution by also considering the relevance of each sentence within each document. However, these techniques are effective only in case a suitable parameter setting could be devised in advance.

In the last years, some attempts to convey the user-generated content into document summarization process have been performed. For instance, Wikipedia content has been exploited to identify relevant concepts in document collections and analyze their semantic correlations (Gong, et al., 2010; Miao & Li, 2010). Query-based summarization systems that rely on UGC to evaluate document terms have been also proposed (Nastase, 2008; Tang, Yao, & Chen, 2009; Sharifi, Hutton, & Kalita, 2010). Knowledge provided by social annotations has been adopted in graph-based summarization to better discriminate among document sentences (Zhu, et al., 2009). Furthermore, supervised summarization approaches focus on training a unified graph-based model that conveys knowledge coming from UGC. For instance, Yang et al. (2011) exploit a Dual-Wing Factor Graph (DWFG) to learn the prediction model from Web documents and social networks. However, results obtained by supervised methods depend on the quality of the training data. Similarly, the system proposed in this Chapter is focused on driving news document summarization by means of UGC. Unlike previously proposed ones, it combines both the use of a well-founded unsupervised data mining technique, i.e., frequent itemset mining, with a statistical term evaluation performed on both user-generated and news content to provide succinct yet appealing news document summaries.

Data Summarization by Means of Frequent Itemsets

An appealing research issue is the summarization of data collections, coming from different application contexts (e.g., market basket data, medical and biological data, network traffic traces), by exploiting frequent itemsets. Frequent itemset mining is widely exploratory data mining technique to discover valuable correlations among data which has been first introduced in Agrawal, Imielinski, and Swami (1993) in the context of market basket analysis.

Selecting a smart subset of frequent itemsets that effectively summarizes the analyzed data is definitely a challenging task. A significant effort has been devoted to transactional data summarization. Preliminary approaches (Brin, Motwami, & Silverstein, 1997; Jaroszewicz & Simovici, 2004) compare the observed frequency of each itemset against some null hypothesis, i.e., measure how its support diverges from its expected value. However, since this approach is static, the discovered summaries are characterized by a high degree of redundancy. Differently, dynamic approaches take the already discovered patterns into account as well. Several previous dynamic approaches (e.g., Kontonasios & De Bie, 2010; Tatti & Heikinheimo, 2008; Tatti & Mampaey, 2010) make use of the maximum entropy model. They commonly perform frequent itemset mining followed by post-pruning. The post-processing phase is driven by one or more analyst-provided thresholds (e.g., the maximum error threshold [Kontonasios & De Bie, 2010] or a minimum significance level [Tatti & Heikinheimo, 2008]). Recently, a novel and parameter-free itemset-based method for succinctly summarizing transactional data has been proposed (Mampaey, Tatti, & Vreeken, 2011). It provides a highly informative and non-redundant data summary by adopting an heuristics to solve the maximum entropy problem. Unlike previous approaches, it allows mining and evaluating on-the-fly itemsets without the need of a post-pruning step. This feature makes it particularly suitable for being applied in text summarization.

A parallel research issue is devoted to summarizing XML data by means of frequent itemsets. For instance, Baralis, Garza, Quintarelli, and Tanca (2007) adopts a preprocessing step performed on tree-based XML data representation to tailor the XML data format to the relational schema. This enables the exploitation of traditional association rule mining algorithms to provide a summarized representation of XML documents. In this Chapter, we adopt a selection of frequent itemsets in news document summarization. To the best of our knowledge, the usage of frequent itemsets in textual document summarization has never been investigated before.

THE NeDocs SYSTEM

NeDocS is a novel summarization system that addresses news document summarization driven by user-generated content. Figure 1 shows the main NeDocS architectural blocks.

NeDocS is based on a three-step process. Each step is accomplished by a specific block. A brief description of each block follows.

- **Data retrieval and preprocessing:** Based on analyst-provided queries, a collection of news documents ranging over a specific topic and a collection of on-topic posts coming from social network sites are retrieved from the Web. The retrieved data is preprocessed to make it suitable for the data mining and knowledge discovery process. To this aim, it exploits both a transactional and a Bag-Of-Word (BOW) representation of the news documents.
- **Web content analysis:** The preprocessed data is analyzed by means of a data mining process that entails: (1) the discovery and selection of the top-k most informative and not redundant frequent itemsets and (2) the evaluation of the news document sentence relevance based on terms occurring in it. The sentence relevance score, based on tf-idf statistics, also considers the term significance in the retrieved social network posts.
- **Sentence selection:** Sentences belonging to the news document collection are selected based on both the top-k selected itemsets and an evaluation score. The problem of sentence selection is modeled as a set covering optimization problem and solved by means of a branch-and-bound algorithm.

In the following, each block and its main functionalities are thoroughly described.

Data Retrieval and Preprocessing

This block focuses on retrieving both news documents and social network posts from the Web as well as making the raw textual data suitable for the data mining and knowledge discovery process.

An analyst-provided query is submitted to a set of Web search engines and social network search Application Programming Interfaces (APIs) to retrieve the following two textual data collections ranging over the same topic: (1) a collection of news documents and (2) a collection of social network posts. The submitted query formats depend on the characteristics of the considered Web Search Engines and Search APIs. The news collection $D=\{d_1, \ldots, d_n\}$ is a set of textual documents stored in the electronic form, where each document d_q is composed of a set of sentences $S_q=\{S_{1q}, \ldots, S_{nq}\}$. Analogously, the post collection $P=\{p_1, \ldots, p_u\}$ is organized as a single textual document where each post p_t corresponds to a sentence.

Figure 1. The NeDocS summarization system

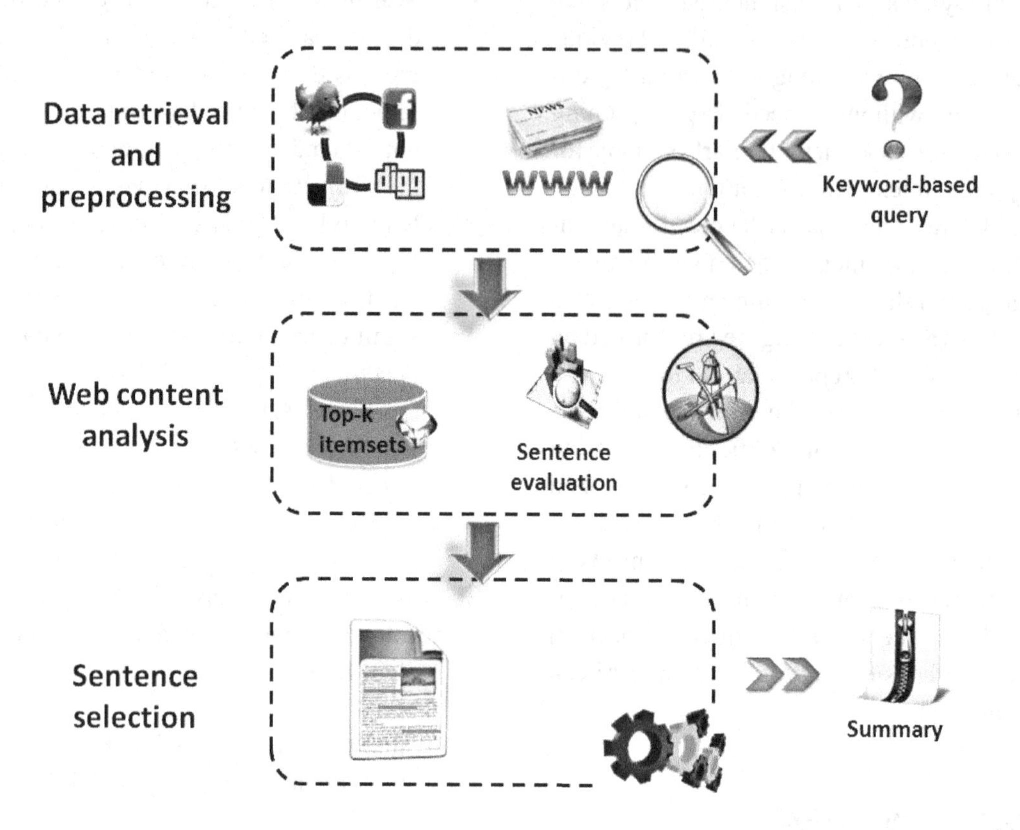

The goal of NeDocS is to summarize the content of the news collection by exploiting the information provided by social network posts. To address this issue, two different document/sentence representations are exploited: (1) the traditional Bag-Of-Word (BOW) sentence representation to evaluate the sentence relevance score and (2) the transactional data format to discover and select the top-k most informative yet non-redundant frequent itemsets from the news collection. The raw textual news and post contents are first pre-processed to make them suitable for the data mining and knowledge discovery process. To avoid noisy information stopwords, numbers, and website URLs are removed, while the Wordnet stemming algorithm (Bird, Klein, & Loper, 2009) is applied to reduce words to their base or root form (i.e., the stem). A formal definition of both the BOW and the transactional sentence representations follows.

The BOW representation of an arbitrary sentence belonging to either a news or the post document is the set of all word stems (i.e., terms) occurring in it. Each term belonging to the BOW representation of a news document sentence is characterized by a tf-idf value (Tan, et al., 2006). A detailed description of the well-founded tf-idf statistics follows. The whole news document content could be represented in a matrix form TC, in which each row represents a distinct term of the document collection while each column corresponds to a document. Each element tc_{iq} of the matrix TC is the tf-idf value associated with a term w_i in the document d_q belonging to the whole collection D. It is computed as follows:

$$tc_{iq} = \frac{n_{iq}}{\sum_r n_{rq}} \cdot \log \frac{|D|}{(\{d_q \in D : w_i \in d_q\})}$$

where n_{iq} is the number of occurrences of i-th term w_i in the q-th document d_q, D is the collection of documents, $\sum_r n_{rq}$ is the sum of the number of occurrences of all terms in the q-th document d_q, and $\log \frac{|D|}{(\{d_q \in D : w_i \in d_q\})}$ represents the inverse document frequency of term w_i.

Differently, we define the transactional sentence representation as follows.

Definition 1 (Transactional sentence representation): Let s_{jq} be an arbitrary sentence of the news collection and $tr_{jq}=\{w_1, _{...}, w_L\}$ the subset of distinct terms occurring in the sentence s_{jq}, i.e., $tr_{jq} \subseteq s_{jq}$ and $w_p \neq w_r$ for any $p \neq r$. The transactional representation of a document sentence s_{jq} is the transaction tr_{jq} whose items are distinct terms taken from its BOW representation.

A transactional representation T of the news document collection D is the union of all transactions tr_{jq} corresponding to each sentence s_{jq} belonging to any document $d_q \in D$.

Web Content Analysis

This block analyzes the relevance and the pertinence of each sentence of the news document collection according to the UGC posted on social networks. To this aim, it performs a two-step process that entails:

- The discovery and selection of a set of top-k most informative and not redundant frequent itemsets from the news document collections tailored to the transactional data format.
- The evaluation of a sentence score, based on the tf-idf statistics associate with single document terms, evaluated on the BOW sentence representation. To take into account the Web user interests in the news topic and not disregard rare but very relevant concepts, the evaluation score also considers the term significance in the post collection.

In the following, we thoroughly discussed the two adopted Web content analysis separately.

Top-k frequent itemset mining and selection: Frequent itemset mining is a well-established data mining technique (Agrawal & Srikant, 1994) that focuses on discovering recurrences, i.e., itemsets, frequently occurring in the source data. This technique is commonly applied to transactional data. A transactional dataset is set of instances, called transactions. Each instance includes an arbitrary set of items (i.e., literals). The most famous examples of transactional datasets are the market basket data, where each transaction consists of set of items bought by a customer at the supermarket (i.e., the market basket).

An itemset I of length k, i.e., a k-itemset, is a set of k distinct items. We denote as *T*(I) the set of transactions covered by I, i.e., $T(I)=\{tr \in T \mid I \subseteq tr\}$. The support of an itemset I is the observed frequency of occurrence of I in T, i.e., sup(I)= $\frac{T(I)}{|T|}$. Since the problem of discovering all itemsets in a transactional dataset is computationally intractable (Agrawal & Srikant, 1994), itemset mining is commonly driven by a minimum support threshold.

Data summarization based on frequent itemsets focuses on discovering the minimal and most informative subset of frequent itemsets, i.e., itemsets that exceed a given support threshold, to succinctly summarize the analyzed data. NeDocS addresses the problem of top-K frequent itemset mining and selection from the transactional representation T of the news collection D. The problem could be formalized as follows.

Definition 2 (Top-K frequent itemset mining and selection problem): Given a news document collection T tailored to the transactional data format, a minimum support threshold *minsup* and a pattern set size *k*, the problem of top-K itemset mining and selection focuses on discovering the most informative yet non-redundant set of *K* frequent itemsets from T.

To maximize the informative content of the summary while minimizing its redundancy, we make use of the maximum entropy model. A maximum entropy model is a compact data representation that optimally exploits the background information. Pattern-based maximum entropy models ensure that the included patterns are characterized by the maximal informative content, according to the analyzed data and the previously selected ones.

Among the set of previously proposed approaches focused on succinctly summarizing transactional data by means of an itemset-based maximum entropy model (e.g., Kontonasios & De Bie, 2010; Tatti & Heikinheimo, 2008; Tatti & Mampaey, 2010 Mampaey, et al., 2011), we adopt an algorithm recently proposed in Mampaey et al. (2011). Unlike previous approaches, it exploits an entropy-based heuristics to drive the mining process and select the most informative yet not redundant itemsets without the need of a post-processing step. More specifically, it adopts the Kullback-Leibler measure (Schwarz, 1978) as a measure of the divergence between the observed itemset support and its expected support, according to the previously selected patterns. This approach allows iteratively selecting the best frequent itemset to include in the model as well as exploiting the measure convexity to early prune a large part of the itemset search space. Its efficiency and effectiveness in discovering succinct transactional data summaries makes it particularly suitable for the application to document summarization.

Sentence relevance score: The sentence relevance score $Sr(s_{jq})$ associated with the j-th sentence of the q-th news document measures the significance of its terms in both the news and post collections. It is computed from the BOW representation of both news and post sentences. To weigh the impact of each term belonging to sentence s_{jq}, a term relevance score $St(w_i, s_{jq})$ for an arbitrary i-th term of sentence s_{jq} is preliminary evaluated as follows.

$$St(w_i s_{jq}) = \lambda \cdot \frac{freq(w_i, P)}{\max_{w_r \in s_{jq}} freq(w_r, P)} + (1-\lambda) \cdot \frac{tc_{iq}}{\max_{r: w_r \in s_{jq}} tc_{rq}}$$

The term evaluation score is composed of two distinct terms weighted by a (analyst-provided) weighting factor $\lambda \in [0,1]$. The former term weighs the term relevance in the post collection by considering its observed absolute frequency of occurrence $freq(w_i, P)$ in the retrieved post collection, while the latter one is based on tf-idf statistics and is evaluated on the news document collection. Both terms are normalized by their maximum observed value achieved in the analyzed sentence s_{jq}. The choice of the weighting factor λ discriminates among the impact of the two terms: the higher is the value of λ the higher is the influence of the social network content on the term evaluation score.

The sentence relevance score $SR(s_{jq})$ is defined as the arithmetic average of the scores of its terms. Its formal expression follows.

$$Sr(s_{jq}) = \frac{\sum_{i: w_i \in t_{jq}} St(w_i, s_{jq})}{|tr_{jq}|}$$

where $|tr_{jq}|$ is the number of distinct terms occurring in sentence s_{jq} and $\sum_{i: w_i \in t_{jq}} St(w_i, s_{jq})$ is the sum of the term score values associated with distinct terms (i.e., word stems) in s_{jq}.

Sentence Selection

The last NeDocS block focuses on selecting most relevant sentences belonging to the news document collection to include in the summary.

The sentence selection process is driven by both the set of generated top-K frequent itemsets and the sentence relevance score. Sentence pertinence to the set of top-K itemsets measures how the analyzed sentence is covered by the most significant correlations among multiple news collection terms. Differently, the previously introduced sentence relevance score evaluates the significance of single sentence terms according to both the news and the social network content. The combination of the two measure allows effectively summarizing the news document content without disregarding the most common Web users' interests.

To evaluate sentence pertinence with respect to the top-k selected itemsets, we introduce the concept of sentence coverage vector.

Definition 3 (Sentence coverage vector): Let T be the collection of news document collection tailored to the transactional data format. Let $I=\{I_1, \ldots, I_{ms}\}$ be the set of top-K frequent itemsets discovered from T. The sentence coverage vector $SC_{jq}=\{sc_{j1}, \ldots, sc_{jp}\}$ associated with sentence $tr_{jq} \in T$ is a K-length binary vector whose i-th element sc_{ji} indicates whether itemset $I_i \in I$ is included or not in tr_{jq}, i.e., $sc_{ji} = 1_{tr_{jq}}(I_i)$ where:

$$1_{tr_{jq}}(I_i) = \begin{cases} 1 & if\ I_i \subseteq tr_{jq} \\ 0 & otherwise \end{cases}$$

The coverage of a sentence s_{jq} with respect to the top-k itemsets is defined as the number of 1's that occur in the corresponding coverage vector SC_{jq}. More generally, the coverage of a set of sentences is given by the number of 1's that occur in the vector obtained from logic OR between the coverage vectors associated with the selected sentences.

We formalize the problem of selecting the most informative and not redundant sentences according to both the top-K frequent itemsets and the sentence relevance score as a set covering optimization problem.

The set covering problem: Given the collection of news collection sentences, each one weighted by its relevance score, and the selected top-k itemsets, the set covering optimization problem focuses on selecting the minimal set of sentences that best covers the top-K itemsets according to the scores, i.e., the subset of sentences that maximizes both coverage and relevance.

The set covering optimization problem is known to be NP-hard. However, since the set covering problem is a min-max problem, it may be converted to a linear programming problem and addressed by using combinatorial optimization strategies. Thus, NeDocS exploits a branch-and-bound algorithm (Ralph & Guzelsoy, 2005) to accomplish the set covering task, whose implementation is available at http://www.coin-or.org/SYMPHONY. Branch-and-bound algorithms enumerate and explore effectively the space of possible solutions to search for the optimal one. The optimal solution provides the subset of most representative yet not redundant sentences to select as news collection summary. More insights on the adopted solution follow.

The adopted approach performs a two-step process in which the set of sentences and their coverage vectors, based on the top-K discovered itemsets, are analyzed. The first step aims at reducing the problem dimensions by pruning less informative elements whereas the second one accomplishes the sentence selection task by means of a linear programming strategy.

1. **Problem dimensionality reduction:** The size of the sentence coverage vectors associated with the news document sentences is minimized by reducing less informative elements. More specifically, each element of the sentence coverage vector that contains all 0's or 1's over all the top-k itemsets is removed because it is uninformative for the searching procedure.

The number of sentences to consider in the sentence selection strategy is minimized based on their eligibility according to both their coverage vector and relevance score. More specifically, sentences whose coverage vector is a subsequence of another one are preventively pruned. If two or more sentences are characterized by the same coverage vector, only the sentence with maximal relevance score is kept.

2. **Sentence selection:** Sentences are selected based on an optimization procedure that searches for the minimum set of sentences needed to cover all the top-k itemsets. Since it is a min-max problem, it can be converted to the following linear programming problem as follows:

$$\min \quad \sum_{i=1}^{|S|} q_i$$

$$\sum_{i=1}^{|S|} sc_{ij} \cdot q_i \quad j \in \{1,2,\ldots,k\}$$

$$q(s_i) \in \{0,1\}$$

where q_i indicates whether the i-th news document sentence is included or not in the summary, K is the size of the sentence coverage vector (i.e., the number of selected itemsets), |S| is the number of sentences in the news document collection, and sc_{ij} is the j-th element of the coverage vector associated with the i-th sentence.

Experimental results, reported in Section "Experimental evaluation," show that the effectiveness of the proposed summarization approach against two of the mostly used open-source text summarizers.

EXPERIMENTAL EVALUATION

We conducted a set of experiments to address the following issues: (1) the effectiveness of NeDocs in summarizing real news documents, (2) a performance comparison between the proposed summarization system and two widely used summarizers, i.e., the Open Text Summarizer (OTS) (Rotem, 2006) and TexLexAn (TexLexAn, 2011), and (3) the impact of the number of most significant itemsets *K*, the weighting factor λ, and the support threshold *minsup* on the performance of NeDocS.

We evaluated all the summarization approaches on a collection of real-life news articles. To this aim, the 10 top-ranked news documents, whose length ranges from 600 to 2,000 words, provided by the Google News Web search engine (http://news.google.com), concerning the following recent news topics have been selected:

- **Afghanistan:** War in Afghanistan 2011.
- **Irene:** The Irene hurricane beats down on the U.S. East Coast.
- **Apple:** Steve Jobs resignation announcement from his role as Apple's CEO.
- **Strauss:** Dominguez Strauss Kahn accused of sexual assault.
- **UK_riots:** The U.K. suffered widespread rioting, looting, and arson in August 2011.
- **US_Open:** The U.S. Open tennis tournament held in New York City (USA) in August 2011.

News are retrieved by the Google search engine from the most authoritative sources (e.g., The Guardian, BBC, Reuters, New York Post, etc.).

To retrieve collections of social network posts ranging over the same news collection topics, we retrieved and analyzed messages posted on Twitter. Twitter (www.twitter.com) posts, i.e., the tweets, are retrieved by means of the Twitter Search Application Programming Interface (API) by specifying the same news queries. On-topic tweet collections composed of 800 tweets each are crawled. From the returned data, provided in the JSON data format, only the textual message content is kept and processed. Both the news collections and the on-topic Twitter posts are made available for research purposes, upon request to the authors.

To compare the results by NeDocS with OTS (Rotem, 2006) and TexLexAn (2011), we used the ROUGE (Lin & Hovy, 2003) toolkit (version

1.5.5), which is widely applied by Document Understanding Conference (DUC) for document summarization performance evaluation[1]. It measures the quality of a summary by counting the unit overlaps between the candidate summary and a set of reference summaries. Intuitively, the summarizer that achieves the highest ROUGE scores could be considered as the most effective one.

Several automatic evaluation scores are implemented in ROUGE. For the sake of brevity, we reported only ROUGE-3 and ROUGE-4 as representative scores. Analogous results were achieved for the other scores. Since a "golden summary" (i.e., the optimal document collection summary) is not available for Web news document, we performed a leave-one-out cross validation. More specifically, for each category, we summarized nine out of ten news documents and we compared the resulting summary with the remaining (not yet considered) document, which has been selected as golden summary at this stage. Next, we tested all other possible combinations by varying the golden summary and we computed the average performance results, in terms of precision, recall, and F-measure, achieved by each summarizer for both ROUGE-3 and ROUGE-4.

Summary Comparison

We compared the summaries generated by NeDocS for the considered news document collections with that generated by two largely used competitors, i.e., OTS (Rotem, 2006) and TexLexAn (2011). Table 1 reports the 3-top sentences included by the summaries generated by NeDocS and the other competitors of the Irene Hurricane news collection, chosen as representative among all the tested ones.

The itemsets {*hurricane, coast, east*}, {*damage, cause, wind*}, and {*agency, emergency, management, federal*} are examples of the itemsets included in the model extracted by our approach. They effectively summarize the main topic of the document collection. The summary provided by our approach captures the main topics of the collection. In particular, the NeDocS

Table 1. Summary examples: hurricane Irene news collection

Method	Summary
NeDocS	New York: hurricane Irene will most likely prove to be one of the 10 costliest catastrophes in the nation's history, and analysts said that much of the damage might not be covered by insurance because it was caused not by winds but by flooding, which is excluded from many standard policies. Homeland Security Secretary Janet Napolitano and Federal Emergency Management Agency Administrator Craig Fugate will tour New York and New Jersey Wednesday to view the damage firsthand. From North Carolina to Vermont, federal officials fanned out along the storm-stricken East Coast Tuesday inspecting damage hurricane Irene inflicted over the weekend.
OTS	As emergency airlift operations brought ready-to-eat meals and water to Vermont residents left isolated and desperate, states along the Eastern Seaboard continued to be battered Tuesday by the after effects of Irene, the destructive hurricane turned tropical storm. Dangerously-damaged infrastructure, 2.5 million people without power and thousands of water-logged homes and businesses continued to overshadow the lives of residents and officials from North Carolina through New England, where the storm has been blamed for at least 44 deaths in 13 states. But new dangers developed in New Jersey and Connecticut, where once benign rivers rose menacingly high.
TexLexAn	As emergency airlift operations brought ready-to-eat meals and water to Vermont residents left isolated and desperate, states along the Eastern Seaboard continued to be battered Tuesday by the after effects of Irene, the destructive hurricane turned tropical storm. Search-and-rescue teams in Paterson have pulled nearly 600 people from flooded homes in the town after the Passaic River rose more than 13 feet above flood stage, the highest level since 1903. It's one of several towns in states such as New Jersey, Connecticut, New York, Vermont and Massachusetts dealing with the damage of torrential rain and flooding spawned by Hurricane Irene (Clevelad hurricane Katrina was a Category 5 hurricane and Hurricane Irene was a Category 1 huriccane).

summary first discusses about the passage of Irene hurricane on the U.S. East coast but it also provides information about the effects generated in Vermont and New Jersey, which is another topic of interest for the online community. Its interestingness is proved by the high document frequency of occurrences of terms like *catastrophe*, *disaster*, *Jersey*, *Vermont*, *York*, *east*, *coast*, and *storm*, in addition to *Irene* and *hurricane*. Differently, OTS and TexLexAn generate less focused summaries, which disregard some of the most relevant topics of interest of online community. For instance, both summarizers select sentences that describe the emergency in Vermont and New Jersey, while disregarding the effects on New York. Moreover, TexLexAn summary also talks about the emergency in Paterson, which may be deemed a topic of less interest for the community.

Performance Evaluation

We evaluated the performance, reported in Tables 2 and 3, in terms of ROUGE-3 and ROUGE-4 precision (Pr), recall (R), and F-measure (F), of NeDocS against OTS and TexLexAn. For both OTS and TexLexAn we adopted the configuration

Table 2. Performance comparison in terms of ROUGE-3 score. Statistically significant improvements between NeDocS and the other approaches are starred.

Dataset	TexLexAn			OTS			NeDocS		
	Recall	Precision	F-score	Recall	Precision	F-score	Recall	Precision	F-score
Afghanistan (K=10)	0.0006*	0.0070*	0.0011*	0.0023*	0.0197*	0.0040*	**0.0028**	**0.0318**	**0.0051**
Irene (K=11)	0.0016*	0.0161*	0.0029*	0.0017*	0.0168*	0.0031*	**0.0030**	**0.0325**	**0.0055**
Apple (K=19)	0.0008*	0.0080*	0.0015*	0.0006*	0.0059*	0.0011*	**0.0017**	**0.0159**	**0.0031**
Strauss (K=7)	0.0037*	0.0307*	0.0065*	0.0020*	0.0199*	0.0036*	**0.0075**	**0.0649**	**0.0134**
UK_riots (K=15)	0.0007*	0.0071*	0.0012*	0.0008*	0.0071*	0.0014*	**0.0018**	**0.0159**	**0.0031**
US_Open (K=8)	0.0020*	0.0168*	0.0036*	0.0038*	0.0322*	0.0067*	**0.0081**	**0.0695**	**0.0145**

Table 3. Performance comparison in terms of ROUGE-4 score. Statistically significant improvements between NeDocS and the other approaches are starred.

Dataset	TexLexAn			OTS			NeDocS		
	Recall	Precision	F-score	Recall	Precision	F-score	Recall	Precision	F-score
Afghanistan (K=10)	0.0001*	0.0010*	0.0001*	0.0002*	0.0020*	0.0004*	**0.0010**	**0.0126**	**0.0018**
Irene (K=11)	0.0002*	0.0020*	0.0003*	0.0003*	0.0030*	0.0005*	**0.0014**	**0.0159**	**0.0025**
Apple (K=19)	0.0001*	0.0001*	0.0001*	0.0001*	0.0010*	0.0002*	**0.0004**	**0.0031**	**0.0007**
Strauss (K=7)	0.0016*	0.0117*	0.0028*	0.0002*	0.0010*	0.0003*	**0.0035**	**0.0299**	**0.0062**
UK_riots (K=15)	0.0001*	0.0010*	0.0002*	0.0001*	0.0001*	0.0001*	**0.0006**	**0.0050**	**0.0011**
US_Open (K=8)	0.0002*	0.0020*	0.0004*	0.0020*	0.0167*	0.0036*	**0.0040**	**0.0335**	**0.0070**

suggested by the respective authors. For NeDocS we enforced a minimum support threshold *minsup*=1%, a weighting factor λ=0.75. To effectively deal with data distributions, the number *k* of selected representative frequent itemsets is set to its best value in the range [5, 30] for each dataset and reported next to the dataset names in Tables 2 and 3. A more detailed discussion on the impact of *minsup*, λ, and *K* on the performance of NeDocS is reported in Section "Impact of the NeDocS parameters."

NeDocS performs better than all the other considered summarizers for every combination of dataset, score, and measure. To validate the statistical significance of the performance improvement achieved by NeDocS against OTS and TexLexAn, we performed the paired t-test of statistical significance (Dietterich, 1998) by setting as significance level p-value = 0.05 for all evaluated datasets and measures. Results, summarized in Tables 2 and 3, confirm the significance of the achieved performance improvements and, thus, the effectiveness of the proposed approach.

Impact of the NeDocS Parameters

We analyzed the impact of both the number of selected itemsets *K* and the weighting factor λ on the performance of the NeDocS summarizer. Since the minimum support threshold *minsup*, enforced during the frequent itemset mining step, relevantly affects the overall quality of the generated itemsets, we analyzed the performance of the NeDocS summarization system by setting three different support thresholds (i.e., 0.5%, 1%, and 1.5%) chosen as representatives of all the tested support threshold values.

In Figures 2 and 3, we reported the F-measure achieved by the NeDocS summarizer by varying, respectively, the number of selected frequent itemsets on the US_Open dataset and the weighting factor on the Apple dataset. In both graphs, the respective curves are associated with different support thresholds. For the sake of brevity, we reported only the results obtained with the ROUGE-3 score. Analogous results have been obtained for the other ROUGE scores, in terms of precision and recall measures, and for all other configurations.

The enforcement of a minimum support threshold in the itemset mining step significantly affects

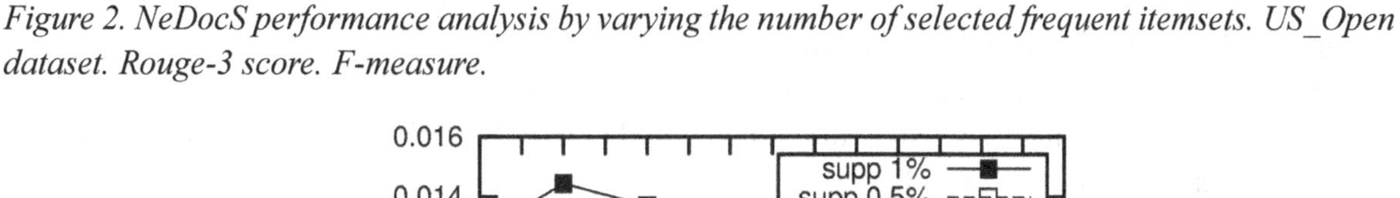

Figure 2. NeDocS performance analysis by varying the number of selected frequent itemsets. US_Open dataset. Rouge-3 score. F-measure.

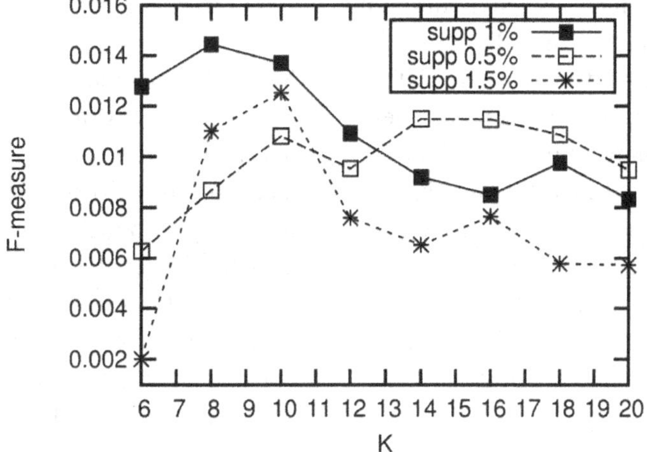

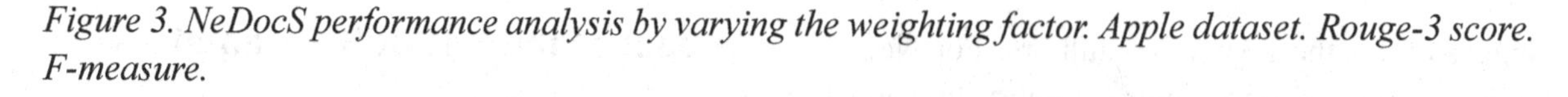

Figure 3. NeDocS performance analysis by varying the weighting factor. Apple dataset. Rouge-3 score. F-measure.

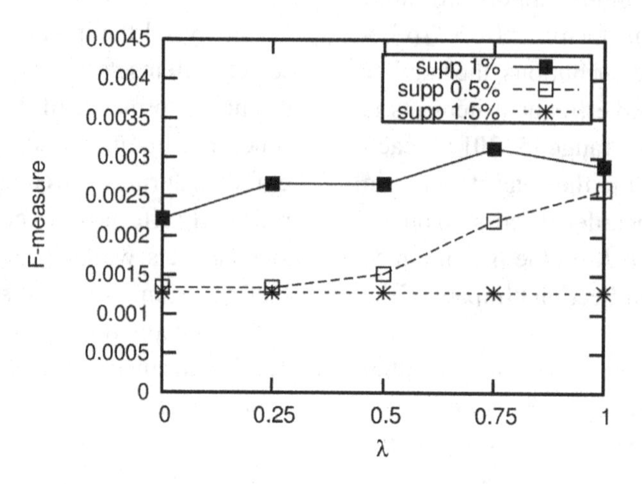

the quality of the selected sentences to include in the summary. When higher support thresholds (e.g., 1.5%) are enforced, the discovered correlations may be too general to succinctly summarize most relevant news document content. Oppositely, when very low support thresholds (e.g., 0.5%) are enforced, the selected itemsets become too much specialized to effectively and concisely summarize the whole news document content. At medium values (e.g., 1%) the best balancing between model specialization and generalization is achieved, thus, on average, the best summarization performance is achieved.

The number of selected itemsets is another important parameter that affects the NeDocS summarization performance. When a limited number of itemsets (e.g., $K = 6$ in Figure 3) is selected, the whole document content may still be partially uncovered by the generated pattern set. Indeed, the average summarization performance is poor. On the contrary, when a large number of itemsets (e.g., $K = 16$) is selected, the coverage of the news content is exhaustive but redundant, thus, the quality of the generated summary is again suboptimal. Finally, the best trade-off is achieved when a medium number of itemsets ($K = 8$) is selected.

Finally, the impact of the UGC content is decisive for summarizing the news document content effectively. When higher weighting factors are enforced, the impact of the UGC published on social networks become more and more relevant. Considering the Web user preferences allows significantly improving the summarization performance on all the tested datasets. For instance, when setting $\lambda = 0.75$ the best summarization performance is achieved. The impact of UGC is less discriminative when the specialization of the selected itemsets is low, e.g., when a high support threshold (*minsup* = 1.5%) is enforced.

FUTURE RESEARCH DIRECTIONS

Multi-document summarization based on frequent itemsets and driven by UGC opens many future research directions. Firstly, summaries generated from Web documents need to be frequently updated due to the continuous evolution and the growth of the content published on the Web. Once a novel and more authoritative set of documents is published, the generated summaries should be updated accordingly. Thus, the extension of the proposed summarization system to also address the problem of incremental summary updating is definitely an interesting future research topic.

Secondly, NeDocS models the set covering problem as a min-max problem and solves it by means of a branch-and-bound implementation. However, discovering the set of sentences that best covers the selected top-K itemsets may be computationally expensive. Indeed, we have a plan to reduce the complexity of the summarization task by using a greedy approach for solving the set-covering problem.

Lastly, NeDocS addresses news summarization by combining the usage of frequent itemsets with a sentence relevance score based on UGC. However, the impact of the two factors is separately evaluated and then jointly considered in the sentence selection step. The pushing of UGC-based constraints into the itemset mining process may effectively reduce the complexity of the mining task while taking into account the community users' opinion as well.

CONCLUSION

This Chapter presents a novel news multi-document summarization system that combines the knowledge provided by a highly informative and not redundant subset of frequent itemsets with a statistical evaluation of the sentence terms' relevance according to on-topic user-generated posts retrieved from social networks. The usage of frequent itemset mining, well established in different research contexts, allows the discovery and selection of the most relevant correlations among terms. Furthermore, the sentence relevance evaluation based on single term significance within the user-generated content coming from social networks allows not disregarding sentences whose terms infrequently occur in the document collection but are deemed appealing by Web users. To the best of knowledge, the combined usage of frequent itemsets and user-generated content in a news summarization system has never been investigated before.

Experiments, conducted on real-life news articles and driven by on-topic messages posted on Twitter, show the effectiveness of the proposed summarization system.

REFERENCES

Abrol, S., & Khan, L. (2010). TWinner: Understanding news queries with geo-content using Twitter. In *Proceedings of the 6th Workshop on Geographic Information Retrieval,* (pp. 1-8). IEEE.

Adomavicius, G., & Tuzhilin, A. (2011). Context-aware recommender systems. *Recommender Systems Handbook.* Retrieved from http://ids.csom.umn.edu/faculty/gedas/NSFCareer/CARS-chapter-2010.pdf

Agrawal, R., Imielinski, T., & Swami, A. (1993). Mining association rules between sets of items in large databases. *SIGMOD Record, 22*(2), 207–216. doi:10.1145/170036.170072

Agrawal, R., & Srikant, R. (1994). Fast algorithms for mining association rules in large databases. In *Proceedings of the 20th VLDB Conference*, (pp. 487-499). VLDB.

Baralis, E., Garza, P., Quintarelli, E., & Tanca, L. (2007). Answering XML queries by means of data summaries. *ACM Transactions on Information Systems, 25*(3), 1–33. doi:10.1145/1247715.1247716

Basile, P., Gendarmi, D., Lanubile, F., & Semeraro, G. (2007). Recommending smart tags in a social bookmarking system. *Bridging the Gap between Semantic Web and Web, 2*, 22-29.

Bender, M., Crecelius, T., Kacimi, M., Michel, S., Neumann, T., & Parreira, J. X. ... Weikum, G. (2008). Exploiting social relations for query expansion and result ranking. In *Proceedings of the IEEE 24th International Conference on Data Engineering Workshop,* (vol. 2, pp. 501-506). IEEE Press.

Bird, S., Klein, E., & Loper, E. (2009). *Natural language processing with Python*. New York, NY: O'Reilly Media.

Brin, S., Motwani, R., & Silverstein, C. (1997). Beyond market baskets: generalizing association rules to correlations. *SIGMOD Record, 26*(2), 265–276. doi:10.1145/253262.253327

Brin, S., & Page, L. (1998). The anatomy of a large-scale hypertextual web search engine. In *Proceedings of the Seventh International Conference on World Wide Web,* (vol. 7, pp. 107-117). IEEE.

Carenini, G., Ng, R. T., & Zhou, X. (2007). Summarizing email conversations with clue words. In *Proceedings of the World Wide Web Conference Series*, (pp. 91–100). IEEE.

Dietterich, T. G. (1998). Approximate statistical test for comparing supervised classification learning algorithms. *Neural Computation, 10*(7), 1895–1923. doi:10.1162/089976698300017197

Dredze, M., Wallach, H. M., Puller, D., & Pereira, F. (2008). Generating summary keywords for emails using topics. In *Proceedings of the 13th International Conference on Intelligent User Interfaces*, (pp. 199-206). IEEE.

Filatova, E., & Hatzivassiloglou, V. (2004). A formal model for information selection in multi-sentence text extraction. In *Proceedings of the 20th International Conference on Computational Linguistics*, (p. 397). IEEE.

Goldstein, J., Mittal, V., Carbonell, J., & Kantrowitz, M. (2000). Multi-document summarization by sentence extraction. In *Proceedings of the ANLP/NAACL Workshop on Automatic Summarization*, (pp. 40-48). ANLP.

Gong, S., Qu, Y., & Tian, S. (2010). Summarization using Wikipedia. In *Proceedings of Text Analysis Conference.* IEEE.

Guha, S., Mishra, N., Motwani, R., & O'Callaghan, L. (2000). Clustering data streams. In *Proceedings of the 41st Annual Symposium on Foundations of Computer Science*, (p. 359). IEEE.

Herlocker, J. L., Konstan, J. A., Terveen, L. G., & Riedl, J. T. (2004). Evaluating collaborative filtering recommender systems. *ACM Transactions on Information Systems, 22*(1), 5–53. doi:10.1145/963770.963772

Heymann, P., Ramage, D., & Garcia-Molina, H. (2008). Social tag prediction. In *Proceedings of the 31st Annual International ACM SIGIR Conference on Research and Development in Information Retrieval*, (pp. 531-538). ACM Press.

Jaroszewicz, S., & Simovici, D. A. (2004). Interestingness of frequent itemsets using Bayesian networks as background knowledge. In *Proceedings of the Tenth ACM SIGKDD International Conference on Knowledge Discovery and Data Mining*, (pp. 178-186). ACM Press.

Kontonasios, K., & De Bie, T. (2010). An information-theoretic approach to finding informative noisy tiles in binary databases. In *Proceedings of the SIAM International Conference on Data Mining.* SIAM.

Lappas, T., Arai, B., Platakis, M., Kotsakos, D., & Gunopulos, D. (2009). On burstiness-aware search for document sequences. In *Proceedings of the 15th ACM SIGKDD International Conference on Knowledge Discovery and Data Mining*, (pp. 477-486). ACM Press.

Li, Q., Wang, J., Chen, Y. P., & Lin, Z. (2010). User comments for news recommendation in forum-based social media. *Information Sciences*, *180*(24), 4929–4939. doi:10.1016/j.ins.2010.08.044

Lin, C., & Hovy, E. (2003). Automatic evaluation of summaries using N-gram co-occurrence statistics. In *Proceedings of the Conference of the North American Chapter of the Association for Computational Linguistics on Human Language Technology*, (vol. 1, pp. 71-78). ACL.

Mampaey, M., Tatti, N., & Vreeken, J. (2011). Tell me what I need to know: Succinctly summarizing data with itemsets. In *Proceedings of the 17th ACM SIGKDD Conference on Knowledge Discovery and Data Mining*. ACM Press.

Mathioudakis, M., & Koudas, N. (2010). Twitter-Monitor: Trend detection over the twitter stream. In *Proceedings of the 2010 International Conference on Management of Data*, (pp. 1155-1158). IEEE.

Miao, Y., & Li, C. (2010). WikiSummarizer - A Wikipedia-based summarization system. In *Proceedings of Text Analysis Conference*. IEEE.

Nastase, V. (2008). Topic-driven multi-document summarization with encyclopedic knowledge and spreading activation. In *Proceedings of Conference on Empirical Methods on Natural Language Processing*, (pp. 763-772). IEEE.

Phelan, O., McCarthy, K., & Smyth, B. (2009). Using Twitter to recommend real-time topical news. In *Proceedings of the Third ACM Conference on Recommender Systems*, (pp. 385-388). ACM Press.

Radev, D. R. (2004). Lexrank: Graph-based lexical centrality as salience in text summarization. *Journal of Artificial Intelligence Research*, 22.

Radev, D. R., Jing, H., Sty, M., & Tam, D. (2004). Centroid-based summarization of multiple documents. *Information Processing & Management*, *40*(6), 919–938. doi:10.1016/j.ipm.2003.10.006

Ralphs, T. K., & Güzelsoy, M. (2005) The symphony callable library for mixed-integer linear programming. In *Proceedings of the Ninth INFORMS Computing Society Conference*, (pp. 61-76). INFORMS.

Rotem, N. (2006). *Open text summarizer (OTS)*. Retrieved in July 2011 from http://libots.sourceforge.net

Schwarz, G. (1978). Estimating the dimension of a model. *Annals of Statistics*, *6*(2), 461–464. doi:10.1214/aos/1176344136

Sharifi, B., Hutton, M. A., & Kalita, J. (2010). Automatic summarization of Twitter topics. In *Proceedings of the National Workshop on Design and Analysis of Algorithms*. IEEE.

Shepitsen, A., Gemmell, J., Mobasher, B., & Burke, R. (2008). Personalized recommendation in social tagging systems using hierarchical clustering. In *Proceedings of the 2008 ACM Conference on Recommender Systems*, (pp. 259-266). ACM Press.

Song, Y., Zhang, L., & Giles, C. L. (2011). Automatic tag recommendation algorithms for social recommender systems. *ACM Transactions on the Web*, *5*(1), 4:1-4:31.

Takamura, H., & Okumura, M. (2009a). Text summarization model based on maximum coverage problem and its variant. In *Proceedings of the 12th Conference of the European Chapter of the Association for Computational Linguistics*, (pp. 781-789). ACL.

Takamura, H., & Okumura, M. (2009b). Text summarization model based on the budgeted median problem. In *Proceeding of the 18th ACM Conference on Information and Knowledge Management*, (pp. 1589-1592). ACM Press.

Tan, P. N., Steinbach, M., & Kumar, V. (2006). *Introduction to data mining*. Boston, MA: Pearson Addison Wesley.

Tang, J., Yao, L., & Chen, D. (2009). Multi-topic based query-oriented summarization. In *Proceedings of the SIAM International Conference Data Mining*. SIAM.

Tatti, N., & Heikinheimo, H. (2008). Decomposable families of itemsets. In *Proceedings of the Machine Learning and Knowledge Discovery in Databases*, (pp. 472-487). IEEE.

Tatti, N., & Mampaey, M. (2010). Using background knowledge to rank itemsets. *ACM Data Mining and Knowledge Discovery, 21*(2), 293–309. doi:10.1007/s10618-010-0188-4

TexLexAn. (2011). *Texlexan: An open-source text summarizer*. Retrieved March 15, 2011, from http://texlexan.sourceforge.net/

Thakkar, K., Dharaskar, R., & Chandak, M. (2010). Graph-based algorithms for text summarization. In *Proceedings of the Third International Conference on Emerging Trends in Engineering and Technology*, (pp. 516–519). IEEE.

Wan, X., & Yang, J. (2006). Improved affinity graph based multi-document summarization. In *Proceedings of the Human Language Technology Conference of the NAACL*, (pp. 181-184). NAACL.

Wang, D., & Li, T. (2010). Document update summarization using incremental hierarchical clustering. In *Proceedings of the 19th ACM International Conference on Information and Knowledge Management*, (pp. 279–288). ACM Press.

Wang, D., Zhu, S., Li, T., Chi, Y., & Gong, Y. (2011). Integrating document clustering and multidocument summarization. *ACM Transactions on Knowledge Discovery from Data, 5*(3), 14. doi:10.1145/1993077.1993078

Wang, J., Li, Q., & Chen, Y. P. (2010). User comments for news recommendation in social media. In *Proceeding of the 33rd International ACM SIGIR Conference on Research and Development in Information Retrieval*, (pp. 881-882). ACM Press.

Yang, Z., Cai, K., Tang, J., Zhang, L., Su, Z., & Li, J. (2011). Social context summarization. In *Proceedings of the International ACM SIGIR Conference on Research and Development in Information Retrieval*. ACM Press.

Yin, Z., Li, R., Mei, Q., & Han, J. (2009). Exploring social tagging graph for web object classification. In *Proceedings of the 15th ACM SIGKDD International Conference on Knowledge Discovery and Data Mining*, (pp. 957-966). ACM Press.

Zhu, J., Wang, C., He, X., Bu, J., Chen, C., & Shang, S. ... Lu, G. (2009). Tag-oriented document summarization. In *Proceedings of the 18th International ACM Conference on World Wide Web*, (pp. 1195-1196). ACM Press.

ADDITIONAL READING

Agarwal, D., Phillips, J. M., & Venkatasubramanian, S. (2006). The hunting of the bump: On maximizing statistical discrepancy. In *Proceedings of the Seventeenth Annual ACM-SIAM Symposium on Discrete Algorithm*, (pp. 1137-1146). ACM Press.

Becker, H., Naaman, M., & Gravano, L. (2011). Selecting quality Twitter content for events. In *Proceedings of the Fifth International AAAI Conference on Weblogs and Social Media*. AAAI.

Bergler, S., Witte, R., Khalife, M., Li, Z., & Rudzicz, F. (2003). Using knowledge-poor coreference resolution for text summarization. In *Proceedings of the Workshop on Text Summarization*, (pp. 85-92). IEEE.

Brunn, M., Chali, Y., & Pinchak, C. J. (2001). Text summarization using lexical chains. In *Proceedings of the Workshop on Text Summarisation in Conjunction with the ACM SIGIR Conference.* ACM Press.

Chakrabarti, D., & Punera, K. (2011). Event summarization using Tweets. In *Proceedings AAAI ICWSM.* AAAI.

Conroy, J. M., Schlesinger, J. D., Goldstein, J., & O'leary, D. P. (2004). Left-brain/right-brain multi-document summarization. In *Proceedings of the Document Understanding Conference*. IEEE.

Diaz, A., & Gervas, P. (2007). User-model based personalized summarization. *Information Processing & Management, 43*(6), 1715–1734. doi:10.1016/j.ipm.2007.01.009

Foong, O. M., Oxley, A., & Sulaiman, S. (2010). Challenges and trends of automatic text summarization. *International Journal of Information and Telecommunication Technology, 1*(1), 34–39.

Ganesan, K., Zhai, C., & Han, J. (2010). Opinosis: A graph-based approach to abstractive summarization of highly redundant opinions. In *Proceedings of the 23rd International Conference on Computational Linguistics*, (pp. 340-348). IEEE.

Girardin, F., Calabrese, F., Fiore, F. D., Ratti, C., & Blat, J. (2008). Digital footprinting: Uncovering tourists with user- generated content. *Pervasive Computing, 7*(4), 36–43. doi:10.1109/MPRV.2008.71

Gong, Y., & Liu, X. (2001). Generic text summarization using relevance measure and latent semantic analysis. In *Proceedings of the 24th Annual International ACM SIGIR Conference on Research and Development in Information Retrieval.* ACM Press.

Gupta, V., & Lehal, G. S. (2010). A survey of text summarization extractive techniques. *Journal of Emerging Technologies in Web Intelligence, 2*(3). doi:10.4304/jetwi.2.3.258-268

Jauua, M., & Hamadou, A. B. (2003). Automatic text summarization of scientific articles based on classification of extract's population. In *Proceedings of the 4th International Conference on Computational Linguistics and Intelligent Text Processing*, (pp. 623-634). IEEE.

Li, L., Wang, D., Shen, C., & Li, T. (2010). Ontology-enriched multi-document summarization in disaster management. In *Proceeding of the 33rd International ACM SIGIR Conference on Research and Development in Information Retrieval*, (pp. 819-820). ACM Press.

Li, X., Guo, L., & Zhao, Y. E. (2008). Tag-based social interest discovery. In *Proceeding of the 17th International Conference on World Wide Web*, (pp. 675-684). IEEE.

Lin, F., & Liang, C. (2008). Storyline-based summarization for news topic retrospection. *Decision Support Systems, 45*(3), 473–490. doi:10.1016/j.dss.2007.06.009

Lloret, E., & Palomar, M. (2010). Challenging issues of automatic summarization: Relevance detection and quality-based evaluation. *Informatica, 34*(1), 29–35.

Nagwani, N. K., & Verma, S. (2011). A frequent term and semantic similarity based single document text summarization algorithm. *International Journal of Computers and Applications, 17*(2), 36–40. doi:10.5120/2190-2778

Nomoto, T., & Matsumoto, Y. (2001). A new approach to unsupervised text summarization. In *Proceedings of the 24th Annual International ACM SIGIR Conference on Research and Development in Information Retrieval*, (pp. 26-34). ACM Press.

Rambow, O., Shrestha, L., Chen, J., & Lauridsen, C. (2004). Summarizing email threads. In *Proceedings of the Human Language Technology Conference of the NAACL*, (pp. 105-108). NAACL.

Rusu, D., Fortuna, B., Grobelnik, M., & Mladenic, D. (2009). Semantic graphs derived from triplets with application in document summarization. *Informatica: An International Journal of Computing and Informatics*, *33*(3), 357–362.

Tatti, N. (2010). Probably the best itemsets. In *Proceedings of the 16th ACM SIGKDD International Conference on Knowledge Discovery and Data Mining*, (pp. 293-302). ACM Press.

Tatti, N., & Mampaey, M. (2010). Using background knowledge to rank itemsets. *Data Mining and Knowledge Discovery*, *21*(2), 293–309. doi:10.1007/s10618-010-0188-4

Vivaldi, J., Cunha, I. D., Moreno, J. M. T., & Velázquez-Morales, P. (2010). Automatic summarization using terminological and semantic resources. In *Proceedings of the International Conference on Language Resources and Evaluation*. IEEE.

Wei, F., Li, W., Lu, Q., & He, Y. (2008). Query-sensitive mutual reinforcement chain and its application in query-oriented multi-document summarization. In *Proceedings of the 31st Annual International ACM SIGIR Conference on Research and Development in Information Retrieval*, (pp. 283-290). ACM Press.

Xue, Y., Zhang, C., Zhou, C., Lin, X., & Li, Q. (2008). An effective news recommendation in social media based on users' preference. In *Proceedings of the 2008 International Workshop on Education Technology and Training & 2008 International Workshop on Geoscience and Remote Sensing*, (vol. 1, pp. 627-631). IEEE.

KEY TERMS AND DEFINITIONS

Document Summarization: The process of conveying the most representative content of either a single document or a document collection to a concise summary.

Frequent Itemset Mining: Frequent itemset mining is a widely exploratory technique to discover relevant recurrences hidden in the analyzed data.

Knowledge Discovery from Data (KDD): The process of extracting hidden information from data. It includes the tasks of data selection, preprocessing, transformation, mining, and evaluation.

Maximum Entropy Model: The maximum entropy model is a model that optimally makes use of the background knowledge to represent a data collection at best.

Set Covering Optimization Problem: Given a universe of elements, the set covering optimization problem is the task of selecting the minimum number of sets whose union still contains all elements in the universe.

Social Network Services: Social network services allow users to define a profile, build a network with other users (friends) and share information or user-generated media content with their friends.

User-Generated Content: User-Generated Content (UGC) refers to various kinds of publicly available media content that are produced by end-users, such as document, photos, and videos.

ENDNOTES

[1] The provided command is: ROUGE-1.5.5.pl -e data -x -m -2 4 -u -c 95 -r 1000 -n 4 -f A -p 0.5 -t 0 -d -a.

Chapter 8
Method for Modeling User Semantics and its Application to Service Navigation on the Web

Munehiko Sasajima
Osaka University, Japan

Yoshinobu Kitamura
Osaka University, Japan

Riichiro Mizoguchi
Osaka University, Japan

ABSTRACT

The value of information accumulated on the Web is enhanced when it is provided to the user who faces a problematic situation that can be solved by the information. The authors have investigated a task-oriented menu that enables users to search for mobile Internet services not by category but by situation. Construction of the task-oriented menu is based on a user modeling method that supports descriptions of user activities, such as task execution and defeating obstacles encountered during the task, which in turn represents the users' situations and/or needs for certain information. They built task models of the mobile users that cover about 97% of the assumed situations of mobile Internet services. Then they reorganized "contexts" in the model and designed a menu hierarchy from the viewpoint of the task. The authors have linked the designed menu to the set of mobile Internet service sites included in the i-mode service operated by NTT docomo, consisting of 5016 services. Among them, 4817 services are properly connected to the menu. This chapter introduces a framework for a real scale task-oriented menu system for mobile service navigation with its relations to the SNS applications as knowledge resources.

DOI: 10.4018/978-1-4666-2806-9.ch008

INTRODUCTION

Today, various kinds of information are accumulated on the Web including SNS. Wikipedia provides generic knowledge about things with hierarchical structure, while Twitter provides real-time information via short messages, for example. Subscribing such services, people are executing many kinds of tasks in daily life.

Providing appropriate information for users in a specific situation should enhance value of information, because value of information is proportional to the necessity of the information for the user. Short messages on Twitter, a late breaking SNS, give latest and dynamic information to users, thus suitable for users who need the latest information to solve certain problem. Messages on Twitter about current train situation support users who seek for the fastest train route to the destination just now, while the service does not work well for users who want to understand whole subway networks of the train-services in Tokyo, for example.

To realize situation-oriented information services, the authors have been investigating a framework to navigate users to the information resources along with the user's situation. For the purpose, we have modeled daily activity of the users who subscribe Japanese mobile Internet service as the first step.

Here we explain characteristics about the Japanese mobile Internet services. While they provide many mobile Internet services via mobile handsets in Japan, such as online shopping, mobile banking, and news services, current methods for mobile service navigation have proven insufficient to guide users efficiently to the mobile Internet services they need. To solve this problem, the authors have been investigating a task-oriented menu, which enables users to search for services by "what they want to do" in certain problem-solving situations, instead of by "name of category." On this first prototype system by Naganuma and Kurakake (2005), it has been proved that the task-oriented menu system has ability to navigating novice users to the mobile services they want faster than conventional domain oriented menu system. The first prototype system by Naganuma, however, was a limited one. In terms of task and domain knowledge, the first prototype assumed only limited situations, thus limited services were built in the menu system.

To extend the first prototype menu system to the real scale one, the authors have been investigating two issues. The first one is how to enhance scalability. The second one is how to develop a menu system with real scale on the basis of the investigation about the scalability. The authors discussed these two issues and developed a new menu system with real scale for navigating users to the mobile services they want, which is linked to a real scale of mobile services consisting of about 9,000 services.

This chapter describes design and development process of the new ontology-based task-oriented menu system, especially from the three viewpoints: (1) how to model people's daily activity to capture their "context" in terms of utilizing information services on the network; (2) how to scale up an application which supports mobile users' daily activity by ontology engineering method; (3) evaluation of the system in terms of coverage and scales with some future directions.

BACKGROUND

Choi, Lee, Kim, and Jeon (2005) investigated which factors of mobile Internet services are important for users. They interviewed people from three countries, Japan, Korea, and Finland, which have mature mobile Internet service markets. According to their analysis, both the "logical order of the menu" and "meaningful classification of the contents" are considered to be important by many subjects from the three countries. The results validate our approach for improving the menu system and classification of the contents by user tasks, which should contribute to user satisfaction.

To satisfy users' needs, many researchers today focus on better composition of existing mobile Internet services. Our modeling method, which focuses on better analysis of users' needs, is able to strengthen the research explained in the following. Hierarchical Task Network planning (a general explanation is given in Kambhampati [1997], and applications for Web services are described in Kuter, Sirin, Nau, Parsia, and Hendler [2005]) supports how to divide and conquer a Web user's "problem," which resembles our task decomposition process in OOPS modeling (Sasajima, Kitamura, Naganuma, Fujii, Kurakake, & Mizoguchi, 2008). In the process of composing Web services, Motahari-Nexhad, Martens, Curbera, and Casati (2007) proposes how to identify mismatches of the interfaces and protocols between two services to be composed. Domingue, Galizia, and Cabral (2005) describes how to cope with heterogeneous interaction patterns with the framework of IRS-III, and Ashri, Denker, Marvin, Surridge, and Payne (2004) discusses the interaction protocols in their experience of IRS-II. In such an organization process, alignment of the ontologies behind the services is necessary. Omelayenko (2003) proposes a method for mapping meta-ontologies among Web services, and Ehrig, Staab, and Sure (2005) describes a machine-learning method for an initial stage of ontology alignment. Au, Kuter, and Nau (2005) points out that it is unrealistic to assume that the information provided by the Web services is static in many cases. They propose another framework to deal with volatile information, taking a ticket reservation service problem as an example.

These studies, however, do not consider the contents of the mobile Internet services. In contrast, our approach starts from analyzing users' activities, including problematic situations, which require mobile Internet services. We then design the menu system for user navigation based upon the user model. Most research on Web services implicitly assumes that Web browsing is done on desktop computers; thus, the time and cost involved in searching and evaluating the answers are not of much concern. On the other hand, in the case of our mobile Internet service problem, users need prompt answers. Thus, we pay attention to navigating users directly from the obstacles, which they face to the proper service which is the source of the answer. We leave evaluation of the answers to the users themselves.

In Masuoka, Parsia, and Labrou (2003), Masuoka proposed a Task Computing framework and built a ubiquitous environment which provides more than 100 Web services. The Web services are described by OWL-S, and the environment changes dynamically. The ubiquitous environment is unique because it deals with dynamic changes such as sudden appearance/ disappearance of clients/services, like the real world.

MIT's Process Handbook Project (Malone, Crowston, & Herman, 2003) deals with knowledge models about businesses. It focuses on modeling business activities and has a taxonomy of basic business activities. However, the method for building the model is implicit, and confusion of task concepts with way concepts occurs with some models. One of the models, "buy in a store," consists of a task concept "buy" and a way concept "in a store," for example. Such confusion lowers the generality of the model, and does not meet our requirements.

In the field of the human-computer interactions, although there are many studies about Web interfaces, there are not so many studies specific to mobile phones. James in James and Reischel (2001) compares the efficiency of two text input methods used on mobile phone: multi-tap and prediction. Kamvar and Baluja (2006) analyzed search patterns of a search engine specifically designed for mobile Internet services on a large scale. The search patterns resembled those of desktop search engines. The results show that mobile Internet services are still not organized well for mobile users. The users rely on search engines, as they do on desktop computers, since they cannot reach the necessary services. As mentioned in INTRODUCTION, basically task-oriented menu system navigates users along with their necessity of the services, thus not so much depends on search technologies.

Task-Oriented Menu for Mobile Service Navigation

Figure 1 shows the process of service selection using a task-oriented menu on the first prototype system (Naganuma & Kurakake, 2005). First, the most abstract task candidates are shown on the mobile phone (see Figure 1 left). A user selects one of them (e.g. "Go to a department store") to solve current problem (e.g., "need to buy clothes"). Then, tasks and/or subtasks associated with each task are unfolded and displayed under the task nodes (see Figure 1 center). Finally, services associated with the task selected by the user are shown, and each of them leads to access to the actual service (see Figure 1 right).

As shown in this example, the task-oriented menu is easy to use for novice users of mobile Internet services. By just selecting what he/she wants to do in the real world from the menu, he/she will be led to a service for solving the current problem. Knowledge about the hierarchy of the domain-oriented menu labeled like "hobbies," "local info," "life," and so on, is not necessary.

Although such a generic task hierarchy looks like the hierarchical structure of the category-based menus of today, there are fundamental differences. In certain cases, it is possible to label a concept with a noun instead of a verb or action. It is acceptable to label a mobile Internet service that sells tickets as "Ticketing" or "Buy a ticket," for example. In the same manner, abstract tasks can be labeled with nouns. Although it seems that any concept can be labeled by both verbs and nouns, it is a hasty generalization. Such a generalization may lead to the misunderstanding that we just followed the process used by the designer of the category-based menu in classifying the mobile Internet services, thus introducing an abstract hierarchy of the tasks.

An important point is that the difference between "Ticketing" and "Buy a ticket" is just the expression of the label. The concepts are the same task. We focus on the concept and essential characteristics of the mobile user's task. Comparing them at the conceptual level, a category-based

Figure 1. Task oriented menu (the first prototype by Naganuma & Kurakake, 2005)

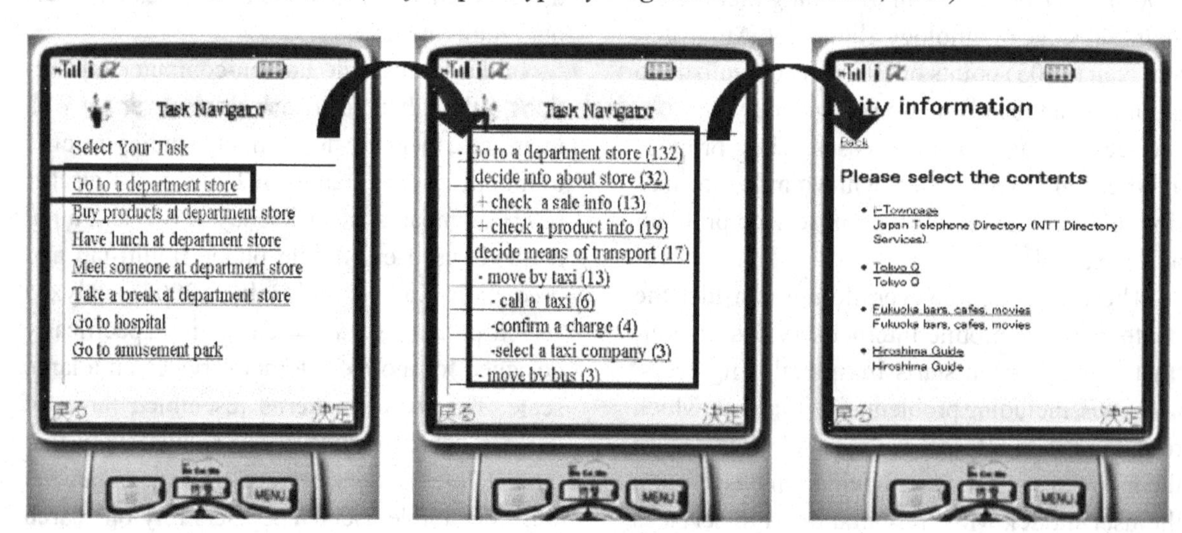

classification of objects is totally different from a task-based one in terms of its structure.

In the case of category-based classifications today, generally speaking, the boundary or definition of each category becomes vague or implicit. Classification of an object or a concept heavily depends on the intention of the designer who developed the menu. The categories "Hobbies" and "Shopping" are both located at the top level of the Japanese i-mode menu, for example. A mobile Internet service that sells cars is classified in the former if the designer considers driving a car as a hobby. On the other hand, the service is classified in the shopping category if the designer focuses on the commercial aspect of the service rather than its object.

On the other hand, in classification of actions from the viewpoint of tasks, the boundary or definition of each category becomes more explicit. Since the criteria for the classification, such as pre-conditions, processes, and effects of the action, appear in both the label of the category and the classified concepts, it is easy to find the location of a new concept in a hierarchy which is classified based on task. For the same reason, it is easy to add a new concept to the task-based classification. A service that sells cars is classified in a sub-category of the task "Buy," whether driving a car is a hobby or not.

For the reasons described so far, task-based categories are more suitable for the classification of mobile services. On this point, Naganuma and Kurakake (2005) conducted a user test involving nine adult subjects to confirm the effectiveness of a task-oriented menu system and evaluate the process used to find services for problem-solving purposes in terms of process functionality. Subjects were divided into three groups according to their experience of mobile Internet services: 1) subjects using mobile Internet services every day, 2) subjects using mobile Internet services a few days a week, and 3) subjects with no experience in using mobile Internet services.

Subjects were asked to retrieve appropriate services to given problems by using the task-oriented menu system, a keyword-type full-text search system newly developed for the experiment, and a major commercial directory-type menu system.

Analyzing the results by the user types, only the task-oriented menu system allowed non-expert users to find the appropriate services with the same success rate as experienced users. The results show that the task-oriented menu system is effective for mobile Internet service navigation.

Issues on Building Real-Scale Task Oriented Menu

The first prototype system, however, was a limited one. In terms of task and domain knowledge, the first prototype assumed only few limited situations, thus limited services were built in the menu system.

For realization of task-oriented menu system in real scale, we have to tackle two issues: (1) Scalability of the system and (2) Building a task-oriented menu system with real scale. For the first issue, the authors have identified four kinds of scalabilities to be satisfied (Sasajima, et al., 2006, 2008): (a) Coverage for domains of mobile services, (b) Granularity of user modeling, (c) Coverage for mobile services in real world, (d) Coverage for mobile users' situations in which they rely on mobile services. For the item (a) and (b), we have already proposed a new ontology-based modeling method, which is named OOPS (abbreviation of "Ontology-based Obstacle, Prevention, and Solution).

Figure 2 represents a process of building an OOPS model. The dotted rectangle labeled (1) corresponds to the basic model of users' activities. It is described by instantiating generic models or ontologies. Description of the OOPS model starts from the task at the level of large granularity.

Next, ways to achieve the task are linked, and each of the ways consists of a sequence of subtasks. Our "way" is similar to the "method" of CommonKADS (Schreiber, et al., 2000) and "how

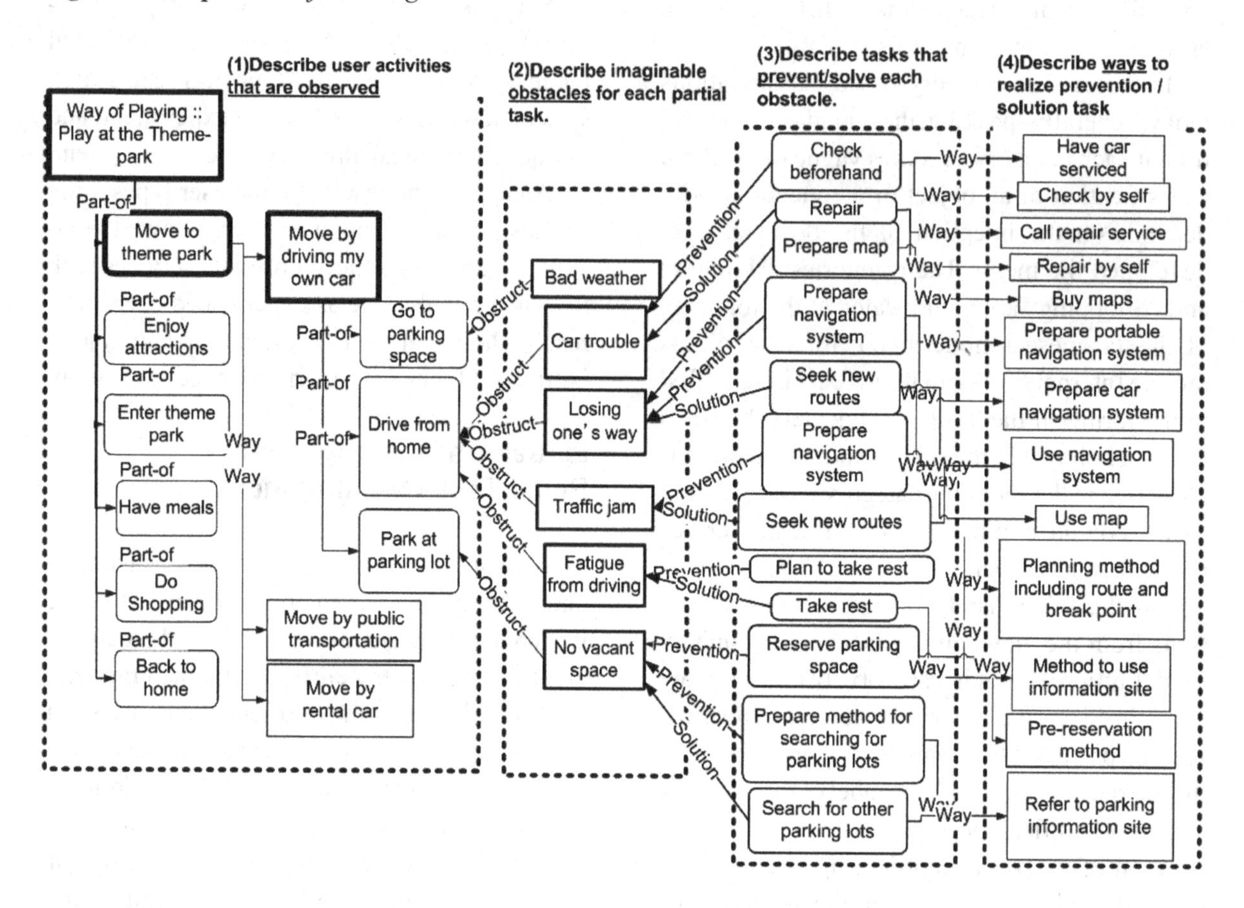

Figure 2. The process of building OOPS models

to bundle" of the Business Process Handbook (Malone, Crowston, & Herman, 2003). Following this process, the task of large granularity is decomposed into sub-tasks via ways. Area (1) in Figure 2 represents that a task "Move to a theme park" is achieved by three ways. Among them, the way "Move by driving my own car" consists of three sub-tasks, "Go to the parking space," "Drive from home," and "Park the car at the parking lot."

We have designed and developed an ontology, which covers users' daily tasks and necessary domain knowledge. Modeling method based on the ontology solves complicated domain modeling (i.e., [a]) and gives guidelines for granularity of the task modeling (i.e., [b]). The modeling method supports descriptions of users' activities and related knowledge, such as how to solve problems that the users encounter and how to prevent or solve them on the spot. By experiments described in Sasajima et al. (2008), OOPS modeling method showed performance that promotes generation of idea for modeling users' daily activities.

Prototyping Real-Scale Task-Oriented Menu

In this research, we concentrated on the issue (2) as well as scalability issues of (c) and (d). For testing the coverage of mobile services and mobile users' situations (item [c] and [d]), a new menu system with real scale is definitely needed.

Analysis of User Activities

To make such a system, analysis of the user activities in a wide range of domains is required. For such analysis, we have applied the OOPS modeling method to "Tourism" domain which covers a broader spectrum of actions from traveling around and consuming money to staying at a hotel. We have evaluated the coverage of the OOPS model by comparing situations assumed and represented in the model which we developed on tourism setting with those situations assumed to be supported by current mobile services.

We have tested coverage of the model by a full set of mobile services, which are available at the official sites of NTT docomo in 2004. Among about 5,000 officially authorized service sites, excluding entertainment services sites (Games, ring-tone downloading, etc.), there are 2,732 sites that consist of 9,162 specific services inside. We analyzed a situation for each of the 9,162 services. Among them, our OOPS user model covered about 98% of the typical situations assumed by the mobile service sites, and just 199 services (2.17% of the 9,162 official services) were not covered by the situations represented by our OOPS model.

Development of New Prototype System

Based on the OOPS user model, we developed a menu system with real scale. Figure 3 depicts our environment for developing the task-oriented menu, which is based on an environment by NTT docomo for building i-appli (applications for i-mode mobile handsets). On the left part of

Figure 3. Environment for developing task-oriented menu

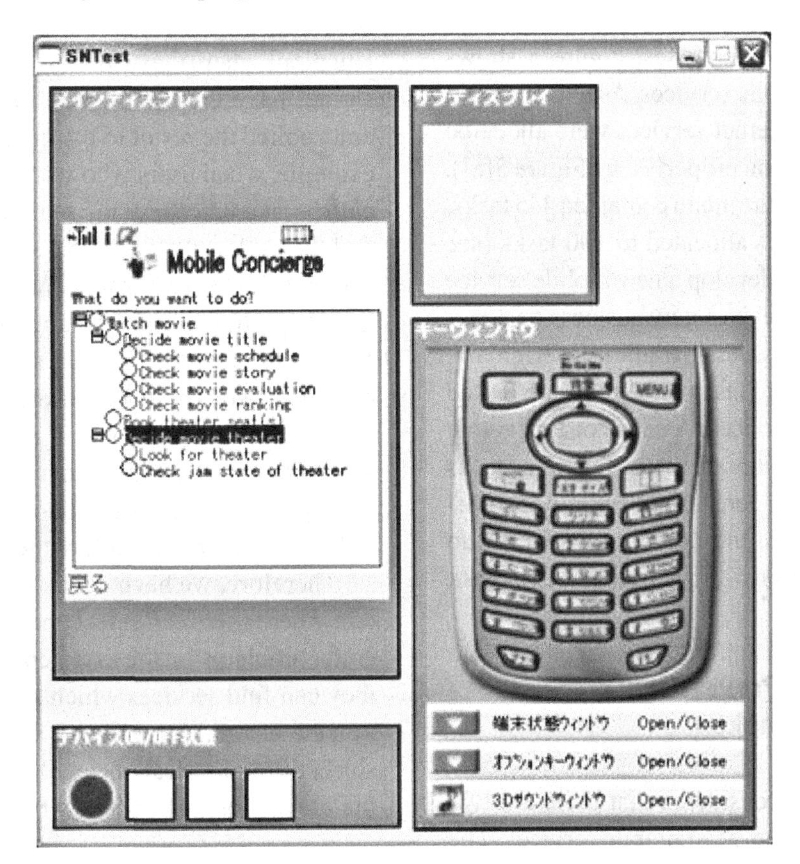

the figure, the menu we designed is displayed hierarchically.

Figure 4 depicts the first two levels of the menu. The OOPS model on tourism domain consists of 5 tasks. At first, we built a menu hierarchy where the 5 tasks are at the top level ("Move," "Have meal," "Have fun," "Buy," and "Stay overnights"). Those at the second level (17 items) have how users achieve the tasks, those at the third level (97 items) have subtasks which consist of methods, those at the fourth level (112 items) have obstacles which can occur when users do subtasks and those at the fifth level (445 tasks) have tasks which can prevent or solve obstacles such as "Go to somewhere," "Have meal," "Draw cash," "Buy things," and so on. As a whole, the menu consists of 5 levels at the deepest level.

The menu hierarchy enables users to search the mobile Internet services they need if they select task, method, subtask, obstacle, and prevention/solution task in order. Then we implemented the menu system and assigned all of the officially authorized service sites. Figure 5 shows statistics about mobile services. As a result, 96% of 5,016 mobile Internet services were allocated to the real-scale menu properly (see Figure 5[a]).

Although the entire menu contained 445 tasks, no mobile service is allocated to 100 tasks (see Figure 5[b]). If we develop a new mobile service for such tasks, it will be a new business opportunity. Furthermore, issues on usability still remain. For example, 11% of task menu items are linked to more than 50 services. A cause of this is that today's mobile services are biased to limited tasks like "know weather forecast," "get movie information," and so on. In addition, we plan to do other usability tests without limitations of task and domain.

Separation of Prevention and Solution Tasks

The authors have considered that there are two situations when users need mobile services. The one is the situations where users want to prevent problems they encountered, and the other is the situations where they want to solve problems. We should have clearly divided the two situations and applied the result to the menu hierarchy. For example, when users who want to move by train cannot take it because the seats are not available and they select the node "No seat available," we can find the prevention task "Make a reservation" and the solution task "Change transportation." When the problem "no seat available"; has occurred already, however, users would be upset because they cannot make a reservation for the seats after they have been fully booked. This means the menu hierarchy should show the node "Make a reservation" before the problem occurs.

Figure 4. First two levels of the menu

Menu top
- Move
    - On foot
    - By public transportation
    - By taxi
    - By car
    - By rent-a-car
- Have meal
    - At a restaurant
    - Take out
    - Cook by self
- Have fun
    - By sight seeing
    - By playing at a theme park
    - By watching sports/play/etc.
- Buy
    - In a town
    - By internet shopping
    - By auction
- Stay over nights
    - Stay at a hotel
    - Stay at friends

Therefore, we have developed the menu system where users can select "before problems" or "after problems" at the first step, following which they can find services which suit their situation. For the example mentioned above, when users select the node "Seat not available" they can find the prevention task "Make a reservation" if they

Figure 5. Statistics about service contents

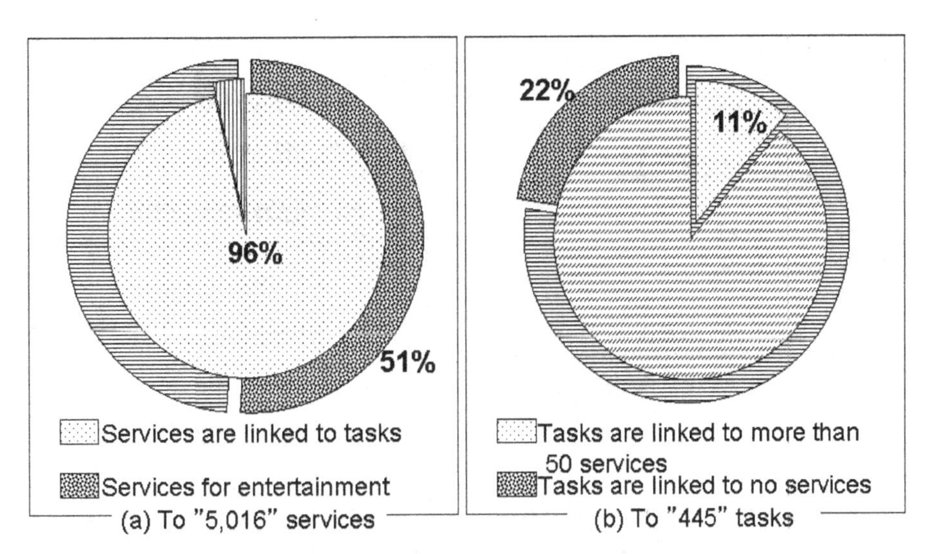

choose "before problems" and the solution task "Change transportation" if they choose "after problems" at the first step.

Process of Mobile Service Navigation

Figure 6 shows screen shots of the developed system. When a user selects one of them (e.g. "Move") to achieve the current goal (e.g., "go to a shop"), methods which can achieve the task are unfolded. By selecting an item among the menu, tasks and subtasks associated with each task are unfolded and displayed under the task nodes. Finally, at the deepest level of the menu, each of the menu items is linked with a URL of an Internet service like "City map service."

Suppose a scenario that a user wants to go to a shop by public transportation system. Figure 6 depicts a sample process of service selection using screen shots of the menu system of the latest version. First, the most abstract task candidates are shown on the mobile phone (see Figure 6, Upper-left). Since the user wants to go to the shop by public transportation system, he/she selects one of them (e.g. "Move") to achieve the current goal (go to a shop). Then, five methods which can achieve the task "Move" are unfolded (see Figure 6 Upper-right). By selecting the second item among the menu (e.g., "By public transportation"), tasks and/or subtasks associated with each task are unfolded and displayed under the task nodes (see Figure 6, Lower-left). Selecting tasks further, plausible obstacles for the subtasks and their solution tasks are unfolded (see Figure 6 Lower-right). The user might lose his/her way to a ticket station, for example. In that case, selecting such a troublesome situation among the menu items, solutions for the trouble are unfolded (e.g., "Find a ticket information" and "Seek for a route map" in Figure 6). Finally, services associated with the task selected by the user are shown, and each of them leads to access to the actual service.

Design Review by Experts

Our new prototype menu system is now under the design review by experts of mobile services. Compared to the original menu system prototyped by Naganuma and Kurakake (2005), they positively point out followings.

1. Granularity of the menu has become uniform. Since original menu was an ad hoc

Figure 6. Sample screen shots

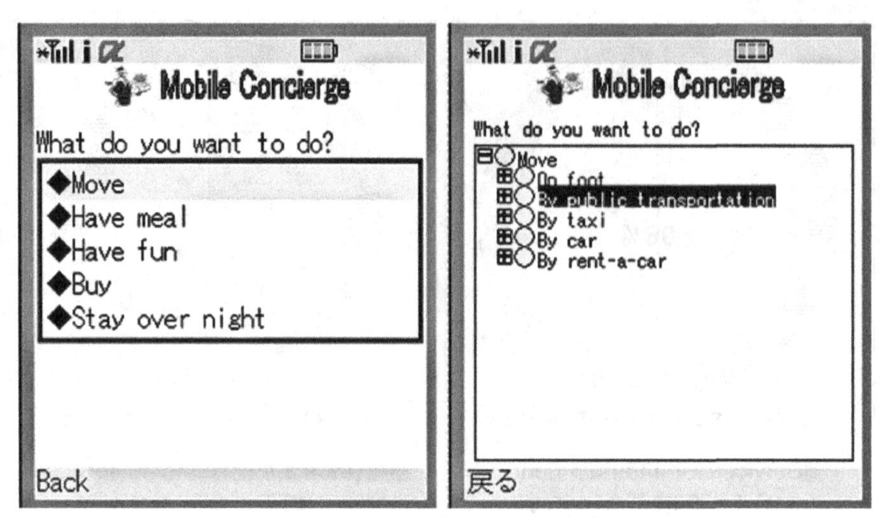

(Left)Top menu: "Move" task is selected by user. (Right) Five methods for "Move" are unfolded and "By public transportation" method is selected.

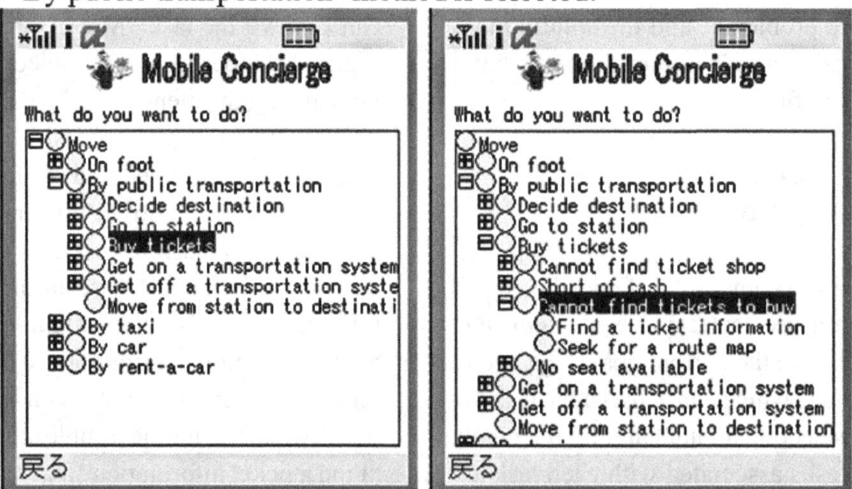

(Left)Subtasks of "move by public transportation" are unfolded. (Right) Plausible obstacles for the subtasks and their solution tasks are unfolded.

one, granularities of some menu items were coarse and others were fine. As a result, understandability of the menu has become better.

2. Since the new menu system is composed of finer grained menu items, users who have definite purpose (e.g., go to a department store to buy cloths) will be guided more easily to the target information services.

3. Those services which are not utilized before are "revealed" and are able to access now. Some television companies, for example, provide information about recipes introduced within their TV programs. Since their sites have been in the "TV" category before, some users should miss the recipes because they cannot imagine TV companies provide such recipes. New menu system guides to

the recipes by "Have meal (task) > Cook by self(method)>Look for recipes (sub-task)." Then links for recipes are listed including those provided by the TV companies.

Social Network Services as a Knowledge Resource

Since valuable knowledge are generated and accumulated on variety of SNS services such as Wikipedia, Twitter, Blog, and so on, appropriate selection of the knowledge source should be done according to the "context" of the users. Suppose that a user plans to do a long business trip on next Monday. He/she has to survey an appropriate route, select and reserve a train, buy tickets, and do the trip. When the user wants to select the route, he/she should refer to the encyclopedia-like knowledge resource from which we can get a quick overview of things. Wikipedia would play the role. On the other hand, if his/her task needs real-time information on the way to the destination, such as train situation of local trains near the destination, Twitter, on which people always tweet the latest information might play the role. Research on mining social media is emerging for such purposes.

Our service navigation framework based on OOPS model has potential to indicate such knowledge resources which are suitable for the users' context. Task model of "Reservation of the seat of the train," for example, is a prevention task to avoid occurrence of the obstacle "The train is full and cannot sit down." Since such preventive tasks are always carried out before the execution of the main task, i.e., goes to the business trip, we can set a heuristic rule reasonably: "For the users seeking information to do preventive task, "static" information services like Wikipedia should be preferred." On the other hand, solution tasks which are carried out when certain problems occur on the spot, dynamic information is should be preferred more to solve the problem on the spot. To solve a problem "Train services are temporally not available," for example, real-time tweets about train situation on Twitter should be helpful for users. Referring to such preferences about the information resources, our task-oriented service navigation system will be able to recommend suitable social network services to users who seek for solution.

FUTURE RESEARCH DIRECTIONS

Now the system is under the design review by experts and we plan to do field test by general users of mobile phones in longer term. We plan to improve user interface for the task-oriented menu system. First, it is unrealistic to replace everything with a task-oriented style; rather, integration of a search engine and/or domain-oriented classification will be necessary for some tasks. For example, the task "buy" deals with millions of items which require conventional search technologies.

Secondly, we are designing "shortcut" menus for some frequently accessible services. The menu hierarchy has some subtasks, which frequently appear under different tasks. Such subtasks are possible to be carried out as not only subtasks associated with each task but also independent tasks. For example, if users who want to draw cash to buy train tickets intend to search services about ATM information, in the current menu system, they must select "Move," "By public transportation," "Buy tickets," "Short of cash," "Draw cash," and "Search ATM" step by step. Although the task oriented menu system can support users to search for the ATM services to solve problems, the shortcut menu should be a good help for users because the services for drawing cash are necessary in many other situations. Therefore, we have been trying to define the problems, which happen frequently and build the shortcut menu for such services.

Lastly, we plan to utilize SNS as resource of knowledge and solution for users. Each SNS services has both strong point and weak point, appropriate recommendation along with user's context should be helpful.

CONCLUSION

This chapter introduced our research on the task-oriented menu system with real-scale mobile Internet services in it, as well as its possibility to work with SNS. We described design and development process of the new ontology-based task-oriented menu system. Through the development process of the real-scale menu system for mobile Internet service navigation, we proved enough scalability of the proposed method for our purpose.

Although this chapter focused on issues about scaling up a system, introduction of ontology for designing and developing knowledge intensive software system is advantageous through the lifecycle of it; design, development, execution, and maintenance of the proposed system, as pointed out by Noy and McGuiness (2001). We plan to evaluate the task-oriented menu system from these points, too.

REFERENCES

Ashri, R., Denker, G., Marvin, D., Surridge, M., & Payne, T. (2004). Semantic web service interaction protocols: An ontological approach. *Lecture Notes in Computer Science*, *3298*, 304–319. doi:10.1007/978-3-540-30475-3_22

Au, T., Kuter, U., & Nau, D. (2005). Web service composition with volatile information. *Lecture Notes in Computer Science*, *3729*, 52–66. doi:10.1007/11574620_7

Choi, B., Lee, I., Kim, J., & Jeon, Y. (2005). A qualitative cross-national study of cultural influences on mobile data service design. In *Proceedings of the SIGCHI Conference on Human Factors in Computing Systems 2005 (CHI 2005)*, (pp. 661-670). ACM Press.

Domingue, J., Galizia, S., & Cabral, L. (2005). Choreography in IRS-III- Coping with heterogeneous interaction patterns in web services. *Lecture Notes in Computer Science*, *3729*, 171–185. doi:10.1007/11574620_15

Ehrig, M., Staab, S., & Sure, Y. (2005). Bootstrapping ontology alignment methods with APFEL. *Lecture Notes in Computer Science*, *3729*, 186–200. doi:10.1007/11574620_16

James, C. L., & Reischel, K. M. (2001). Text input for mobile devices: Comparing model prediction to actual performance. In *Proceedings of CHI 2001*, (pp. 365-371). ACM.

Kambhampati, S. (1997, Summer). Refinement planning as a unifying framework for plan synthesis. *AI Magazine*, 67–97.

Kamvar, M., & Baluja, S. (2006). A large scale study of wireless search behavior: Google mobile search. In *Proceedings of CHI 2006*, (pp. 701-709). ACM Press.

Kuter, U., Sirin, E., Nau, D., Parsia, B., & Hendler, J. (2005). Information gathering during planning for web service composition. *Journal of Web Semantics*, *3*(2-3), 183–205. doi:10.1016/j.websem.2005.07.001

Malone, T. W., Crowston, K., & Herman, G. A. (2003). *Organizing business knowledge - The MIT process hand book*. Cambridge, MA: MIT Press.

Masuoka, R., Parsia, B., & Labrou, Y. (2003). Task computing - The semantic web meets pervasive computing. *Lecture Notes in Computer Science*, *2870*, 866–881. doi:10.1007/978-3-540-39718-2_55

Motahari-Nexhad, H. R., Martens, A., Curbera, F., & Casati, F. (2007). Semi-automated adaptation of service interactions. [IEEE.]. *Proceedings of WWW, 2007*, 993–1002.

Naganuma, T., & Kurakake, S. (2005). Task knowledge based retrieval for services relevant to mobile user's activity. [ISWC.]. *Proceedings of the ISWC, 2005*, 959–973.

Noy, N. F., & McGuiness, D. (2001). *Ontology development 101: A guide to creating your first ontology*. New York, NY: Knowledge Systems Laboratory.

Omelayenko, B. (2003). RDFT: A mapping meta-ontology for web service integration. In Omelayenko, B., & Klein, M. (Eds.), *Knowledge Transformation for the Semantic Web* (pp. 137–153). Boca Raton, FL: IOS Press.

Sasajima, M., Kitamura, Y., Naganuma, T., Fujii, K., Kurakake, S., & Mizoguchi, R. (2008). Obstacles reveal the needs of mobile internet services -OOPS: Ontology-based obstacle, prevention, and solution modeling framework. *Journal of Web Engineering*, *7*(2), 133–157.

Sasajima, M., Kitamura, Y., Naganuma, T., Kurakake, S., & Mizoguchi, R. (2006). Task ontology-based framework for modeling users' activities for mobile service navigation. [ESWC.]. *Proceedings of Posters and Demos of the ESWC, 2006*, 71–72.

Schreiber, G., Akkermans, H., Anjewierden, A., de Hoog, R., Shadbolt, N. V., de Velde, W., & Wielinga, B. (2000). *Knowledge engineering and management - The CommonKADS methodology*. Cambridge, MA: MIT Press.

Chapter 9
Tag Cloud Reorganization:
Finding Groups of Related Tags on Delicious

Alberto Pérez García-Plaza
UNED, Spain

Arkaitz Zubiaga
City University of New York, USA

Víctor Fresno
UNED, Spain

Raquel Martínez
UNED, Spain

ABSTRACT

Tag clouds have become an appealing way of navigating through Web pages on social tagging systems. Recent research has focused on finding relations among tags to improve visualization and access to Web documents from tag clouds. Reorganizing tag clouds according to tag relatedness has been suggested as an effective solution to ease navigation. Most of the approaches either rely on co-occurrences or rely on textual content to represent tags. In this chapter, the authors explore tag cloud reorganization based on both of them. They compare these clouds from a qualitative point of view, analyzing pros and cons of each approach. The authors show encouraging results suggesting that co-occurrences produce more compelling reorganization of tag clouds than textual content, being computationally less expensive.

INTRODUCTION

Social bookmarking sites allow to collaboratively annotate Web pages, relying on the social tagging philosophy. Social tagging is a Web 2.0 application based phenomenon where users can describe Web contents by adding tags or keywords as metadata in an open, collaborative, and non-hierarchical way (Smith, 2008). Social bookmarking is a popular way to store, organize, comment on and search links to any Web page, and it has emerged as one of the most important Web applications that eases information sharing. Popular collaborative

DOI: 10.4018/978-1-4666-2806-9.ch009

tagging sites have aggregated a vast amount of user-created metadata in the form of tags, providing valuable information about the interests and expertise of the users. Because of this, it becomes a fertile area to scientific research on social media (Gupta, et al., 2010).

In order to facilitate access to tagged resources, and to enable visual browsing, social bookmarking tools typically provide an interface model known as the tag cloud. A tag cloud is an appealing way to enable users to navigate through the most popular tags of a social bookmarking site. When users access the information contained in these structures, it is presented in the form of a cloud consisting of the most popular tags, where the bigger is the font size of a tag, the more popular it is on the site. Typical tag clouds include between 100 and 200 tags, and tag weights are represented by different font sizes, or other visual clues. In addition, tags can be sorted in alphabetical, size-based, or random order, and users can sometimes customize clouds with different fonts, layouts, and color schemes. These structures are particularly useful for browsing and for information discovery, because they provide a visual summary of the content in the collection. However, related tags do not appear in nearby spaces of the tag cloud, and it is not easy to find the tags of one's interest. To solve this problem, research in the field has pointed out that grouping related tags, and showing them close to each other can help enhance navigation through tag clouds.

In order to enhance browsing phase in a tag cloud, an effective way is to identify inter-related tags and relations among contents. This book chapter aims to discuss the tag grouping task so that it enables an enhanced visualization and improved navigation through the tag cloud. To this end, several methods of representing tags have been proposed in earlier research. Most of them consider co-occurrences among tags to group related tags into clusters, but do not pay special attention on the algorithm employed to weight such co-occurrences. In this work, we focus on the reorganization of a tag cloud based on the identification of groups of inter-related tags, and compare different methods for weighting tag co-occurrences. We rely on a well-known clustering algorithm for this purpose.

Recently, there has been an increasing interest on tag clustering tasks; most of them tackle the problem from the point of view of tag co-occurrences (Specia & Motta, 2007; Mika, 2007; Sbodio & Simpson, 2009). Other works have followed a content-based approach, such as Zubiaga et al. (2009). All of them performed a qualitative evaluation of their results, finding appealing groupings for human users. Nevertheless, these works did not compare content-based methods with those based on tag co-occurrences, widely used in the literature.

In this book chapter we further explore several state-of-the-art weighting functions to represent co-occurrences among tags. After clustering tags with these weightings, we compare the results with those obtained by the content-based approach. Going further, we analyze and discuss the appropriateness and performance of each approach.

Next, in Section Background we cover some basic ideas about social tagging systems and present the related work. In Section Tag Cloud Reorganization, we explain the settings of our experiments, including dataset, tag representation approaches, and tag clustering algorithm. To conclude the section, we analyze the resulting clouds and discuss some possible applications. Finally, we summarize the future research directions and conclusion.

BACKGROUND

Tagging is an open way to assign tags or keywords to resources or items (e.g., a Web page), in order to describe the characteristics of them. This enables later retrieval of resources in an easier way. As opposed to a classical taxonomy-based categorization system, they are usually non-hierarchical, and the vocabulary is open, so it tends to grow indefinitely. For instance, a user could tag this chapter as *social-tagging*, *clustering*, and *delicious* whereas another user could use *chapter*, *research*, and *tagging* tags to annotate it.

A tagging system becomes social when tags are publicly visible, and so profitable for anyone. The fact of a tagging system being social implies that a user could take advantage of tags defined by others to retrieve a resource. As a result, the collection of tags defined by the community creates a tag-based organization, so-called folksonomy. A folksonomy is also known as a community-based taxonomy, where the classification scheme is plain, there are no predefined tags, and therefore users can freely choose new words as tags.

Depending how users assign tags, two types of social tagging systems can be distinguished (Smith, 2008):

- **Simple Tagging**: users describe their own resources or items, such as photos on Flickr.com, news on Digg.com or videos on Youtube.com, but nobody else tags another user's resources. Usually, the author of the resource is who tags it. This means no more than one user tags a concrete item.
- **Collaborative Tagging**: many users tag the same item, and every person can tag it with his own tags in his own vocabulary. As a result, several users tend to post the same item. For instance, CiteULike.org, LibraryThing.com and Delicious.com are based on collaborative tagging, where each resource (papers, books and URLs, respectively) could be tagged by all the users who considered it interesting.

Figure 1 shows an illustrative example of how each of these systems work.

Within the collaborative tagging systems, this work is focused on social bookmarking. In social bookmarking sites, people can post and tag their favorite Web pages by using the tags they consider representative. These tags represent the keywords a user would use to look for it, and tend to differ from user to user. Thus, the more users describe an item, the more precise its tag set is. This is one of the main hypotheses of social tagging.

Formally, in a social bookmarking site each user u_i can post an item i_j with a set of tags $T_{ij} = \{t_1,...,t_p\}$, with a variable number p of tags. After k users posted i_j, it is described as a weighted set of tags $T_j = \{w_1 t_1,...,w_n t_n\}$, where $w_1,...,w_n \leq k$.

The nature of social bookmarking sites offers us multiple positive aspects:

- Collaborative tagging of the same item by different users allows to create a weighted list of tags established by general consent. This leads to a wide set of tags, where a few tags are high-weighted and many of

Figure 1. Simple tagging vs. collaborative tagging

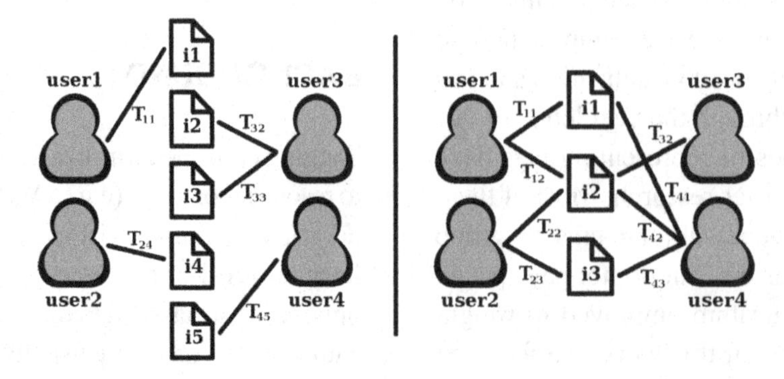

them are low-weighted, following a power law distribution.

- The open vocabulary allows users to create non-existing tags required by the current affairs or personal needs.

As far as the research on social tagging is concerned, there are some unhanded aspects that interfere in the performance of social bookmarking sites:

- Different tags can be synonymous (e.g., photo and photography), or have similar or related meanings.
- Different hypernym/hyponym relations can be found (e.g., programming and java), with different levels of specificity for related tags.
- No distinction is made for polysemous tags, such as *library*, which could mean both a place containing books and a collection of sub-programs, amongst others.
- The purpose of users is different when tagging a page from Youtube or a tutorial about movie editing by using a common tag like *video*.

As regards to the nature and purpose of tags, different classification schemes have been proposed. In Golder and Huberman (2006), the authors consider that tags may be categorized in a list of 7 types: (1) identifying what (or who) it is about, (2) identifying what it is, (3) identifying who owns it, (4) refining categories, (5) identifying qualities or characteristics, (6) self reference, and (7) task organizing. These 7 types may be reduced into the following three more general types, according to Sen et al. (2006):

- **Factual tags:** Describes item topics, kinds of item or category refinements being objective tags, e.g., *design, video*.
- **Subjective tags:** Describes item quality. e.g., *cool, interesting*.
- **Personal tags:** Describes item ownership, self-reference or tasks organization, e.g., *toread*.

Even though these classifications of tags include most of the cases, it is not so easy to classify all of them, as the open vocabulary allows to assign new kinds of them. For instance, we can also find temporal tags, such as *2008*, which are not considered in the above lists.

In summary, social bookmarking sites allow us to get large annotated datasets, but some kind of preprocessing could be required in order to access the needed information in a suitable way.

Related Work

With the emergence and popularity of social tagging systems, several researchers have shown their concern about improving the navigation through tag clouds. The lack of meaningful spatial interpretations in tag clouds has already been addressed by several authors. Lately, there has been an increasing interest on the discovery of semantic relations among social tags Garcia-Silva et al. (2011). With the aim of getting a reorganized tag cloud, a representation of tags must be performed first. Many of the works represent tags consider co-occurrences among tags, whereas a few rely on the textual content of the tagged documents.

For instance, Dattolo et al. (2011) present an approach to discover tag semantics using clustering techniques to find different categories of related tags. They present an approach that considers distributional measures of tags, besides intersection (co-occurrences) and Jaccard (normalized co-occurrences). They perform a qualitative evaluation over a reduced set of top 20 tags, a group of tags known to be ambiguous, and a set of subjective tags. Vandic et al. (2011) propose a method to improve search on social tagging systems by clustering syntactic variations of tags with the same meaning. They use the cosine similarity based on co-occurrence vectors for measuring semantic relatedness. In a similar approach, Specia and Motta (2007) perform the clustering process based on the similarity among tags given by their co-occurrence, where each tag is represented using the intersection with each other tag in the whole tag set.

In Begelman et al. (2006), they build an undirected graph representing the tag space, where the vertices correspond to tags, and the edges between them represent their co-occurrence frequency. The tag space is built with the pairs of tags that co-occur more frequently than expected, by looking for a cut-off point above which a pair of tags is considered strongly related. The authors obtain clusters of related tags using a clustering algorithm based on the spectral bisection.

Textual content has also been considered to represent and find inter-related tags. In Brooks and Montanez (2006), the authors analyzed document similarity based on weighted word frequency using the TF-IDF term weighting function. They grouped documents with sharing tags into clusters, and then compared the similarity of all documents within a cluster, by means of the average pairwise cosine similarity and an agglomerative algorithm. Zubiaga et al. (2009) present a methodology to obtain and visualize a cloud of related tags based on the use of Self-Organizing Maps, where relations among tags are established taking into account the textual content of the tagged documents. Although the resultant tag cloud was promising, they did not compare the content based representation with any other co-occurrence based representation.

In this context, in addition to other clustering algorithms, Self-Organizing Maps have been used to cluster related tags. An advantage of SOMs over other methods is that the clustering step itself produces a graphical map of the folksonomy. Graph-based clustering methods such as that by Simpson (2008) have also been used to produce a visual graph of tags, but these graphs are often more complex, with many edges, and require more expensive layout algorithms. The visualization capabilities of SOMs provide an intuitive way of representing the distribution of data as well as the object similarities. In Sbodio (2009), tags were clustered by means of a SOM, using a tag representation based on co-occurrence among tags. Once the map was trained, the authors used it to classify new tagged documents. Other works that use Self Organizing Maps to find related tags are Li and Zhu (2008) and Gabrielsson and Gabrielsson (2006).

There are works in clustering tags on the social bookmarking site Last.fm. In Chen et al. (2009), a clustered tag cloud for the social bookmarking site Last.fm—a popular social bookmarking site, where music-related resources like artists and songs can be tagged by users—is presented, and Lehwark et al. (2008) use a SOM and U-Map techniques to visualize and cluster tagged music data from Last.fm. These two works rely on tagging patterns to discover relations among them, e.g., they group tags containing the string 'rock.' Afterwards, they calculate semantic similarities between tags by means of co-occurrences and other measures like the TF-IDF weighting function, euclidean distance and cosine similarity. Nonetheless, their approaches are not automated but manual, and they manually define similar words to be clustered, so their systems do not allow to easily update the clustered tag cloud.

Different from those above, we perform a qualitative comparison of two tag representations: tag co-occurrences and document textual content.

TAG CLOUD REORGANIZATION

Dataset

First of all, we need a dataset to conduct the experiments. We decided to use the DeliciousT140 dataset released by Zubiaga et al. (2009). This dataset is made up by 144,574 unique URLs, all of them with their corresponding social tags retrieved from Delicious on June 2008. This set of documents is annotated with 67,104 different tags.

Going into further details about this collection, it was created starting from the 140 most popular tags of the site, that is, the whole tag cloud (in the following T140). Each URL is attached to an amount k of annotators, and a list of weighted tags $T = \{w_1\ t_1,...,w_n\ t_n\}$, where n is at most 25, limited by the social bookmarking site at the time of the dataset generation. Along with the social tags from Delicious, it includes the HTML content corresponding to that URLs. Moreover, the dataset contains only English-written documents. This dataset is available on the Web for research purposes.

Tag Representations

The great majority of approaches to represent tags are based on co-occurrences among tags. As far as we know, there is not any comparison between the co-occurrence representation and any other representation based on the textual content of the annotated Web documents. At the heart of both approaches is the same kind of information, but they stress in a different way. On the one hand, both take into account the document content, one in an explicit way (content based) and the other in an implicit way (co-occurrence based), since considering tag co-occurrences assumes relations among contents from the tagged documents. On the other hand, both use tag co-occurrence data: one in an explicit way (co-occurrence based) and the other in an implicit way (content based), since the content of a document can take part in the representation of more than one tag we take into account co-occurrence information in an implicit way.

In this chapter, we tried these two approaches described above to represent tags in order to reorganize a tag cloud: representation by tag co-occurrence, and by textual document content. In both cases, we use the VSM.

Representation by Tag Co-Occurrence

User posts present interesting features to represent tags. When a user tags a document, the implicit semantics of the tag is assigned to document content. Since we considered only popular tags (only the 140 tags in the tag cloud are taken into account) we can expect a reasonable user agreement, and these tags will fit the documents they represent quite well. Therefore, we are taking into account information provided by user classification, in such a way that we could say we are building a tag representation based on human knowledge. Moreover, this classification was performed by a large number of users. Therefore, if we find two highly posted tags labeling the same document, we can assume the document content is related to both of them. Thus, if this tag co-occurrence is found in several documents, being the number of documents large enough to be representative in our dataset, we can conclude that a relation between those tags exists, due to the fact that system users posted the same documents with both of them. From this assumption we formulate our main hypothesis for co-occurrence based tag representation: *the greater the number of documents tagged by the same tags, the greater is the similarity among these tags.*

Based on these ideas, we propose four different tag weighting functions. For each tag we build a vector representing its degree of co-occurrence with every tag within T140. Therefore, we obtained 140 vectors with 140 dimensions each, one per tag. Hence, each vector component corresponds to a different T140 tag, and the value set for this component, hereafter tag weight, measures the degree of co-occurrence between the tag corresponding to that component and the tag represented by the vector. Equation (1.1) shows how a tag vector is organized:

$$Tag_i = (W_{tag_i,tag_1}, \ldots, W_{tag_i,tag_j}, \ldots, W_{tag_i,tag_{140}}) \quad \forall \quad tag_i \in T140 \qquad (1.1)$$

being Tag_i the vector representation of tag *i*, and W_{tag_i,tag_j} the weight between Tag_i and Tag_j.

Thus, the tag vectors corresponding to the whole collection (140 vectors) make up the matrix (1.2).

$$\begin{pmatrix} W_{tag_1,tag_1} & \cdots & W_{tag_1,tag_{140}} \\ \vdots & \ddots & \vdots \\ W_{tag_{140},tag_1} & \cdots & W_{tag_{140},tag_{140}} \end{pmatrix} \quad (1.2)$$

So far, we have defined the vector space used to represent tags. We have also talked about the weighting functions used to build the vectors and the main ideas we took into account to choose them. Now, we will define in detail each of the weighting functions. We consider three main features to be combined with the number of documents tagged with both tags: (1) the minimum document frequency between tags, (2) the maximum tag document frequency between tags, and (3) the number of documents tagged with at least one of the tags. We combine these 3 weights to define 4 different weighting functions:

Document frequency of the intersection of two tags (Equation [1.3]): The absolute number of documents in the dataset tagged with both tags. In this case, we make use of the main hypothesis previously formulated directly. This function is not normalized to the dataset dimension and so, its values will not be relative but absolute within the dataset.

$$W_{tag_i,tag_j} = df(tag_i \cap tag_j) \quad (1.3)$$

Document frequency of the intersection of two tags over document frequency of the union of those tags (Equation [1.4]): This function represents the Jaccard similarity coefficient. If two tags have a high Jaccard score, then they almost always occur in the dataset as a pair, and one will almost never occur in the absence of the other. This function also assumes the main hypothesis, but in this case, the values are scaled-down by the number of documents tagged with one of the tags.

$$W^{union}_{tag_i,tag_j} = \frac{df(tag_i \cap tag_j)}{df(tag_i \cup tag_j)} \quad (1.4)$$

The Jaccard similarity coefficient has been assumed by several tag clustering studies like Simpson (2008) and Sbodio (2009), reason why we consider this tag weighting function within our baseline. However, its appropriateness as compared to other measures has not yet been shown. In this work, we also aim to show whether or not Jaccard is suitable for the task. We consider this tag weighting function within our baseline because it is one of the most used function in the literature.

Document frequency of the intersection of two tags over the minimum tag document frequency between them (Equation [1.5]): In this case we adjust the value using the minimum tag document frequency of both tags in the dataset, in such a way that the greater the number of documents tagged with the least common tag in connection with the intersection value, the lower the weight is. This function also assumes the previous hypothesis, but in this case, the values are scaled-down by the number of documents tagged with the least common tag.

$$W^{min}_{tag_i,tag_j} = \frac{df(tag_i \cap tag_j)}{min\{df(tag_i), df(tag_j)\}} \quad (1.5)$$

Document frequency of the intersection of two tags over the maximum tag document frequency between them (Equation [1.6]): The weight is adjusted with the maximum tag document frequency of both tags in the dataset. In this weighting function, we assume again the initial hypothesis, but unlike the preceding one, the values are scaled-down by the number of documents tagged with the most common tag.

$$W^{max}_{tag_i,tag_j} = \frac{df(tag_i \cap tag_j)}{max\{df(tag_i), df(tag_j)\}} \quad (1.6)$$

Representation by Document Content

In order to represent a tag by content, we consider the documents that were annotated with that tag. Specifically, we limit to the textual content. Since each tag has many documents annotated, we merged the textual content of all those underlying documents. This approach was first introduced in Zubiaga et al. (2009).

However, we think we should not include all the tags in the same way in the document representation, as some of them may be hardly important because they have lower post count, and because of the associated computational cost. In order to decide which tags to consider relevant for a document, we need to set a threshold; in this manner, only tags with a higher post count than the threshold are selected. We consider the average post count (26) like our threshold, extrapolating the average in the collection to each and every single document (see Figure 2). Hence, working only with the top ranked tags could be more precise in order to discover document content semantics and to find relations among the tags in T140 set.

Then, each of the T140 tags is represented by its corresponding documents and instead of representing each and every document as a vector, we merge all the documents corresponding to a particular tag (hereafter super-documents). Thus, we obtain 140 super-documents representing the tags in T140. Since a document can belong to more than one super-document if it has been tagged with more than one of the 140 tags, then documents might represent more than one tag, and so we would be taking into account co-occurrence information in an implicit way, as we introduced above at the beginning of this section.

The next step is to represent each super-document into the vector space model. At this stage we follow the process described by Zubiaga et al. (2009) in order to represent tags using *TF-IDF* weighting function. The final result of this process

Figure 2. X axis represents the rank of a tag in the top list of tags of the annotated resources, whereas Y axis represents the average post count for each of the positions in the ranking. Note that the tag ranked first could be different from resource to resource. The dashed line means the average post count for every tag positions from first to 25th. In consequence, only the tags with a higher post count than the average (above dashed line) were selected in representations by document content.

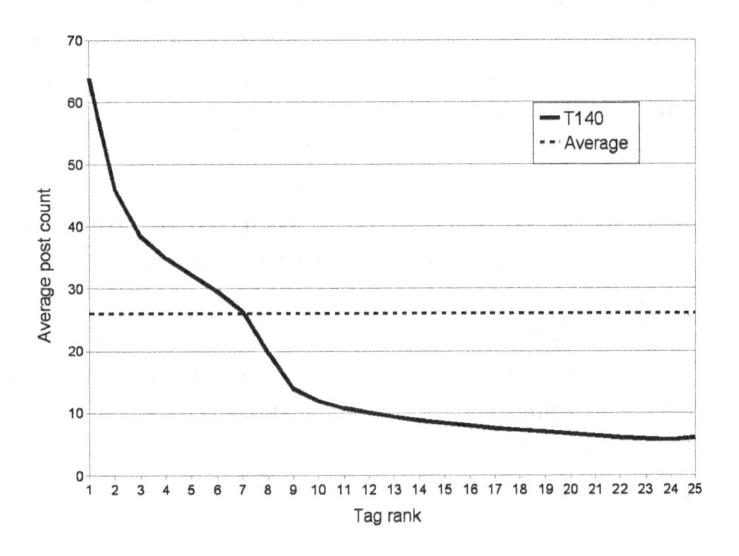

are 140 term vectors, corresponding to each of the T140 tags, composed by 17,518 features each.

Tag Clustering

As a state-of-the-art clustering algorithm, we use Self-Organizing Maps (SOM) (Kohonen, 1990, 2001) to carry out the experimentation. SOM has proven to be an effective way not only to organize information, but also to visualize it, and even to allow content addressable searches (Rauber, et al., 2002; Vesanto & Alhoniemi, 2000; Russell, et al., 2002; Perelomov, et al., 2002; Roh, et al., 2003).

Kohonen's Self-Organizing Maps are unsupervised neural network architectures that use competitive learning in order to produce a spatial-topological relationship between the reference vectors of each neuron in a Vector Space Model (VSM); after a training process, and depending on high dimensional input vectors. The neurons are arranged as a regular node grid, usually with 2 dimensions. Thus, after the training phase, similar inputs to the map will produce nearby outputs into the node grid.

The SOM size was set to 12x12, in order to obtain a square map with a number of neurons close to the number of tags (144 neurons, and 140 tags) with a rectangular lattice. In this way, we have at least one neuron per tag. We did not want to force tag grouping due to map size, that is, if the number of tags is greater than the number of neurons, then multiple tags must share the same neuron because there is no space enough to allocate them in separate ones.

During map training the initial learning rate was set to 0.1, the initial neighborhood was set to 12, equal to map width, and the number of training iterations was 50,000. These values were chosen measuring map quality with the Average Quantization Error (AQE) after several tests with different configurations. AQE measures the average distance between input vectors and their associated reference vectors in the map. Other issues about the SOM are the same as in the standard implementation SOMlib (Rauber, et al., 2002).

Analysis

The different weighting functions based on co-occurrences produced very similar maps from the point of view of tag groups, but W function showed better grouping than the others from a qualitative point of view. As our main goal in this chapter is comparing approaches based on co-occurrences with a content based one, we chose the map generated using W function to represent the co-occurrences approach, simplifying the comparison process.

In order to analyze our results, content-based map is shown in Figure 3, while co-occurrences based map is shown in Figure 4. In these maps, each table cell is a SOM neuron, which may contain tags. Each tag is formatted in such a way that the bigger is the font, the higher the popularity of that tag in the dataset. As it was stated before, proximity on the map implies relatedness among the tags. Finally, throughout this section we will refer to neurons using these positions in the table, e.g. *(x/y)*, being *x* the row number and *y* the column number.

Before analyzing our results, it is worth to note that this dataset has a clear bias towards computer related topics. Because of this, we analyzed the maps in a deeper level of detail, assuming that some tags having different meanings are biased to their computer related one, e.g. tool and tools are related to programming and not to diy (*do it yourself*).

Therefore, analyzing and comparing the maps, several issues emerge:

- Tags sharing the same stem not always share the same neuron in the content approach, e.g. blog (11/8) and blogging (11/4), fun (11/11) and funny (5/11), photo (3/11) and photos (11/0), tool (4/5) and tools (10/1), article (11/9) and articles

Figure 3. Content-based tag cloud reorganization

	0	1	2	3	4	5	6	7	8	9	10	11
0	apple osx mac	**software** freeware windows	howto ubuntu linux	java performance	tech computer hardware		email library	flex flash		web wordpress	css javascript jquery webdesign	portfolio inspiration illustration
1	**programming** python _net development	tips	security					reference	tutorials	ajax	typography	**design** art
2	3d graphics opensource			audio			rails ruby					fashion
3					videos **video** youtube					images	photoshop	photo shopping shop photography
4	games game mobile					tool		tv movies	cool			diy
5	science database			google search politics	news				interesting			funny
6	visualization math				seo jobs	work		economics finance	travel	green home		fic au humor
7	architecture					2008		articles		environment		language english
8	lifehacks					community		culture				health
9	iphone			research	online	internet						food cooking recipes recipe
10	collaboration socialnetworking php	tools		resources		twitter	advertising					
11	flickr productivity firefox wiki photos	tutorial		education teaching learning	technology blogging web2.0	social business socialmedia	marketing media		music mp3 download blogs **blog**	article	free toread writing	books book fun history webdev

(7/7). In the co-occurrences based map this fact only affects to one case, though they are in adjoining neurons: tool (1/1) and tools (1/2), appearing the others together.

- There are very related terms which appear together in the co-occurrences map, while they are separated in the content based, as: flickr (11/0), images (3/9), photo (3/11), photography (3/11) and photos (11/0); iphone (9/0) and mobile (4/0); google (5/3), search (5/3), and seo (6/4).
- In the content approach there are some strange groups, like: music, mp3, download, blog, and blogs (11/8); book, books, fun, history, and Webdev (11/11). In the co-occurrences map some of these tags appear in groups that make more sense, e.g. css and Webdev (6/0); audio, mp3, and music (11/8); fun, funny, and humor (11/6).
- Some tags, which seem to be correctly grouped in the co-occurrences approach are located far away in the content side. This is the case of, for instance: fun (11/11), funny (5/11) and humor (6/11); blog (11/8), blogs (11/8), and blogging (11/4); Webdev (11/11) and css (0/10); music (11/8), mp3 (11/8), and audio (2/3).

This analysis suggests that co-occurrences produce better groupings and therefore a better reorganized tag cloud. Our results show the high efficiency and accurate performance of approaches based on tag co-occurrences, which qualitatively outperform the grouping based on content. Furthermore, approaches based on co-occurrences greatly reduce the computational cost. Computing the tag co-occurrence values is feasible for tag clouds that contain about 100-200 tags. Accordingly, tag cloud reorganization is an affordable task that provides useful features to the browsing. This reorganization represents a good complement to traditional tag clouds, easing user

Figure 4. Co-occurrences-based tag cloud reorganization

	0	1	2	3	4	5	6	7	8	9	10	11
0	linux ubuntu	download free freeware windows	apple mac osx	computer hardware tech	iphone mobile	email		jobs work	google search seo	internet	advertising business marketing	collaboration community social socialmedia socialnetworking twitter web2.0
1	opensource security	firefox tool	**software** tools									**blog** blogging blogs
2							wiki				media	2008 news
3	.net java performance python	database				lifehacks productivity	article articles toread					online technology
4	rails ruby	library	development programming			howto tips	reference				resources	research
5	ajax javascript jquery	php									writing	education english language learning teaching
6	css webdev		tutorial tutorials								math science	book books
7	wordpress		flex				diy					culture history interesting
8	web webdesign											economics finance politics
9	**design**	architecture									travel	environment green health
10	graphics typography	photoshop	3d visualization	flash				movies tv				fashion home shop shopping
11	art illustration inspiration portfolio	flickr images photo photography photos		cool		game games	fun funny humor	video videos youtube	audio mp3 music		au fic	cooking food recipe recipes

navigation through big document collections. In consequence, relying on social data provided by end users shows to be a reliable source to find tag relations.

Applications

A reorganized tag cloud presents different applications as compared to the original one:

- Feed subscription. Within the new cloud, users not only could subscribe to a unique tag, but also they can subscribe to a neuron or even a group of neurons, which contains a set of related tags.
- Finding collection-specific relations among tags allows to discover user communities, or even temporal trends; the new visualization improves the way in which users can explore the whole document collection.
- Analyzing the evolution of tag relations over time could show interesting characteristics of each tag, e.g., whether a tag is temporarily popular.

FUTURE RESEARCH DIRECTIONS

In this work, we performed a qualitative evaluation of tag cloud reorganization based on two different approaches for tag representation. An interesting work to corroborate our findings would be to carry out a quantitative evaluation relying on external evaluation measures. In order to use this kind of measures, a gold standard would be needed. In this sense, another interesting work would be to develop a benchmark to be used as a gold standard for evaluation purposes, because as far as we know, there is not any available for the research community at the moment.

Besides, an analysis on tag evolution throughout time could be done based on the progressive map updates, e.g., a tag like *news* may vary its neighborhood due to the trends of the news in a specific period.

CONCLUSION

In this book chapter, we have motivated the need of a tag cloud reorganization process, which applied to the traditional tag clouds can help the user navigate through related content easily. We have also covered two different approaches to represent tags: one based on document content and another based on tag co-occurrences. In the case of co-occurrences, we used four different tag weighting functions in order to obtain a value representing the degree of co-occurrence between tag pairs, building a set of input vectors, one per tag, containing the similarity values between the vector tag and the rest of the T140 tags. These four functions were chosen in order to establish a baseline for tag co-occurrence representation facing the comparison with the content-based representation, which is based on the TF-IDF term weighting function. We have shown that representing tags by co-occurrences yields more accurate clusters than representing them by content.

Summarizing, we have shown that relying on social data provided by end users is a reliable source to find tag relations. These relations allow the composition of a reorganized tag cloud where each tag is surrounded by other related tags, enhancing users experience in social tagging systems.

ACKNOWLEDGMENT

The authors would like to thank the MA2VICMR consortium (S2009/TIC-1542, http://www.mavir.net) for the financial support for this research, a network of excellence funded by the Madrid Regional Government, and the Spanish research project Holopedia funded by the Ministerio de Ciencia e Innovación under grant TIN2010-21128-C02.

REFERENCES

Begelman, G., Keller, P., & Smadja, F. (2006). Automated tag clustering: Improving search and exploration in the tag space. In *Proceedings of WWW 2006 Collaborative Web Tagging Workshop*, (pp. 15-33). IEEE.

Brooks, C. H., & Montanez, N. (2006). Improved annotation of the blogosphere via autotagging and hierarchical clustering. *In Proceedings of the 15th International Conference on World Wide Web*, (pp. 625-632). ACM.

Chen, Y.-X., Santaía, R., Butz, A., & Therón, R. (2009). Tagclusters: Semantic aggregation of collaborative tags beyond tagclouds. In Butz, A., Fisher, B. D., Christie, M., Krüger, A., Olivier, P., & Therón, R. (Eds.), *Smart Graphics* (pp. 56–67). Berlin, Germany: Springer. doi:10.1007/978-3-642-02115-2_5

Dattolo, A., Eynard, D., & Mazzola, L. (2011). An integrated approach to discover tag semantics. In *Proceedings of the 2011 ACM Symposium on Applied Computing SAC 2011*, (pp. 814-820). ACM.

Gabrielsson, S., & Gabrielsson, S. (2006). *The use of self-organizing maps in recommender systems*. (Master's Thesis). Uppsala University. Uppsala, Sweden.

Garcia-Silva, A., Corcho, O., Alani, H., & Gomez-Perez, A. (2011). Review of the state of the art: Discovering and associating semantics to tags in folksonomies. *The Knowledge Engineering Review*, *26*(4).

Golder, S. A., & Huberman, B. A. (2006). The structure of collaborative tagging systems. *Journal of Information Science*, *32*(2), 198–208. doi:10.1177/0165551506062337

Gupta, M., Li, R., Yin, Z., & Han, J. (2010). Survey on social tagging techniques. *SIGKDD Explorations*, *12*(1), 58–72. doi:10.1145/1882471.1882480

Kohonen, T. (1990). The self-organizing map. *Proceedings of the IEEE*, *78*(9), 1464–1480. doi:10.1109/5.58325

Kohonen, T. (2001). *Self-organizing maps*. Berlin, Germany: Springer. doi:10.1007/978-3-642-56927-2

Lehwark, P., Risi, S., & Ultsch, A. (2007). Visualization and clustering of tagged music data. In *Proceedings 31st Annual Conference of the German Classification Society*, (pp. 673-680). Berlin, Germany: Springer.

Li, B., & Zhu, Q. (2008). The determination of semantic dimension in social tagging system based on some model. In *Proceedings of the Second International Symposium on Intelligent Information Technology Application, IITA 2008*, (pp. 909-913). IEEE.

Mika, P. (2007). Ontologies are us: A unified model of social networks and semantics. *Journal of Web Semantics*, *5*(1), 5–15. doi:10.1016/j.websem.2006.11.002

Perelomov, I., Azcarraga, A. P., Tan, J., & Chua, T. S. (2002). Using structured self-organizing maps in news integration websites. In *Proceedings of the 11th International World Wide Web Conference*. ACM.

Rauber, A., Merkl, D., & Dittenbach, M. (2002). The growing hierarchical self-organizing map: Exploratory analysis of high-dimensional data. *IEEE Transactions on Neural Networks*, *13*(6), 1331–1341. doi:10.1109/TNN.2002.804221

Roh, T. H., Oh, K. J., & Han, I. (2003). The collaborative filtering recommendation based on SOM cluster-indexing CBR. *Expert Systems with Applications*, *25*(3), 413–423. doi:10.1016/S0957-4174(03)00067-8

Russell, B., Yin, H., & Allinson, N. M. (2002). Document clustering using the 1 + 1 dimensional self-organising map. In *Proceedings of the Intelligent Data Engineering and Automated Learning—IDEAL 2002*, (pp. 167-174). Berlin, Germany: Springer.

Sbodio, M. L., & Simpson, E. (2009). *Tag clustering with self organizing maps. HP Labs Technical Reports*. New York, NY: HP Labs.

Sen, S., Lam, S. K., Rashid, A. M., Cosley, D., Frankowski, D., & Osterhouse, J. … Riedl, J. (2006). Tagging, communities, vocabulary, evolution. In *Proceedings of the 2006 20th Anniversary Conference on Computer Supported Cooperative Work*, (pp. 181-190). ACM.

Simpson, E. (2008). *Clustering tags in enterprise and web folksonomies. HP Labs Technical Reports*. New York, NY: HP Labs.

Smith, G. (2008). *Tagging: People-powered metadata for the social web*. Berkeley, CA: New Riders.

Specia, L., & Motta, E. (2007). Integrating folksonomies with the semantic web. In Franconi, E., Kifer, M., & May, W. (Eds.), *The Semantic Web: Research and Applications* (pp. 624–639). Berlin, Germany: Springer. doi:10.1007/978-3-540-72667-8_44

Vandic, D., van Dam, J.-W., Hogenboom, F., & Frasincar, F. (2011). A semantic clustering-based approach for searching and browsing tag spaces. In *Proceedings of the 2011 ACM Symposium on Applied Computing, SAC 2011*, (pp. 1693-1699). ACM Press.

Vesanto, J., & Alhoniemi, E. (2000). Clustering of the self-organizing map. *IEEE Transactions on Neural Networks*, *11*(3), 586–600. doi:10.1109/72.846731

Zubiaga, A., García-Plaza, A. P., Fresno, V., & Martínez, R. (2009). Content-based clustering for tag cloud visualization. In *Proceedings of the 2009 International Conference on Advances in Social Network Analysis and Mining*, (pp. 316-319). IEEE Computer Society.

ADDITIONAL READING

Babinec, M. S., & Mercer, H. (2009). *Metadata and open access repositories*. Philadelphia, PA: Taylor & Francis.

Baca, M., & Getty Research Institute. (2008). *Introduction to metadata*. Los Angeles, CA: Getty Research Institute.

Bonino, S. (2009). *Social tagging as a classification and search strategy: A smart way to label and find web resources*. Saarbrcken, Germany: VDM Verlag.

Caplan, P. (2009). *Metadata fundamentals for all librarians*. New Delhi, India: Indiana Pub House.

Foulonneau, M., & Riley, J. (2008). *Metadata for digital resources: Implementation, systems design and interoperability*. Oxford, UK: Chandos.

Gartner, R., L'Hours, H., & Young, G. (2008). *Metadata for digital libraries: State of the art and future directions*. Bristol, UK: JISC.

Golder, S., & Huberman, B. A. (2006). The structure of collaborative tagging systems. *Journal of Information Science*, *32*(2), 198–208. doi:10.1177/0165551506062337

Granitzer, M., Lux, M., & Spaniol, M. (2008). *Multimedia semantics: The role of metadata*. Berlin, Germany: Springer. doi:10.1007/978-3-540-77473-0

Heymann, P., Koutrika, G., & Garcia-Molina, H. (2008). Can social bookmarking improve web search? In *Proceedings of the International Conference on Web Search and Web Data Mining*, (pp. 195-206). New York, NY: ACM.

Hider, P. (2009). *Information resource description: Creating and managing metadata*. London, UK: Facet.

Hillmann, D. I., Guenther, R., & Hayes, A. Library of Congress, & Association for Library Collections and Technical Services. (2008). *Metadata standards & applications*. Washington, DC: Library of Congress.

Lanius, L., & Vermont. (2009). *Embracing metadata: Understanding MARC and Dublin core [workshop]*. Montpelier, VT: Vermont Dept of Libraries.

Mittal, A. C. (2009). *Metadata management*. Delhi, India: Vista International Pub House.

Noll, M. G., & Meinel, C. (2008a). Exploring social annotations for web document classification. In *Proceedings of the 2008 ACM Symposium on Applied Computing*, (pp. 2315-2320). Fortaleza, Brazil: ACM.

Paukkeri, M., Pérez García-Plaza, A., Fresno, V., Martínez, R., & Honkela, T. (2012). Learning a taxonomy from a set of text documents. *Applied Soft Computing*, *12*(3), 1138–1148. doi:10.1016/j.asoc.2011.11.009

Pérez García-Plaza, A., Fresno, V., & Martínez, R. (2008). Web page clustering using a fuzzy logic based representation and self-organizing maps. In *Proceedings of the IEEE/WIC/ACM International Conference on Web Intelligence and Intelligent Agent Technology*, (Vol. 1, pp. 851-854). Sydney, Australia: IEEE Press.

Peters, I. (2009). *Folksonomies: Indexing and retrieval in web 2.0*. Berlin, Germany: De Gruyter/Saur. doi:10.1515/9783598441851

Ramage, D., Heymann, P., Manning, C. D., & Garcia-Molina, H. (2009). Clustering the tagged web. In *Proceedings of the Second ACM International Conference on Web Search and Data Mining,* (pp. 54-63). Barcelona, Spain: ACM.

Sartori, F., Sicilia, M.-A., & Manouselis, N. (2009). *Metadata and semantic research: Third international conference*. Milan, Italy: Springer.

Smith, G. (2008). *Tagging: People-powered metadata for the social web*. Berkeley, CA: New Riders.

Taylor, A. G., & Joudrey, D. N. (2009). *The organization of information*. Westport, CT: Libraries Unlimited.

Turrell, A. (2008). *Augmenting classifications and search with tags to create usable content- and product-based websites*. Baltimore, MD: University of Baltimore.

Wu, X., Zhang, L., & Yu, Y. (2006). Exploring social annotations for the semantic web. In *Proceedings of the 15th International Conference on World Wide Web,* (pp. 417-426). New York, NY: ACM.

Zhou, D., Bian, J., Zheng, S., Zha, H., & Giles, C. L. (2008). Exploring social annotations for information retrieval. In *Proceedings of the 17th International Conference on World Wide Web,* (pp. 715-724). Beijing, China: ACM.

Zubiaga, A., Martínez, R., & Fresno, V. (2009). Getting the most out of social annotations for web page classification. In *Proceedings of the 9th ACM Symposium on Document Engineering,* (pp. 74-83). New York, NY: ACM.

KEY TERMS AND DEFINITIONS

Collaborative Tagging: Many users tag the same item, and every person can tag it with their own tags in their own vocabulary. The collection of tags assigned by a single user creates a smaller folksonomy, also known as personomy. As a result, several users tend to post the same item. For instance, CiteULike, LibraryThing and Delicious are based on collaborative tagging, where each resource (papers, books, and URLs, respectively) could be tagged (therefore annotated) by all the users who considered it interesting.

Folksonomy: As a result of a community tagging resources, the collection of tags defined by them creates a tag-based organization, so-called folksonomy. A folksonomy is also known as a community-based taxonomy, where the classification scheme is plain, there are no predefined tags, and therefore users can freely choose new words as tags. A folksonomy is basically known as weighted set of tags, and may refer to a whole collection/site, a resource or a user. A summary of a folksonomy is usually presented in the form of a tag cloud.

Simple Tagging: Users describe their own resources or items, such as photos on Flickr, news on Digg or videos on Youtube, but nobody else tags another user's resources. Usually, the author of the resource is who tags it. This means no more than one user tags an item. In many cases, like in Flickr and Youtube, simple tagging systems include an attachment to the resource, and not just a reference to it.

Social Bookmarking: Delicious, StumbleUpon, and Diigo, amongst others, are known as social bookmarking sites. They provide a social means to save Web pages (or other online resources like images or videos) as bookmarks, in order to retrieve them later on. In contrast to saving bookmarks in user's local browser, posting them to social bookmarking sites allows the community to discover others' links and, besides, to access the bookmarks from any computer to the user itself. In these systems, bookmarks represent references to Web resources, and do not attach a copy of them, but just a link. Note that social bookmarking sites do not always rely on social tags to organize resources, e.g., Reddit is a social bookmarking approach to add comments on Web pages instead of tags. The use of social tags in social bookmarking systems is a common approach, though.

Social Tagging: A tagging system becomes social when its tag annotations are publicly visible, and so profitable for anyone. The fact of a tagging system being social implies that a user could take advantage of tags defined by others to retrieve a resource.

Tag Cloud: In order to enable visual browsing, social bookmarking systems typically provide an interface model known as tag cloud. These clouds are one of the main ways of browsing and discovering Web documents on social bookmarking systems, as a structure that provides a visual summary of the most popular topics in the collection. Tag clouds comprise between 50 and 200 of the most popular tags on the site, where the more popular is a tag, the bigger it is shown. Sometimes, tags are sorted alphabetically, randomly, or using other non-semantic orderings.

Tagging: Tagging is an open way to assign tags or keywords to resources or items (e.g., Web pages, movies, or books), in order to describe them. This enables the later retrieval of the resources in an easier way, using tags as resource metadata. As opposed to a classical taxonomy-based categorization system, they are usually non-hierarchical, and the vocabulary is open, so it tends to grow indefinitely. For instance, a user could tag this chapter as social-tagging, research and chapter, whereas another user could use Web 2.0, social-bookmarking and tagging tags to annotate it.

Chapter 10
Global Community Extraction in Social Network Analysis

Xianchao Zhang
Dalian University of Technology, China

Liang Wang
Dalian University of Technology, China

Yueting Li
Dalian University of Technology, China

Wenxin Liang
Dalian University of Technology, China

ABSTRACT

Great efforts have been made in retrieving the structure of social networks, in which one of the most relevant features is community extraction. A community in social networks presents a group of people focusing on a common topic or interest. Extracting all communities in the whole network, one can easily classify and analyze a specified group of people, which yields amazing results. Global community extraction is due to this demand. In global community extraction (also global clustering), each person of the input network is assigned to a community in the output of the method. This chapter focuses on global community extraction in social network analysis, previous methods proposed by some outstanding researchers, future directions, and so on.

INTRODUCTION

Social network analysis has been focused by great many researchers. The community structure, containing huge information of the network, is one of the most important directions in social network analysis.

DOI: 10.4018/978-1-4666-2806-9.ch010

A social network is usually organized as a graph. A vertex indicates a person in the network, and edges present the relation between persons. Therefore, communities in social networks are groups of vertices which probably share common properties. As another popular definition, communities can also be a dense sub-graph, that is, edges between vertices in the sub-graph are much more than those between vertices in and out of the sub-graph. Figure 1 shows a schematic example of a graph with three communities.

In order to extract the whole structure of the network, researchers propose the global community extraction methods, in which each vertex of the input graph is assigned to a community in the output of the method. There are mainly three kinds of methods at present to extract global communities: divisive algorithms, modularity-based algorithms, and dense sub-graph algorithms.

Divisive clustering algorithms are a class of hierarchical methods that work top-down, recursively partitioning the graph into clusters (communities), during which the split is typically into two sets. Modularity-based algorithms are to optimize the modularity criterion, defined by Newman and Girvan (Newman, 2004a). For dense sub-graph algorithms, they extract the dense regions in the graph to indicate the whole network structure, which is already acceptable.

In this chapter, we will detailed describe the global community extraction methods. Firstly, the background for global community extraction is presented. Section 1 discusses divisive algorithms, such as, related works and potential problems. Then modularity-based algorithms and dense sub-graph algorithms are shown in section 2 and section 3, respectively, followed with conclusion and future work.

Figure 1. A simple graph with three communities, enclosed by the gray circles

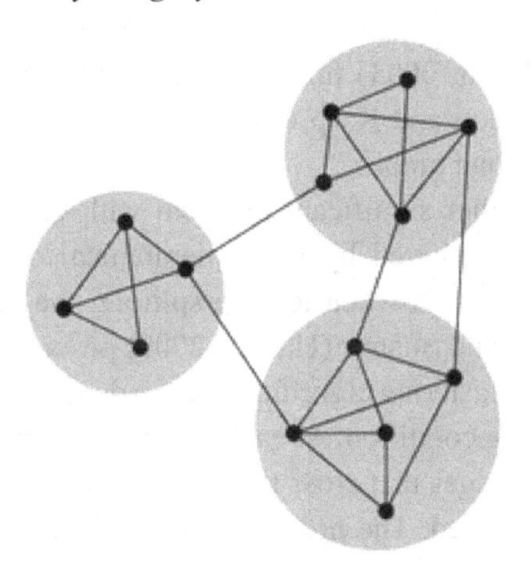

BACKGROUND

As for the huge information, community structure extraction is an essential part in social network analysis. With this structure, we can analyze the group feature of the people more precisely, and then deal with these people, respectively. Plenty of researchers have been conducting research on this direction and done great efforts, for the variety of networks and communities, however, achievement has not been as perfect as we expect.

At the risk of oversimplifying the large and often intricate body of work on community detection in complex networks, the following five-part story describes the general methodology:

1. Data are modeled by an interaction graph. In particular, part of the world gets mapped to a graph, in which vertices represent entities and edges represent some type of interaction between pairs of those entities.
2. Hypothesis is made that the world contains groups of entities that interact more strongly amongst themselves than with the outside world.
3. An algorithm is then selected to find sets of nodes that exactly or approximately optimize this or some other related metric.
4. The clusters or communities or modules are evaluated in some way.

To extract global community structure, one also needs to choose or design the extracting algorithm to be used within a certain "framework" of defining what global community extraction of a graph is. One decision is that all clusters $C_1, \ldots, C_\ell$ are independent with each other, namely partition:

$$C_i \cap C_j = \varnothing \, if \, i \neq j. \quad (1)$$

The other decision allows for each datum to belong to more than one community, namely overlap or cover.

According to these strategies, a number of algorithms have been put forward in recent years. At the rest of this chapter, some popular global community extraction methods will be discussed.

DIVISIVE ALGORITHMS

A simple way to identify communities in a graph is to detect the edges that connect vertices of different communities and remove them, so that the clusters get disconnected from each other. This is the philosophy of divisive algorithms. Divisive clustering algorithms are a class of hierarchical methods that work top-down, recursively partitioning the graph into clusters, during which the split is typically into two sets. The graph bisection is NP-hard, meanwhile, ℓ-partition where the graph is to be partitioned to ℓ equal-sized graph is also NP-hard.

There are some useful methods proposed to deal with the partition problem, cut-based methods, maximum flow, spectral methods, betweenness, and so on.

Cut-Based Algorithms

One intuitive approach is to look for the small cuts. Given a undirected graph $G = (V, E)$, the cut size is the number of edges that connect vertices in C to vertices in $V \setminus C$:

$$Cut(C, V \setminus C) = \left|\left\{(v,u) \in E \,\middle|\, u \in C, v \in V \setminus C\right\}\right|. \quad (2)$$

Minimizing the cut size, plenty of approaches are proposed. One of the earliest methods is the Kernighan-Lin algorithm (Kernighan, 1970), which is still frequently used in combination with other techniques. It starts from a previous partition, and iteratively averages the *gain* metric that swap vertices with negative gains. The *gain* for vertex v is defined as:

$$Gain_v = \sum_{(v,u) \in E \vee P[v] \neq P[u]} w(v,u) - \sum_{(v,u) \in E \vee P[v] = P[u]} w(v,u). \quad (3)$$

The KL method, scaling as $O(n^2 \log n)$, keeps the two clusters with equal size and the gain for every vertex with negative value. It is preferable to start with good guess about the sought partition; otherwise the results are quite poor. Therefore, the method is typically used to improve on the partitions found through other techniques, by using them as starting configurations for the algorithm.

The KL algorithm has been extended to extract partitions in any number of parts (Suaris, 2002); however, it does not have good scalability. Cooperate with KL algorithm, Karypis et al. (Karypis, 1998) propose a multi-level partitioning algorithm, namely Metis, to solve ℓ-partition problem. It mainly contains three phases: coarsening phase, partition phase, and uncoarsening phase.

In addition, Condon and Karp (Condon, 2004) present an ℓ-bisection algorithm that finds in linear time the optimal partition. Their algorithm greedily classifies the vertices into two groups, C_1 and C_2, minimizing the total number of edges crossing the various cuts. Dubhashi et al. (Dubhashi, 2003) further develop the approach of Condon and Karp to cluster categorical data rather than graphs.

Another significant problem with cut-based methods, shared by most hierarchical divisive algorithms, is when to stop splitting the graph. Hartuv and Shamir (Hartuv, 2000) propose a divisive clustering algorithm using a density-based stopping condition. At each iteration, to split the graph, edges that cross the current minimum cut are removed. The procedure stops if each component is highly connected, that is, the edge-connectivity of the graph of order n is above $\frac{n}{2}$.

Instead of simple cut size, a popular criterion for partitioning is that of low conductance (Carrasco, 2003; Shi, 2002; Sima, 2006), believed to be in general superior to simple minimum cuts when used in graph clustering (Flake, 2004; Kannan, 2004). It is defined as:

$$\Phi(C) = \frac{Cut(C, V \setminus C)}{\min(\deg(C), \deg(V \setminus C))}. \quad (4)$$

where $\deg(C)$ is defined as the sum of degree of all vertices in C.

Conductance also takes into account the orders of the sets that are being cut apart, yielding often in more significant separations. It is NP-hard to find a cut with minimum conductance (Sima, 2006); however, a number of techniques have been put forward based on variants of conductance (Kannan, 2004; He, 2002; Flake, 2004; Brandes, 2003; Cheng, 2006; Gkantsidis, 2003), generally iteratively finding a cut with low conductance, ratio cut or normalized cut size and splitting the graph. The ratio cut of C is defined as:

$$\Phi_R(C) = \frac{Cut(C, V \setminus C)}{| V_C || V_{V \setminus C} |}. \quad (5)$$

where $| V |$ denotes the num of edges in C.

The normalized cut of C is

$$\Phi_N(C) = \frac{Cut(C, V \setminus C)}{\deg(C)}. \quad (6)$$

The general idea is that conductance and other similar measures tend to reach optimal values at community boundaries and not within communities, as each community should be internally dense while being sparsely connected to the rest of the graph.

Max-Flow Methods

Referring to the maximum-flow and minimum-cut problems, plenty of algorithms (Brandes, 2003; Carrasco, 2003; Flake, 2000, 2002; Andersen, 2008; Khandekar, 2009; Lang, 2004) based on flow are proposed to extract community structure. The well known maximum-flow and minimum-cut theorem by Ford and Fulkerson (Ford, 1956) state that the minimum cut between any two vertices s and t of a graph, i.e. any minimal subset of edges whose deletion would topologically separate s from t, carries the maximum flow that can be transported from s to t across the graph.

Most flow-based methods take edge weights readily into account. Gomory and Hu defines a *minimum-cut tree*, built with minimum cuts, to define the cluster of a given vertex v with respect to the artificial sink vertex t. Andersen and Lang (Andersen, 2008) present an algorithm called *Improve* that improves a proposed partition of a graph, inputting a subset of vertices and returning a new subset of vertices with a smaller *quotient cut* score. It finds such a set by solving a sequence of polynomially many $s-t$ minimum cut problems, a sequence that cannot be cast as a single parametric flow problem, done by previous known methods. The *quotient cut* is defined as:

$$\Phi_Q(C) = \frac{Cut(C, V \setminus C)}{\min(w(C), w(V \setminus C))}. \quad (7)$$

Khandekar et al. (Khandekar, 2009) present a method using single commodity flows that iteratively route flows across $O(\log^2 n)$ cuts in the graph and embed an expander with congestion with $O(\log^2 n)$ times the optimum.

In addition, the edge weights can also be assigned more precisely. Gallo et al. (Gallo,

1989) use single-parameter functions instead of constants to decide edge capacities. Zhang et al (Zhang, 2010a) define a similarity function to automatically decide the degree of each edge in the graph, which can extract communities with more accurate boundary.

Flow-based approaches can well split graph to subparts, and great efforts have been made to reduce the computational complexity. However, they are not scalable enough to deal with enormous data yet. They are often used to co-operate with other algorithms.

Betweenness Approaches

Betweenness is some measure that favors edges that lie between communities and disfavors those that lie inside communities. In order to cluster graph $G = (V, E)$, one technique is that repeatedly removes edges with highest betweenness. It is supposed that edges with high betweenness are most possible those of between communities. The general steps are:

1. Computation of the centrality for all edges.
2. Removal of edge with largest centrality: in case of ties with other edges, one of them is picked at random.
3. Recalculation of centralities on the running graphs.
4. Iteration of the cycle from step 2.

The most popular algorithm is that proposed by Girvan and Newman (Girvan, 2002; Newman, 2004a), which is based on the node-betweenness defined by Freeman (1977) for use in sociological studies. They assume edges with high betweenness to be links between clusters instead of internal links within a cluster. They consider three alternative definitions: geodesic edge betweenness, random-walk edge betweenness and current-flow betweenness.

Edge betweenness is the number of shortest paths between all pairs of vertices, which can be calculated in a time that scales as $O(mn)$, or $O(n^2)$ on sparse graph, with techniques based on breadth-first-search (Newman, 2004a; Tao, 2006). m is the number of edges, and n is the number of vertices. The random-walk edge betweenness is calculated through a random walker that gets the expected net number of times that a random walker between a particular pair of vertices will pass down a particular edge and sum over all vertex pairs. Current-flow betweenness, defined by considering the graph a resistor network, with edges having unit resistance, is the average value of the current carried by the edge after several voltage differences.

Calculating edge betweenness is much faster than the other two. Moreover, edge betweenness gives much better results according to the Girvan-Newman algorithm by applying many centrality measures. However, the algorithm requires an additional factor m as the running time, scaling as $O(m^2 n)$, which is quite slow.

Spectral Methods

Based on the properties of the spectrum of the *Laplacian* matrix L ($L = I - A$), spectral bisection methods have became another popular technique for graph partitioning. *Spectral clustering* includes all methods and techniques that partition the set into clusters by using the eigenvectors of matrices. It is typically based on computing the eigenvectors corresponding to the second-smallest eigenvalue of the normalized *Laplacian* or some eigenvector of some other matrix representing the graph structure. The normalized Laplacian is:

$$L_N = D^{-\frac{1}{2}} L D^{-\frac{1}{2}} = I - D^{-\frac{1}{2}} A D^{-\frac{1}{2}} \quad (8)$$

In 1973, Donath and Hoffman (Donath, 1973) firstly proposed a spectral clustering algorithm that used the eigenvectors of the *adjacency matrix* for graph partitions. In the same year, Fiedler (1973) realized that from the eigenvector of the second smallest eigenvalue of the *Laplacian* matrix it was possible to obtain a bipartition of the graph with very low cut size. Based on this point, Qiu and Hancock (Qiu, 2006) present a spectral method for clustering graphs. For weighted graphs, Capoccia et al. (Capoccia, 2005) give a spectral clustering approach that is to compute eigenvectors and uses the correlations between the elements to determine the cluster structure. In addition, Pothen et al. (Pothen, 1990) present a heuristic algorithm for minimum bisection using the eigenvectors of the *Laplacian* matrix.

Spectral clustering runs fast and the *Laplacian* matrix is particularly suitable for spectral clustering. How well spectral measures work as separators in clustering is studied by Guattery and Miller (Guattery, 1998), who find that using the Fiedler vector to partition a graph into two equal-sized vertex sets work poorly for a family of bounded-degree planar graphs and in that there exists a family of graphs on which spectral methods in general work poorly.

Markov Chains

Markov chains and random walks are techniques based on the probability. If a vertex is in a certain cluster, another vertex in the cluster is significantly the next random visited one. Therefore, before a random walker walks out the cluster, all vertices can be visited. Then, we can get the cluster. For example, in Figure 2, if a random walk is currently in one of the two clusters, it is more likely to remain in that cluster than to move over to the other cluster.

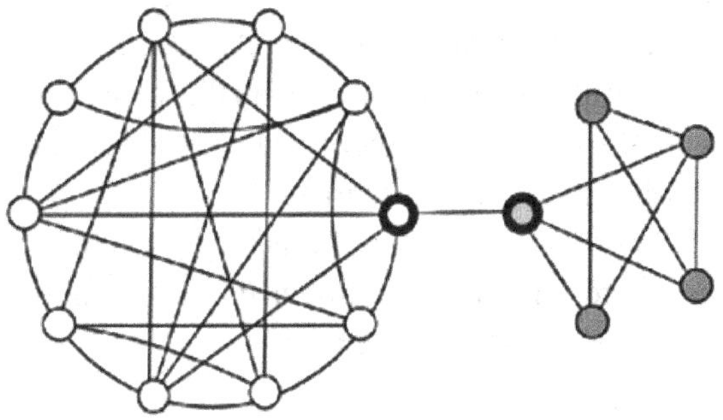

The thesis of van Dongen (2000) discusses graph clustering using *Markov* chains, to apply a sequence algebraic matrix operations on the *Markov* chain corresponding to the input graph such that performing the operations will eliminate inter-cluster interactions and only leave the intra-cluster parts.

Figure 2. A two-cluster graph. If a blind random walk is currently in the white thick-bordered vertex, it will remain in the white cluster with probability $\frac{5}{6}$ and switch over to the grey cluster with probability $\frac{1}{6}$

Spielman and Teng (Spielman, 2004) approximate the distributions of several random walkers on the input graph, and give an algorithm for approximate graph partitioning. In addition, some modifications are used to local clustering (Orponen, 2008, 2005).

MODULARITY-BASED ALGORITHMS

For a long time, there is no good and general criterion to evaluate the results of clustering methods. *Conductance* is a considerable measurement to evaluate the performance of algorithms; however, it, as the ratio of inter-edges and intra-edges, ignores the density inside the cluster. Therefore, if an algorithm extracts sparsely connected clusters, it may also give good values, as soon as the total networks are sparse enough. To cope with this problem, Newman and Girvan (Newman, 2004a) define a measure of the quality of a particular division of a network, namely *modularity*. This measure is based on a previous measure of assortative mixing proposed by Newman (2003).

He assumed that simply counting edges, such as *Conductance*, is not a good way to quantify the intuitive concept of community structure. A good division of a network into communities is not merely one in which there are fewer than expected edges between communities. If the number of edges between two groups is only what one would expect on the basis of random chance, then few thoughtful observers would claim this constitutes evidence of meaningful community structure. This measure is called *modularity*.

Consider a particular division of a network into k communities. Define a $k \times k$ symmetric matrix e whose element $\mathbf{e}_{ij}$ is the fraction of all edges in the network that link vertices in community i to vertices in community j. Then the modularity measure is defined as:

$$Q = \sum_i (\mathbf{e}_{ii} - \mathbf{a}_i^2) = Tr\mathbf{e} - \left\|\mathbf{e}^2\right\| \quad (9)$$

where $\mathbf{a}_i = \sum_j \mathbf{e}_{ij}$, which represents the fraction of edges that connect to vertices in community i; $Tr\mathbf{e} = \sum_i e_{ii}$ gives the fraction of edges in the network that connect vertices in the same community; $\left\|\mathbf{e}\right\|$ indicates the sum of the elements of the matrix $\mathbf{x}$.

Given the adjacency matrix $\mathbf{A}$ of $\mathbf{G}$, the modularity can also be expressed as:

$$Q = \frac{1}{2m}\sum_{w}[A_{vu} - \frac{k_v k_u}{2m}]\delta(C_v, C_u) \quad (10)$$

where k_v is the degree of vertex v and $m = \frac{1}{2}\sum_{v \in \mathbf{V}} k_v$.

Modularity is by far the most used and best known quality function, representing one of the first attempts to achieve a first principle understanding of the clustering problem, and embedding in its compact form all essential ingredients and questions, from the definition of community, to the choice of a null model, to the expression of the "strength" of communities and partitions. By far, great techniques have been put forward to detecting community structure by maximizing modularity.

In this section, we will review some outstanding algorithms that extract community structure based on *modularity optimization*, and discuss some extensions/modifications of it.

Modularity Optimization

According to Newman and Girman's assumption, a higher value of modularity gives a better partition. So, there is an obvious strategy for clustering to maximize modularity of the partition. However, recently, it has been proved that modularity optimization is an NP-complete problem (Brandes, 2006). However, there are many approximate approaches to maximizing modularity. Details are as follows.

Greedy Techniques

Newman (2004b), achieving the first step, proposes a greedy algorithm based on modularity, which is an agglomerative hierarchical clustering method. It starts with a state in which each vertex is the sole member of one of n communities, and repeatedly joins communities together in pairs, choosing at each step the join that results in the greatest increase (or smallest decrease) in Q. The progress of the algorithm can be represented as a "dendrogram" (see Figure 3), a tree that shows the order of the joins.

Clauset et al. (Clauset, 2004) propose a different algorithm that performs the same greedy optimization as the algorithm of Newman (2004b) and therefore gives identical results, however, runs more quickly, in time $O(md \log n)$, where d is the depth of the "dendrogram."

Figure 3. Dendrogram of the communities found by Newman's greedy algorithm based on modularity

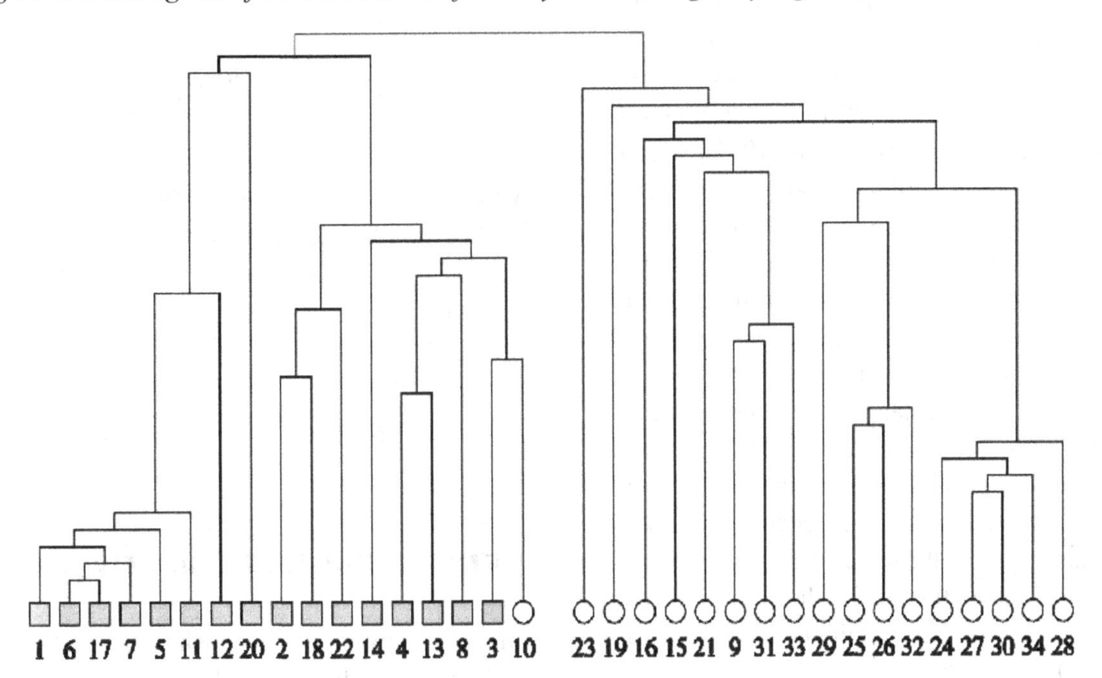

To deal with this problem of modularity optimization, Danon et al. (Danon, 2006) suggest to normalize the modularity variation ΔQ produced by the merger of two communities by the fraction of edges incident to one of the two communities, in order to favor small clusters. In this approach, they take the inhomogeneities in community size into account and modify the algorithm proposed by Newman (Newman, 2004b) by calculating the normalized ΔQ. They normalize ΔQ by the number of links:

$$\Delta Q' = \frac{\Delta Q}{k_i} \tag{11}$$

Spectral Techniques

Modularity can also be optimized using the eigenvalues and eigenvectors of a special *modularity matrix* $\mathbf{B}$ with elements:

$$\mathbf{B}_{ij} = \mathbf{A}_{ij} - \frac{k_i k_j}{2m} \tag{12}$$

According to Equation (10), Q can also be defined as:

$$\begin{aligned} Q &= \frac{1}{2m}\sum_{ij}(A_{ij} - \frac{k_i k_j}{2m})\delta(C_i, C_j) \\ &= \frac{1}{4m}\mathbf{s}^{\mathrm{T}}\mathbf{B}\mathbf{s} \end{aligned} \tag{13}$$

where $\mathbf{s}$ is the column vector whose elements $\mathbf{s}_i = 1$ if vertex i belongs to group 1 and $\mathbf{s}_i = -1$ if it belongs to group 2. Writing $\mathbf{s}$ as a linear combination of the normalized eigenvectors $\mathbf{u}_i$ of $\mathbf{B}$ so that $\mathbf{s} = \sum_{i=1}^{n} \mathbf{a}_i \mathbf{u}_i$ with $\mathbf{a}_i = \mathbf{u}_i^{\mathrm{T}} \bullet \mathbf{s}$, then Q can also be presented as:

$$Q = \frac{1}{4m}\sum_i \mathbf{a}_i \mathbf{u}_u^{\mathrm{T}} \mathbf{B} \sum_j \mathbf{a}_j \mathbf{u}_j = \frac{1}{4m}\sum_{i=1}^{n} (\mathbf{u}_i^{\mathrm{T}} \bullet \mathbf{s})^2 \beta_i \tag{14}$$

Note that the elements of each of $\mathbf{B}$'s rows and columns sum to zero, so that it always has an eigenvector $(1,1,1,\ldots)$ with eigenvalue zero. This observation is reminiscent of the matrix known as the graph *Laplacian* (Chung, 1997), which is the basis for one of the best-known methods of graph partitioning, spectral partitioning (Fiedler, 1973; Pothen, 1990), and has the same property. This suggests that one can optimize modularity on bipartitions via spectral bisection, by replacing the *Laplacian* matrix with $\mathbf{B}$.

As the first outstanding achievement, Newman (2006) proposes a spectral algorithm for community detection based on modularity matrix. This method chooses the leading eigenvector corresponding to the largest eigenvalue to do the bipartition that all vertices whose corresponding elements are positive go in one group and all of the rest in the other.

White and Smyth (White, 2005) put forward a different spectral approach for weighted graph. Given a partition of g into k clusters described trough an $n \times k$ assignment matrix $\mathbf{X}$, where $\mathbf{X}_{ic} = 1$ if vertex i belongs to cluster C, otherwise $\mathbf{X}_{ic} = 0$, they rewrite modularity as:

$$Q \propto Tr\mathbf{X}^{\mathrm{T}}(\mathbf{W}-\mathbf{D})\mathbf{X} = -Tr\mathbf{X}^{\mathrm{T}}\mathbf{L}_{\mathbf{Q}}\mathbf{X} \tag{15}$$

where $\mathbf{W}$ is a diagonal matrix with identical elements, equal to the sum of all edge weights, and the entries of $\mathbf{D}$ are $\mathbf{D}_{ij} = k_i k_j$. $\mathbf{L}_Q = \mathbf{D} - \mathbf{W}$ is called the Q*-Laplacian,* which is used to perform spectral partitioning.

Modifications of Modularity

Ever since the birth of modularity, it has been focused by many researchers, and it has many modifications proposed.

Modularity can be easily extended to graphs with weighted edges (Newman, 2004c). The same procedure as in the case of unweighted graphs can be used to define weighted modularity:

$$Qw = \frac{1}{2\|\mathbf{W}\|}\sum_{ij}(\mathbf{W}_{ij} - \frac{s_i s_j}{2\|\mathbf{W}\|})\delta(C_i, C_j) \tag{16}$$

where s_i is the *strength* of vertex i.

For *overlapping* clustering, the modularity, produced by Nicosia et al. (Nicosia, 2009) is defined as:

$$Q_{ov} = \frac{1}{m}\sum_{c=1}^{n_c}\sum_{ij}\left[r_{ijc}\mathbf{A}_{ij} - s_{ijc}\frac{k_i^{out}k_j^{in}}{m}\right] \tag{17}$$

DENSE SUB-GRAPH DISCOVERY

In almost any network, density is an indication of importance. It is defined as:

$$d(S) = \frac{2\,|\,E(S)\,|}{|\,S\,|(|\,S\,|-1)} \tag{18}$$

Depending on what properties are being modeled by the graph's vertices and edges, dense regions may indicate high degrees of interaction, mutual similarity and hence collective characteristic, attractive forces, favorable environments, or critical mass.

From a theoretical perspective, dense regions have many interesting properties. And for a long period, dense components are thought the same as the definition of communities, that is, detecting community structure is to find all the dense sub-graphs of the networks. That is absolutely true to some extent, as the dense component satisfies the basic rules of community, high density inside. However, with density constraint, dense components always prefer small modules to community, and cannot maintain that they are sparsely link to outside.

In this section, we will outline representative approaches for dense sub-graph extraction in more detail.

Clique

In graph theory, a *clique* is a complete sub-graph, in which every vertex connects to any other vertex. Obviously, it is the densest part among the graph, that is, the most satisfying community structure. However, it is hard to extract all the cliques, for not only the computing complexity, but also the density of the network. Not all social networks contain cliques, or enough cliques, at least.

In order to extract the community structure, we need to enumerate all the cliques in the graph. However, the clique problem is NP-hard. Plenty of heuristic methods are proposed to do this job.

When mining for cliques, a few questions must be addressed:

1. **Minimum size:** Obviously, a clique is usually smaller than a community. To achieve satisfying results, the minimum number of vertices in cliques should be decided. Only cliques with enough vertices can be enumerated as a dense sub-graph.
2. **Must be clique:** A clique is a most dense community, however, not all networks contain cliques. Clique-like components can be much better to generate the community structure. The problem that needs to be decided is how density the component should be.
3. **The whole structure:** We can enumerate all the cliques, meeting the size, density, degree, or discover the N highest-ranking components.

According to the different selection when mining dense components, there are various methods.

The most natural way to extract dense sub-graph is to enumerate all possible subsets of vertices, and then check if they are dense enough. Bron and Kerbosch (Bron, 1973) propose a famous approach for extracting cliques, applying the branch-and-bound technique to prune and generating a clique. It is shown that, without theoretical performance, the computing complexity is $O(3.14^n)$. Makino et al. (Makino, 2004) put forward another enumerating algorithm, using of efficient matrix multiplication, which is suitable for sparse networks.

Other than exactly enumerate cliques, Liu and Wong (Liu, 2008) proposed an approximate algorithm to quickly discover *quasi-cliques*. A quasi-clique S is defined as a clique-like component that each vertex in S has at least γ percent of the possible connections to other vertices. The author employs some novel pruning techniques based on diameter, minimum size threshold, and vertex degree. Another quasi-clique method, proposed by Abello et al. (Abello, 2002), is based on the density. The quasi-clique here is recognized as a dense component S with at least $\gamma \mid S \mid (\mid S \mid - 1) / 2$ edges. By using an existing framework, a Greedy Randomized Adaptive Search Procedure (GRASP), this approach owns a new evaluation measure to choose potential vertices into current quasi-clique, with time complexity $O(\mid S \mid \mid V \mid^2)$.

Bipartite Core

There are some special networks that are not suitable for previous clique-based algorithms. One typical graph is the bipartite graph (see Figure 4), a graph whose node set can be partitioned into two sets, F and C. Every directed edge in the graph is directed from a vertex u of F to a vertex v of C. We would better not transform these bipartite graphs into undirected, as the direction of links shows great many information. Actually, there should be more efficient algorithm proposed for discovering dense bipartite sub-graphs, namely, *bipartite cores*, as the signature of communities of bipartite networks.

Figure 4. Bipartite graph

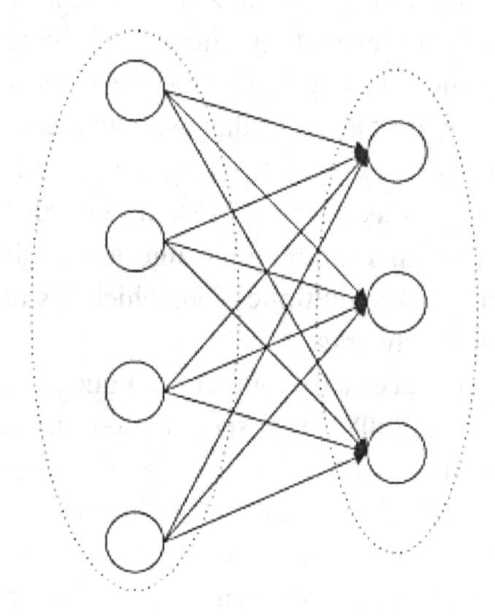

One of the most important and ground-breaking is due to Kumar et al. (Kumer, 1999), proposing the *Trawling* algorithm to enumerate the *Complete Bipartite Graphs* (CBGs), that is, cores. To discover all the complete bipartite cores, it runs in three phases:

1. **Iterative pruning:** Remove vertices that have out-links less than i or in-links less than j and corresponding edges, as these vertices does not benefit the extracted CBGs.
2. **Inclusion-exclusion pruning:** At every step, the method either eliminate a page from contention, or discover (and output) an (i, j) core.
3. **Finishing it off:** Fix a value for j, and note that every proper subset of the fans in any (i, j) core forms a core of smaller size. Here the method uses a priori algorithm of Agrawal and Srikant (Agrawal, 1994), that stats with all $(1, j)$ cores, and then $(2, j)$, and so on.

Trawling algorithm gives another way to extracting community structure, and as soon as it is proposed, it is focused by large number of researchers.

Reddy et al. (Reddy, 2001) proposed a different strategy to enumerate cores. Instead of the CBG, they indicate that the *Dense Bipartite Graph* (DBG) is much suitable to illustrate the Web community. They constrain that, in the DBG, each fan vertex $u \in F$ has at least p links to vertices of C, meanwhile, each center vertex $v \in C$ has at least q links from vertices in F.

Figure 5. Cores extraction algorithm of C&C

Algorithm 1 Cores extraction

Input: *seeds*: a map of nodes and their corresponding *outdegree*
G: a web graph $< V, E >$
i: number of fans
j: number of centers

Output: all existing $C_{3,3}$

1: **for** $r \in seeds$, compute $C^3_{T(r)}$ **do**
2:   **for** each $C^3_{T(r)}$ of T(r) **do**
3:     compute the intersection size of fans pointed to these 3 centers;
4:     **if** the intersection size ≥ 3 **then**
5:       output these fans and 3 centers;
6:     **end if**
7:   **end for**
8:   delete r and iteratively modify seeds and G;
9: **end for**

Gibson et al. (Gibson, 2005) propose a new algorithm based on shingling for discovering large DBGs in massive graphs. Since this algorithm utilizes the shingling technique, firstly introduced in Broder (1997), to convert each dense component with arbitrary size into shingles with constant size, it is very efficient and practical for single large graphs and can be easily extended for streaming graph data.

In addition, Yang and Lin (Dourisboure, 2007) propose an exhaustive and edge removal algorithm to find DBGs. Zhang et al. (Zhang, 2010b) introduces a more effective modification of Trawling algorithm. Their method is based on *Combination and Consolidation* (C&C algorithm for short) using several heuristic methods (see Figure 5 and Figure 6). In the C&C algorithm, a number of unit cores are firstly extracted by the technique of combination, and then the cores that belong to larger complete bipartite graphs are automatically combined by the algorithm of consolidation. The C&C algorithm is much more efficient and effective than the Trawling algorithm because: 1) C&C does not need any user-determined parameters; 2) Identifying all the embedded cores with different sizes only requires one-pass execution of C&C, while it needs many times of iterative execution of Trawling; 3) C&C is capable of extracting all the largest complete bipartite sub-graphs emerged in the Web graph, which is more valuable than the fixed-size cores extracted by Trawling.

FUTURE RESEARCH DIRECTIONS

It has been a long time since community structure extraction appeared in social network analysis. However, there is no common definition about what a community is. According to different definition, there are various algorithms proposed. Giving a generally definition for community will definitely contribute a lot.

Figure 6. Cores consolidation algorithm of C&C

Algorithm 2 Cores consolidation

Input: $Cores$: a map of cores output by Algorithm 1
T: a set of fans
I: a set of centers
Output: merged cores

```
1: set Cores to ∅;
2: for each core output by algorithm 1 do
3:    set T to ∅;
4:    set I to ∅;
5:    read fans of core to set T;
6:    read centers of core to set I;
7:    if T has exist in Cores then
8:       insert ⟨T, I ∪ Cores[T]⟩ into Cores;
9:    end if
10:   if I has exist in Cores then
11:      insert ⟨T ∪ Cores[I], I⟩ into Cores;
12:   else
13:      insert⟨T, I⟩ into Cores;
14:   end if
15: end for
```

For global community extraction in social networks, experiment results are always evaluated manually, by reading the detailed information of people in the community. Obviously, it is not dependable. Conductance and modularity are two mainly strategy to show the results of community extraction algorithms, which are feasible for some specific networks. Work should be done to find a general criterion for global community extraction.

CONCLUSION

As the traditional methods, divisive algorithms are widely used for community extraction in social network analysis. It is the most direct way to use graph theory. Modularity based algorithms and dense sub-graph algorithms are new emerging methods, which are for exactly extraction and large graphs, respectively.

Global community extraction is a pretty hot point in social network analysis. More and more researchers are paying attention on it. Definitely, it deserves all the efforts, and more and more achievement will be made.

REFERENCES

Abello, J., Resende, M., & Sudarsky, S. (2002). Massive quasi-clique detection. In *Proceedings of the 5th Latin American Symposium on Theoretical Informatics*, (pp. 598-612). Springer-Verlag.

Agrawal, R., & Srikant, R. (1994). Fast algorithms for mining association rules. In *Proceedings of VLDB*. Citeseer.

Andersen, R., & Lang, K. J. (2008). An algorithm for improving graph partitions. In *Proceedings of the Nineteenth Annual ACM-SIAM Symposium on Discrete Algorithms*, (pp 651–660). Society for Industrial and Applied Mathematics.

Brandes, U., & Delling, D. (2006). *On modularity-np-completeness and beyond.* Bibliothek.

Brandes, U., Gaertler, M., & Wagner, D. (2003). Experiments on graph clustering algorithms. [ESA.]. *Proceedings of Algorithms-ESA, 2003*, 568–579. doi:10.1007/978-3-540-39658-1_52

Broder, A. Z., Glassman, S. C., Manasse, M. S., & Zwerg, G. (1997). Syntactic clustering of the web. *Computer Networks and ISDN Systems, 29*(8-13), 1157-1166.

Bron, C., & Kerbosch, J. (1973). Algorithm 457: Finding all cliques of an undirected graph. *Communications of the ACM, 16*(9), 575–577. doi:10.1145/362342.362367

Carrasco, J. J., Fain, D. C., Lang, K. J., & Zhukov, L. (2003). Clustering of bipartite advertiser-keyword graph. In *Proceedings of the Third IEEE International Conference on Data Mining, Workshop on Clustering Large Data Sets*. Melbourne, FL: IEEE Press.

Cheng, D., Kannan, R., Vempala, S., & Wang, G. (2006). A divide-and-merge methodology for clustering. *ACM Transactions on Database Systems, 31*(4), 1499–1525. doi:10.1145/1189769.1189779

Chung, F. R. K. (1997). *CBMS conference on recent advances in spectral graph theory, and conference board of the mathematical sciences: Spectral graph theory*. Academic Press.

Clauset, A., Newman, M. E. J., & Moore, C. (2004). Finding community structure in very large networks. *Physical Review E: Statistical, Nonlinear, and Soft Matter Physics, 70*(6). doi:10.1103/PhysRevE.70.066111

Condon, A., & Karp, R. M. (2004). Algorithms for graph partitioning on the planted partition model: Randomization, approximation, and combinatorial optimization. *Algorithms and Techniques, 18*(2), 221–232.

Danon, L., Diaz-Guilera, A., & Arenas, A. (2006). The effect of size heterogeneity on community identification in complex networks. *Journal of Statistical Mechanics*, 11.

Donath, W. E., & Hoffman, A. J. (1973). Lower bounds for the partitioning of graphs. *IBM Journal of Research and Development*, *17*(5), 420–425. doi:10.1147/rd.175.0420

Dongen, S. V. (2000). Graph clustering by flow simulation. *Computer Science Review*, *1*(1), 27–64.

Dourisboure, Y., Geraci, F., & Pellegrini, M. (2007). Extraction and classification of dense communities in the web. In *Proceedings of 16th International Conference on World Wide Web*, (pp. 461-470). ACM.

Dubhashi, D., Laura, L., & Panconesi, A. (2003). Analysis and experimental evaluation of a simple algorithm for collaborative filtering in planted partition models. In *Proceedings of Foundations of Software Technology and Theoretical Computer Science* (pp. 168–182). IEEE. doi:10.1007/978-3-540-24597-1_15

Fiedler, M. (1973). Algebraic connectivity of graphs. *Czechoslovak Mathematical Journal*, *23*(2), 298–305.

Flake, G. W., Lawrence, S., & Giles, C. L. (2000). Efficient in social and biological networks. In *Proceedings of the Sixth ACM SIGKDD International Conference on Knowledge Discovery and Data Mining*, (p. 160). ACM Press.

Flake, G. W., Lawrence, S., Giles, C. L., & Coetzee, F. M. (2002). Self-organization and identification of web communities. *Computer*, *35*(3), 66–70. doi:10.1109/2.989932

Flake, G. W., Tarjan, R. E., & Tsioutsiouliklis, K. (2004). Graph clustering and minimum cut trees. *Internet Mathematics*, *1*(4), 385–408. doi:10.1080/15427951.2004.10129093

Ford, L. R., & Fulkerson, D. R. (2008). Maximal flow through a network. *Canadian Journal of Mathematics*, *8*(3), 399–404.

Freeman, L. C. (1977). A set of measures of centrality based on betweenness. *Sociometry*, *40*(1), 35–41. doi:10.2307/3033543

Gallo, G., Grigoriadis, M. D., & Tarjan, R. E. (1989). A fast parametric maximum flow algorithm and applications. *SIAM Journal on Computing*, *18*(1), 30–55. doi:10.1137/0218003

Gibson, D., Kumar, R., & Tomkins, A. (2005). Discovering large dense sub-graphs in massive graphs. In *Proceedings of 31st International Conference on very Large Data Bases*, (pp. 721-732). ACM.

Girvan, M., & Newman, M. E. J. (2002). Community structure in social and biological networks. *Proceedings of the National Academy of Sciences of the United States of America*, *99*(12). doi:10.1073/pnas.122653799

Gkantsidis, C., Mihail, M., & Saberi, A. (2003). Conductance and congestion in power law graphs. In *Proceedings of the 2003 ACM SIGMETRICS International Conference on Measurement and Modeling of Computer Systems*, (pp. 148–159). ACM Press.

Guattery, S., & Miller, G. L. (1998). On the quality of spectral separators. *SIAM Journal on Matrix Analysis and Applications*, *19*(3), 701–719. doi:10.1137/S0895479896312262

Hartuv, E., & Shamir, R. (2000). A clustering algorithm based on graph connectivity. *Information Processing Letters*, *76*(4-6), 175–181. doi:10.1016/S0020-0190(00)00142-3

He, X., & Zha, H. (2002). Web document clustering using hyperlink structures. *Computational Statistics & Data Analysis*, *41*(1), 19–45. doi:10.1016/S0167-9473(02)00070-1

Kannan, R., Vempala, S., & Vetta, A. (2004). On clustering: Good bad and spectral. *Journal of the ACM*, *51*(3), 497–515. doi:10.1145/990308.990313

Karypis, G., & Kumar, V. (1998). A fast and high quality multilevel scheme for partitioning irregular graphs. *SIAM Journal on Scientific Computing*, *20*(1), 359–392. doi:10.1137/S1064827595287997

Kernighan, B. W., & Lin, S. (1970). An efficient heuristic procedure for partitioning graphs. *The Bell System Technical Journal*, *49*(2), 291–307.

Khandekar, R., Rao, S., & Vazirani, U. (2009). Graph partitioning using single commodity flows. *Journal of the ACM*, *56*(4), 1–15. doi:10.1145/1538902.1538903

Kumar, R., Raghavan, P., Rajagopalan, S., & Tomkins, A. (1999). Trawling the web for emerging cyber-communities. *Computer Networks*, *31*(11-16), 1481-1493.

Lang, K., & Rao, S. (2004). A flow-based method for improving the expansion or conductance of graph cuts. *Integer Programming and Combinatorial Optimization*, 383–400.

Liu, G., & Wong, L. (2008). Effective pruning techniques for mining quasi-cliques. *Lecture Notes in Computer Science*, *5222*, 33–49. doi:10.1007/978-3-540-87481-2_3

Makino, K., & Uno, T. (2004). New algorithms for enumerating all maximal cliques. *Algorithm Theory – SWAT 2004*, (pp. 260-272). Retrieved from http://research.nii.ac.jp/~uno/papers/04swat.pdf

Newman, M. E. J. (2003). Mixing patterns in networks. *Physical Review E: Statistical, Nonlinear, and Soft Matter Physics*, *67*(2). doi:10.1103/PhysRevE.67.026126

Newman, M. E. J. (2004b). Fast algorithm for detecting community structure in networks. *Physical Review E: Statistical, Nonlinear, and Soft Matter Physics*, *69*(6). doi:10.1103/PhysRevE.69.066133

Newman, M. E. J. (2004c). Analysis of weighted networks. *Physical Review E: Statistical, Nonlinear, and Soft Matter Physics*, *70*(5). doi:10.1103/PhysRevE.70.056131

Newman, M. E. J. (2006). Modularity and community structure in networks. *Proceedings of the National Academy of Sciences of the United States of America*, *103*(23). doi:10.1073/pnas.0601602103

Newman, M. E. J., & Girvan, M. (2004a). Finding and evaluating community structure in networks. *Physical Review E: Statistical, Nonlinear, and Soft Matter Physics*, *69*(2). doi:10.1103/PhysRevE.69.026113

Nicosia, V., Mangioni, G., Carchiolo, V., & Malgeri, M. (2009). Extending the definition of modularity to directed graphs with overlapping communities. *Journal of Statistical Mechanics: Theory and Experiment*. Retrieved from arxiv.org

Orponen, P., & Schaeffer, S. E. (2005). Local clustering of large graphs by approximate Fiedler vectors. *Experimental and Efficient Algorithms*, *3503*, 524–533. doi:10.1007/11427186_45

Orponen, P., Schaeffer, S. E., & Gaytan, V. A. (2008). *Locally computable approximations for spectral clustering and apsorption times of random walks*. Retrieved from arxiv.org

Pothen, A., Simon, H. D., & Liou, K. (1990). Partitioning sparse matrices with eigenvectors of graphs. *SLAM Journal on Metrix Analysis and Applications*, *11*(3).

Qiu, H., & Hancock, E. R. (2006). Graph matching and clustering using spectral partitions. *Pattern Recognition*, *39*(1), 22–34. doi:10.1016/j.patcog.2005.06.014

Reddy, P. K., & Kitsuregawa, M. (2001). An approach to relate the web communities through bipartite graphs. In *Proceedings of WISE* (p. 301). IEEE Press.

Shi, J., & Malik, J. (2002). Normalized cuts and image segmentation. *IEEE Transactions on Pattern Analysis and Machine Intelligence*, *22*(8), 888–905.

Sima, J., & Schaeffer, S. (2006). On the NP-completeness of some graph cluster measures. In *Proceedings of the SOFSEM 2006: Theory and Practice of Computer Science,* (pp. 530-537). SOFSEM.

Spielman, D. A., & Teng, S. H. (2004). Nearly-linear time algorithms for graph partitioning, graph sparsification, and solving linear systems. In *Proceedings of the Thirty-Sixth Annual ACM Symposium on Theory of Computing*, (pp. 81–90). ACM Press.

Suaris, P. R., & Kedem, G. (2002). An algorithm for quadrisection and its application to standard cell placement. *IEEE Transactions on Circuits and Systems*, *35*(3), 294–303. doi:10.1109/31.1742

White, S., & Smyth, P. (2005). A spectral clustering approach to finding communities in graphs. In *Proceedings of the Fifth SIAM International Conference on Data Mining*, (p. 274). Society for Industrial Mathematics.

Zhang, X., Li, Y., & Liang, W. (2010b). C&C: An effective algorithm for extracting web community cores. In *Database Systems for Advanced Applications* (pp. 316–326). Berlin, Germany: Springer. doi:10.1007/978-3-642-14589-6_32

Zhang, X., Xu, W., & Liang, W. (2010a). Extracting local web communities using lexical similarity. In *Database Systems for Advanced Applications* (pp. 327–337). Berlin, Germany: Springer. doi:10.1007/978-3-642-14589-6_33

Chapter 11
Local Community Extraction in Social Network Analysis

Xianchao Zhang
Dalian University of Technology, China

Liang Wang
Dalian University of Technology, China

Yueting Li
Dalian University of Technology, China

Wenxin Liang
Dalian University of Technology, China

ABSTRACT

To identify global community structures in networks is a great challenge that requires complete information of graphs, which is infeasible for some large networks, e.g. large social networks. Recently, local algorithms have been proposed to extract communities for social networks in nearly linear time, which only require a small part of the graphs. In local community extraction, the community extracting assignments are only done for a certain subset of vertices, i.e., identifying one community at a time. Typically, local community detecting techniques randomly start from a vertex v and gradually merge neighboring vertices one-at-a-time by optimizing a measure metric. In this chapter, plenty of popular methods are presented that are designed to obtain a local community for a given graph.

INTRODUCTION

Community structure is one of the most relevant features of social networks, and with the growth of interest in the study of social networks, great efforts have been made on the area of community detection.

DOI: 10.4018/978-1-4666-2806-9.ch011

Popularly, a community is a tightly knit subgraph in a network, in which the within-group links are stronger or denser than between-group links (Wasserman, 1994). Directly optimizing the density inside or the sparseness outside, a number of algorithms have been proposed (Gibson, 1998; Newman, 2004a; Andersen, 2008a; Sharan, 2009), which work well in special cases. Similarly, several approaches come forth based on the concept of flow (Flake, 2000; Flake, 2002; Khandekar, 2009). Recently, researchers have been focusing on algorithms based on the *modularity* metric (Newman, 2004b; Neman, 2006).

However, these techniques require information of the whole graphs, which might be problematic for large graphs, as the constraint of running time and memory consumption. To overcome the limitation, local community detecting methods emerged, which only need a small subset of vertices instead of the whole graphs.

In general, local community extraction techniques need a seed vertex to create the initial community structure, and then gradually merge neighboring vertices one-at-a-time by optimizing a fitness function. Computing the desired answer by a clustering algorithm for many applications only requires a small subset of vertices to be clustered instead of the whole graph. The process takes time polynomial in k (the size of extracted community), which is independent to the size of the whole graph. For these two characteristics, local clustering technique is able to extract community structure of huge scale networks. To extract the entire input graph, we can iteratively run the local method, which may be simultaneously obtained by parallel computation.

In this chapter, we will first define the locality for local clustering. Then some famous fitness functions are illustrated, followed with local clustering algorithms presentation.

BACKGROUND

Extracting the full community structure, e.g. global community extraction, requires information of the whole graphs, which might be problematic for large graphs like large social networks. For massive data sets, the running time of a clustering algorithm should not grow faster than $O(n)$ in order to be scalable; sub-linearity is strongly preferable. On the other hand, storing the complete edge set ($O(n^2)$ for dense graphs) is also often infeasible.

However, if the graph is stored in a format that allows access to connected sub-graphs or adjacency lists of nearby vertices, ideas similar to agglomerative clustering can be applied: clusters can be computed one at a time based on only partial views of the graph topology. This is called local clustering.

DEFINITION OF LOCALITY

In order to distinguish local computation from global computation, we must define what information about an input graph we consider to be locally available. Given a single vertex v, it is assumed that at least the adjacency list containing the identifiers of vertices in $N(v)$ are known.

For a wider definition, it also allows direct access from a vertex u to its second neighbors:

$$\bigcup_{u \in N(v)} N(u) \tag{1}$$

Optionally we can allow knowledge of the degrees of the vertices in $N(v)$.

To extract a local community, we should at least store sufficient partial of network. According to the definition of the locality, the part of network should contain the local community and the immediate neighbors of the local community. As shown in Figure 1, C denotes the local community; Neighbor set $N = \{v \mid (u,v) \in E_C, u \in V_C, u \in V_C \wedge v \notin V_C\}$; vertices like u constitute boundary set. The portion of local network should at least have the tree parts: C, B and N.

FITNESS FUNCTIONS

In this section, we present some widely used fitness functions for local community optimization.

Figure 1. Portion graph G for identifying local community C

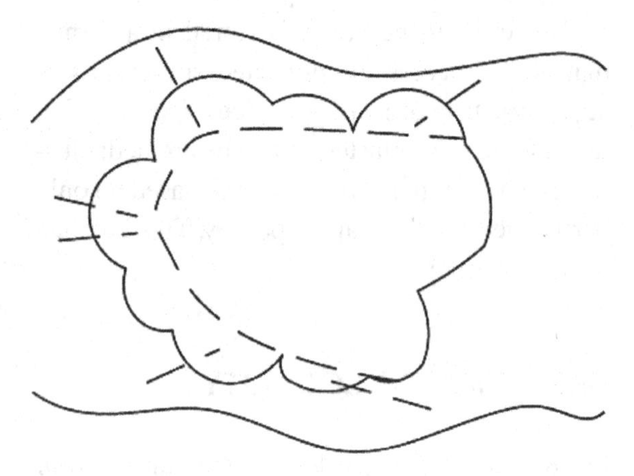

Totally, possible fitness functions for such local clustering are numerous. According to the definition of community, almost all existing metrics directly use the internal and external degrees to represent these two significant factors.

Cheeger Ratio

Cheeger ratio (Cheeger, 1970) is one of the most famous fitness functions, which measures the ratio of the cut size of the cluster to the minimum of the sums of degrees either inside the cluster or outside it:

$$h(C) = \frac{|\{(u,v)\} \mid u \in C, v \in V \setminus C\}|}{\min\left\{\sum_{u \in C} \deg(u), \sum_{v \in V \setminus C} \deg(v)\right\}} \quad (2)$$

where $\deg(u)$ is the degree of vertex u.

Optimizing the *Cheeger ratio* is equivalent to a graph partition problem (He, 2002).

Density

Density can also be used to extract local community. Each of the vertices has a contribution to the density and *vertex density* can be measured as:

$$den(v,C) = \frac{\deg_{\text{int}}(v,C)}{|C|-1} \quad (3)$$

where $\deg_{\text{int}}(v,C)$ measures the number of edges between vertex v and other vertices inside C; $|C|$ donates the num of vertices of C.

The quality of a given community can be evaluated on the basis of the suitability of the included vertices. According to Equation (3), the *cluster density* would be a scaled sum:

$$den(C) = \frac{1}{|C|}\sum_{v \in C} den(v,C) = \frac{1}{|C|(|C|-1)}\sum_{v \in C} \deg_{\text{int}}(v,\dot{C}) \quad (4)$$

As the sum of the internal degrees of vertices in C is twice the internal degree (edges) of the cluster, so Equation (4) can also be written as:

$$den(C) = \frac{1}{|C|(|C|-1) \cdot 2\deg_{\text{int}}(C)} = \frac{\deg_{\text{int}}(C)}{\binom{|C|}{2}} \quad (5)$$

This is equal to the global definition for graph density.

Another metric, *introversion*, measures the ratio of internal edges to all related edges. The *vertex introversion* is defined as:

$$\rho(v,C) = \frac{\deg_{\text{int}}(v,C)}{\deg(v)} \quad (6)$$

The *cluster introversion* can be similarity characterized by summing the suitability measures of Equation (6) and scaling with the cluster order to obtain a measure in $[0,1]$:

$$\rho(C) = \frac{1}{|C|}\sum_{v \in C} \rho(v,C) = \frac{1}{|C|}\sum_{v \in C} \frac{\deg_{\text{int}}(v,C)}{\deg(v)} \quad (7)$$

Similarly, there is an "independence" measure used in local community extraction: the *relative density* den_r (Mihail, 2002). *Relative density* is the ratio of the internal degree to the number of incident edges:

$$den_r(C) = \frac{\deg_{\text{int}}(C)}{\deg_{\text{int}}(C) + \deg_{ext}(C)} = \frac{\sum_{v \in C} \deg_{\text{int}}(v, C)}{\sum_{v \in C} \deg_{\text{int}}(v, C) + 2\deg_{ext}(v, C)} \quad (8)$$

where $\deg_{ext}(v, C)$ denotes the number of edges between vertex v and vertices outside C.

Cut

As to the definition of community, a community is properly introverted if Equation (7) has a high value and the capacity of the *cut* is low. The *cut capacity* is obtained as:

$$Q(C, V \setminus C) = \sum_{v \in C, u \in V \setminus C} \frac{1}{\deg(v)} = \sum_{v \in C} \frac{\deg_{ext}(v, C)}{\deg(v)} \quad (9)$$

Modularity

The metric of *modularity* proposed by Newman and Girvan (Newman, 2004a), does a good job in global clustering:

$$Q = \frac{1}{2m} \sum_{vu} \left[\mathbf{A}_{vu} - \frac{k_v k_u}{2m} \right] \delta(C_v, C_u) \quad (10)$$

where k_v is the degree of vertex v and $m = \frac{1}{2} \sum_{v \in V} k_v$. $\delta = 1$ if v and u are in the same community; otherwise, $\delta = 0$.

In 2005, Clauset (2005) defines a local fitness function based on *modularity*. The *local modularity* metric is defined as:

$$Q_l = \frac{\sum_{ij} \mathbf{A}_{ij} \delta(i, j)}{\sum_{ij} \mathbf{A}_{ij}} = \frac{1}{2m} \sum_{ij} \mathbf{A}_{ij} \delta(i, j) \quad (11)$$

where $\mathbf{A}$ is the adjacency matrix.

ALGORITHMS OF LOCAL COMMUNITY EXTRACTION

Great efforts have been made to extract local community from social networks and many of them do a good job, in spite of some predefined parameters. In this section, we will present some outstanding local algorithms.

Modularity-Based Algorithms

Modularity, proposed by Newman and Girvan (Newman, 2004a), has been an essential metric to evaluate clustering algorithms. Also, it can be used to identify global community structure. Clauset (2005) recognizes that modularity can also be used for local clustering, and proposes the *local modularity* metric (see Equation [11]). In addition, Clauset restricts their consideration to those vertices of the subset of C that have at least one neighbor in $\mathbf{N}$, i.e., vertices of $\mathbf{B}$.

Intuitively, they expect that a community with a sharp boundary will have few connections from its boundary to the unknown portion of the graph, while having a greater proportion of connections from the boundary back into the local community. So *local modularity* metric is modified to be:

$$Q_R = \frac{\sum_{ij} \mathbf{B}_{ij} \delta(i, j)}{\sum_{ij} \mathbf{B}_{ij}} = \frac{\mathbf{I}}{\mathbf{T}} \quad (12)$$

where matrix $\mathbf{B}$ is the boundary-adjacency matrix with elements as: $\mathbf{B}_{ij} = 1$ if vertex i and vertex

j are connected, and either vertex is in $\mathbf{B}$; otherwise, $\mathbf{B}_{ij} = 0$.

Initially, this method places the random source vertex v_0 into C, and its neighbors in $\mathbf{N}$. At each step, it adds to C (and to $\mathbf{B}$, if necessary) the neighbor that results in the largest increase (or smallest decrease) in Q_R. That is, for each vertex $v_j \in \mathbf{N}$, they calculate the ΔQ_R that corresponds to the change in local modularity as a result of joining v_j to C. This process continues until it has agglomerated either a given number of vertices k, or it has discovered the entire enclosing component. The value of ΔQ_R can be quickly computed by using an expression derived from Equation (12):

$$\Delta Q_R = \frac{x - Q_R y - z(1 - Q_R)}{\mathbf{T} - z + y} \tag{13}$$

where x is the number of edges in $\mathbf{T}$ that terminated at v_j, y is the number of edges that will be added to $\mathbf{T}$ by the agglomeration of v_j (i.e., the degree of v_j is $x + y$), and z is the number of edges that will be removed from $\mathbf{T}$ by the agglomeration. The pseudocode for this process is given in Figure 2.

Clauset's local modularity method firstly applied modularity to extracting local communities. However, it cannot do satisfactorily, as it stops depending on the given community size which is hard to decide. In addition, it ignores "outliers," which decrease the inner density of local community.

Similarly, Luo et al. (Luo, 2008) produce another local modularity metric Q_M:

$$Q_M = \frac{\deg_{\text{int}}(C)}{\deg_{ext}(C)} \tag{14}$$

which measures the ratio of internal and external edges.

This method adds the *deletion* step after each *addition* step, in which the algorithm iteratively removes vertices in C that result in the increase of Q_M of C rather than separating C. So it has considered the "outliers." However, it also lacks a good stopping criterion. The detailed steps of the algorithm is shown in Figure 3.

To tackle the problem of "outliers" and stopping criteria, Zhang et al. (Zhang, 2011) propose a three-phase algorithm based on local cores that takes time polynomial in k (the size of extracted community):

1. Instead of choosing a random starting vertex, it firstly automatically extracts a local core as the initial state for local community extraction.
2. Starting from the local core, the corresponding community is expanded by maximizing local modularity metric (Equation [12]).
3. To prune potential outliers.

A local core is a dense sub-graph within a community, generated by a vertex and its neighbors. The core, centric of certain communities, lying in

Figure 2. The general algorithm for the greedy maximization of local modularity of Clauset

add v_0 **to** $\mathcal{C}$
add all neighbors of v_0 **to** $\mathcal{U}$
set $\mathcal{B}=v_0$
while $|\mathcal{C}|<k$ **do**
for each $v_j \in \mathcal{U}$ **do**
compute ΔR_j **from Eq. (5)**
end for
find v_j **such that its** ΔR_j **is maximum**
add that v_j **to** $\mathcal{C}$
add all new neighbors of that v_j **to** $\mathcal{U}$
update R **and** $\mathcal{B}$
end while

Figure 3. The local community discovery algorithm of Luo et al. (2008)

```
1 /* find a community for a give source v */
2 /* initializing*/
3         create a new sub-graph S with v
4         do
5                 create a new neighbor set N with adjacent vertices of v.
6                 /*addition step*/
7                 Do
8                         create a new list Q to store new adding vertices
9                         sort the vertex in N based on their degree incrementing
10                        for each vertex uj in N
11                                compute ΔM;
12                                if ΔM > 0
13                                        add uj to sub-graph S
14                                        add uj list Q;
15                                end if
16                        end for
17                while Q is empty
18                /*deletion step*/
19                do
20                        create a new list deleteQ
21                        for each vertex ui in VS
22                                compute ΔM;
23                                if ΔM > 0 and removing of ui does not disconnect sub-graph S
24                                        remove vi form subgraph
25                                add vi to deleteQ
26                                end if
27                        end for
28                while deleteQ is empty
29                Create a new neighbor set newN of S
30        while newN is the same as N
31        if modularity of S > 1 and S contain v
32                return S
33        else
34                Print out ("no community found for v")
35 /* end */
```

the densest region of the community, is capable of locating the position of the community by rule and line. Instead of starting from a randomly vertex, local core approach is able to detect much more stable and meaningful community with initial reliable core.

Density-Based Algorithms

Xu (2009) proposed a dense-similar algorithm (see Figure 4) for detecting local communities based on the *group cohesion* (Wasserman, 1994), defined as:

$$Cohesion(C) = \frac{den_{\text{int}}(C)}{den_{ext}(C)} = \frac{\dfrac{2m_{\text{int}}(C)}{n_c(n_c-1)}}{\dfrac{m_{ext}(C)}{n_c(n-n_c)}} = 2\frac{(n-n_c)m_{\text{int}}(C)}{(n_c-1)m_{ext}(C)} \quad (15)$$

where $n_c = |C|$. They believe a sub-graph is a cohesive community if its internal link density is

Figure 4. The pseudo code for the local cohesion based algorithm of Xu

```
C = {v0}
add all neighbors of v0 to H

/*The merging phase*/
while H is not empty
  for each vi in H do
    calculate ΔCohesion if vi is merged into C
  end for

  find vj that provides the maximal positive ΔCohesion
  if no such vj is found
    exit the while loop
  else
    add vj to C
    add all external neighbors of vj to H
    update Cohesion
    update the internal and external degrees of vj and its neighbors
end while

/*The pruning phase*/
for each vi in C do
  calculate ΔCohesion if vi is removed from C
  if ΔCohesion > 0
    remove vi from C and update Cohesion
    if vi = v0
      print "No local community found for v0."
      exit the for loop
end for
```

greater than its external link density, or $Cohesion(C) > 1$.

This method also contains a "pruning" phase. But, different from Luo et al. (Luo, 2008) and Zhang et al. (Zhang, 2011), it does pruning after the initial local community has been decided. It stops as soon as $Cohesion(C) > 1$ cannot increase by merging. However, it always extracts much too small modules.

Considering "outliers," Chen et al. (Chen, 2009) propose another local algorithm based on density. They define a local community metric L that calculates the ratio of intra-density and inter-density of cluster C:

$$L = \frac{L_{\text{int}}}{L_{ext}} \tag{16}$$

However, the inter-density is not similar to Xu (2009). Here the *internal degree* and *external degree* are defined as:

$$L_{\text{int}} = \frac{\sum_{v \in C} \deg(v, C)}{|C|} \tag{17}$$

$$L_{ext} = \frac{\sum_{v \in \mathbf{B}} \deg(v, \mathbf{N})}{|\mathrm{B}|} \tag{18}$$

Analyzing the property of merging vertex, Chen et al. indicate that, after merging v into C, there are total three cases that will probably result in $L' > L$ (L' is the value after merging) are:

- $L'_{\text{int}} > L_{\text{int}}$ and $L'_{ext} < L_{ext}$

- $L'_{\text{int}} < L_{\text{int}}$ and $L'_{ext} < L_{ext}$
- $L'_{\text{int}} > L_{\text{int}}$ and $L'_{ext} > L_{ext}$

They only maintain vertices in the final local community that satisfy the first case, which is used to handle the "outliers" (see pseudocode as Figure 5). This is too strict to detect integrated community.

Schaeffer (2005) wants to optimize the density of the community and the sparseness between communities. Therefore, they propose a method by combining Equation (5) and Equation (8):

$$F(C) = den(C)\bullet den_r(C) = \frac{2\deg_{\text{int}}(C)^2}{|C|(|C|-1)(\deg_{\text{int}}(C)+\deg_{ext}(C))} \quad (19)$$

Besides, they guide the local clustering with simulated annealing (Kirkpatrick, 1983).

For bipartite networks, Andersen (Andersen, 2008a, 2008b) proposes a density-based local method. The density is the bipartite density:

$$den(S,T) = \frac{\deg(S,T)}{\sqrt{|S|}\sqrt{|T|}} \quad (20)$$

FUTURE RESEARCH DIRECTIONS

A number of local algorithms have been advanced to identify a local community at a time, which cost less on storage and running time. However, the quality of the local communities is not as perfect as we demand, especially in real social networks.

Efforts should be made to improve the performance of local methods in some aspects as follows:

- It is a challenge for previous methods to stop automatically while reaching the boundary of the community. Most of them depend on a predefined threshold. If the fitness functions satisfy the threshold, they stop and output the current community.

Figure 5. The pseudocode of Chen et al.'s algorithm

Input: A social network G and a start node n_0.
Output: A local community with its quality score L.
1. Discovery Phase:
  Add n_0 to D and B, add all n_0's neighbours to S.
  do
    for each $n_i \in S$ **do**
      compute L'_i
    end for
    Find n_i with the maximum L'_i, breaking ties randomly
    Add n_i to D if it belongs to the first or third case
    Otherwise remove n_i from S.
    Update B, S, C, L.
  While $(L' > L)$
2. Examination Phase:
  for each $n_i \in D$ **do**
    Compute L'_i, keep n_i only when it is the first case
  end for
3. If $n_0 \in D$, return D, otherwise there is no local community for n_0.

The threshold is hard to decide and cannot always suit to all networks with varied of size and property. A good stopping criteria need to be found to better optimize the community boundary.

- To some extent, "outlier" can decrease the density of local community. In the future research, outliers should be paid more attention to make the corresponding community dense inside and sparse outside.
- For large dynamic social networks, there should be more efficient local methods to satisfy the time-variant characteristic.

CONCLUSION

As mentioned above, local community extraction techniques are perfect solutions to extract the structure of large-scale social networks. Recently, this technology has become a hot research point, as the information explosion is essential to be coped with. However, most of previous approaches cannot do an ideal job for extracting community structures. It is really demanded for extra work.

REFERENCES

Andersen, R. (2008b). A local algorithm for finding dense sub-graphs. In *Proceedings of the Nineteenth Annual ACM-SIAM Symposium on Discrete Algorithms*, (pp. 1003-1009). Society for Industrial and Applied Mathematics.

Andersen, R., & Lang, K. J. (2008a). An algorithm for improving graph partitions. In *Proceedings of the Nineteenth Annual ACM-SIAM Symposium on Discrete Algorithms*, (pp. 651-660). Society for Industrial and Applied Mathematics.

Cheeger, J. (1970). A lower bound for the smallest eigenvalue of the Laplacian. In *Problems in Analysis* (pp. 195–199). Princeton, NJ: Princeton University Press.

Chen, J., Zaizne, O., & Goebel, R. (2009). Local community identification in social networks. In *Proceedings of the International Conference on Advances in Social Network Analysis and Mining*, (pp. 237-242). IEEE.

Clauset, A. (2005). Finding local community structure in networks. *Physical Review E: Statistical, Nonlinear, and Soft Matter Physics*, *72*(2). doi:10.1103/PhysRevE.72.026132

Flake, G. W., Lawrence, S., et al. (2000). Efficient identification of web communities. In *Proceedings of the 6th ACM SIGKDD International Conference on Knowledge Discovery and Data Mining*, (pp. 150-160). ACM Press.

Flake, G. W., & Lawrence, S. (2002). Self-organization and Identification of web communities. *Computer*, *35*, 66–70. doi:10.1109/2.989932

Gibson, D., Kleinberg, J., & Raghavan, J. (1998). Inferring web communities from link topology. In *Proceedings of the 9th ACM Conference on Hypertext and Hypermedia: Links, Objects, Time and Space*, (pp. 225-234). New York, NY: ACM.

He, X., & Zha, H. (2002). Web document clustering using hyperlink structures. *Computational Statistics & Data Analysis*, *41*(1), 19–45. doi:10.1016/S0167-9473(02)00070-1

Khandekar, R., Rao, S., & Vazirani, S. (2009). Graph partitioning using commodity flows. *Journal of the ACM*, *56*, 1–15. doi:10.1145/1538902.1538903

Kirkpatrick, S., Gelatt, Jr., et al. (1983). Optimization by simulated annealing. *Science*, *220*(4598), 671–679. doi:10.1126/science.220.4598.671

Luo, F., Wang, J. Z., & Promislow, E. (2008). Exploring local community structures in large networks. *Web Intelligence and Agent Systems*, *6*(4), 387–400.

Mihail, M., Gkantisidis, C., & Saberi, A. (2002). *On the semantics of internet topologies*. Atlanta, GA: Georgia Institute of Technology.

Newman, M. E. J. (2004b). Fast algorithm for detecting community structure in networks. *Physical Review E: Statistical, Nonlinear, and Soft Matter Physics*, 69.

Newman, M. E. J. (2006). Modularity and community structure in networks. *Proceedings of the National Academy of Sciences of the United States of America*, *103*, 8577–8582. doi:10.1073/pnas.0601602103

Newman, M. E. J., & Grivan, M. (2004a). Finding and evaluating community strucuture in networks. *Physical Review E: Statistical, Nonlinear, and Soft Matter Physics*, *69*(2). doi:10.1103/PhysRevE.69.026113

Schaeffer, S. E. (2005). Stochastic local clustering for massive graphs. *Lecture Notes in Computer Science*, *3518*, 413–424. doi:10.1007/11430919_42

Sharan, A., & Gupta, S. L. (2009). Identification of web communities through link based approaches. In *Proceedings of the International Conference on Information Management and Engineering*, (pp. 703-708). IEEE.

Wasserman, S., & Faust, K. (1994). *Social network analysis: Methods and applications*. Cambridge, UK: Cambridge University Press. doi:10.1017/CBO9780511815478

Xu, J. (2009). Identifying cohesive local community structures in networks. In *Proceedings of ICIS 2009*, (p. 112). ICIS.

Zhang, X., Wang, L., Li, Y., & Liang, W. (2011). Extracting local community structure from local cores. [Springer.]. *Proceedings of Database Systems for Advanced Applications: SNSMW, 2011*, 287–298.

Chapter 12
Supporting Social Interaction in Campus-Scale Environments by Embracing Mobile Social Networking

Zhiwen Yu
Northwestern Polytechnical University, China

Yunji Liang
Northwestern Polytechnical University, China

Yue Yang
Northwestern Polytechnical University, China

Bin Guo
Northwestern Polytechnical University, China

ABSTRACT

With the popularity of smart phones, the warm embrace of social networking services, and the perfection of wireless communication, mobile social networking has become a hot research topic. The characteristics of mobile devices and requirements of services in social environments pose challenges to the construction of a social platform. In this chapter, the authors elaborate a flexible system architecture based on the service-oriented specification to support social interaction in a university campus. For the client side, they designed a mobile middleware to collect social contexts such as proximity, acceleration, and cell phone logs, etc. The server backend aggregates such contexts, analyzes social connections among users, and provides social services to facilitate social interaction. A prototype of mobile social networking system is deployed on campus, and several applications are implemented to demonstrate the effectiveness of the proposed architecture. Experiments were carried out to evaluate the performance (in terms of response time and energy consumption) of our system. A user study was also conducted to investigate user acceptance of our prototype. The experimental results show that the proposed architecture provides real-time response to users. Furthermore, the user study demonstrates that the applications are useful to enhance social interaction in campus environments.

DOI: 10.4018/978-1-4666-2806-9.ch012

INTRODUCTION

With the development of wireless communication and pervasive computing technology, smart campuses are built to benefit the faculty members and students, manage the available resources, and enhance user experience with proactive services. A smart campus ranges from a smart classroom (Shi, 2003; Suo, 2009; Yau, 2003; Griswold, 2004; Murray, 2003; Barkhuus, 2005), which benefits the teaching process in an enclosed environment, to an intelligent campus (Halawani, 2003; Rohs, 2003; Dong, 2009) that provides lots of proactive services in a campus-scale environment. Many contexts such as location, activity and user profiles are widely adopted by existing work to provide proactive services, social contexts (e.g., the proximity and the communication history), however, are rarely concerned in their systems. Campus is a social environment where college students have lots of social interaction with each other. However, little research has been conducted on social aspects, e.g., supporting social interactions.

Social interaction with others is important for college students. On one hand, social interaction is vital to human health, both mentally and physically. The presence of social interaction with others significantly enhances the individual's abilities to cope with stress and reduces the probability of depression and suicide in campus. On the other hand, social interaction may help college students acquire knowledge and skills from which they will benefit throughout their life, especially for freshman. Amount of knowledge about public infrastructures or arrangements of campus life is shared through interactions with seniors, which benefits the quick adjustment for the freshmen. Several examples of frequent questions asked in a university campus are shown as follows:

- Whether the study lounge is available?
- Who is available to play basketball with me?
- Whether my friend is in the cafeteria?

Nowadays, Social Networking Sits (SNS) such as Facebook, MySpace, and Twitter are very popular for college students. Tens of millions of college students spend numerous hours logging on such sites, communicating with their friends, and sharing media in their community. With the development of wireless communication and smart phones, social networking sits are going mobile, which promotes the emergence of mobile social networking. IDC (International Data Center) predicts that the user number of smart phones will reach 2 billion in 2014. The explosive growth of user number of smart phones makes it valuable to perform research on mobile devices. Different from Web-based social networks, Mobile Social Network (MSN) is capable of continuous, seamless sensing, which allows us to obtain contexts from the physical world. As shown in Figure 1, MSN collects various contexts through mobile phone sensing, such as location, text message, phone records, etc. All those contexts present social interaction among people. By analyzing those contexts, we can get a number of high-level semantic information, such as social relationship, transportation mode, user activity, etc. Those semantic information benefits the design of proactive services. The mobility of MSN endows data with spatiotemporal information, which benefits the understanding of contexts where the user situates. Furthermore, MSN makes it possible to provide ego-centric services for users anywhere, anytime to enhance user experience. The MSN outperforms Web-based social networks with the support to social interactions in the real world. It makes full use of the smart phones to extract social contexts by analyzing sensing data and provides social services. Thus we introduce MSN into the built of smart campuses to support social interaction among faculty members and students.

However, the support to social interaction in campus is still discontented. In comparison with conventional face-to-face social interactions, smart campus should additionally support intelligent social services, such as friend search and

Figure 1. Illustration of a mobile social networking system

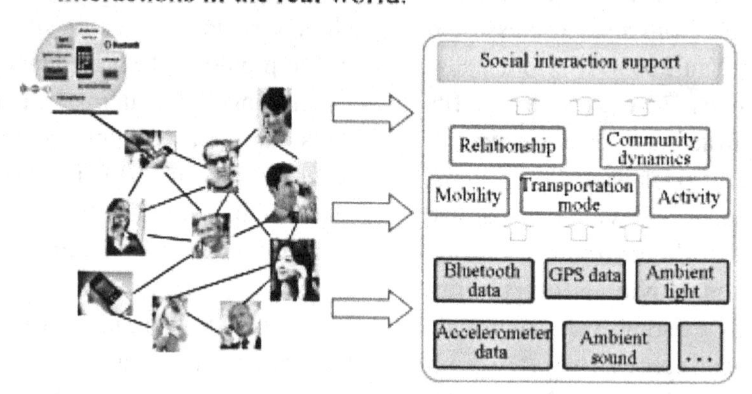

recommendation, significant location detection, etc. In this chapter, we propose a flexible mobile social networking architecture in a campus to support social interaction. It is built from the service-oriented perspective, with a centric server and mobile clients. In the client side, a mobile middleware is introduced to collect social contexts such as proximity, communication history, etc. The server aggregates those contexts, analyses social connections among users and provides social services to facilitate social interactions. Based on the proposed architecture, several applications are implemented to demonstrate the usefulness of the architecture.

The rest of this chapter is organized as follows. In Section 2, the state of art of the research on smart campus and mobile social networking are described. The system requirements are presented in Section 3. Section 4 elaborates the system architecture and illustrates details of both the server backend and the client. Three applications based on the proposed architecture are described in Section 5. In Section 6, we test the system performance in terms of response time and energy consumption. Finally, we conclude the chapter in Section 7.

RELATED WORK

A number of studies have been conducted in building smart campus. Context-awareness has been introduced to provide proactive services for faculty members and students in a campus. A campus is essentially a social environment, where human communication plays an important role in daily life. The emergence of the mobile social networking makes it possible to support social interactions in a campus.

Smart Campus

The smart classroom takes a key step in the development of smart campuses. As teaching is one of the most important activities in colleges, much attention has been paid to smart classrooms (Shi, 2003; Suo, 2009; Yau, 2003; Griswold, 2004; Murray, 2003; Barkhuus, 2005). In Shi's work, a smart classroom prototype was designed to provide seamless tele-education. The prototype integrates voice recognition, computer vision, and other technologies to make a physical classroom a natural user interface, which provides tele-education experience like in a real classroom (Shi, 2003). An architecture based on the Web service technology was proposed to overcome the weakness in extensibility and scalability (Suo, 2009). Stephen et al. presented a reconfigurable context-sensitive middleware to achieve collaborative learning (Yau, 2003). In the prototype, everyone is attached with a PDA to recognize situation. The system provides different services to users based on their current situation, ultimately to enhance the collaboration among students and teachers.

With the evolvement of pervasive computing, several smart campus prototypes were developed to offer assistive services for users situated in a smart space. Talal et al. (2003) proposed to efficiently integrate services in a campus with the smart card. Rohs et al. (2003) presented the ETHOC system, which focuses on the integration of virtual and physical elements in the campus environment. The system supports the interaction with virtual counterparts of printed document using a variety of devices, such as mobile phones or PDAs. Dong et al. (2009) designed a location based service system to support the interaction between the 3D virtual world and the physical world. A prototype system was deployed on a campus to make user experience the school life in the virtual world.

Although lots of context-aware applications on the campus have been implemented, most of them focus on the interaction between users and services, which provides relevant services according to user situation. In other words, previous works mainly aim to support individual users with ego-centric services. However, a campus is essentially a social environment, where social interaction is very important. The social contexts should be introduced to promote social connection among campus-users.

Social Interaction

Social interaction is very important for human-centric intelligent system. To date, numerous artificial systems were constructed to enhance the social interaction between both socially capable agents and between artificial systems and humans.

FreeWalk is a social interaction platform where people and agents can socially and spatially interact with one another. FreeWalk has evolved to support heterogeneous interaction styles including meetings, cross-cultural encounters, and evacuation drills (Nakanishi, 2004). Dragone et al. proposed a framework, named Social Robot Architecture (SRA), to support coherent social interaction between real and artificial systems with the developments of heterogeneous robots. SRA is a bridge between real-world environments and artificial systems whether through robot-robot, robot-avatar, or robot-human social interaction (Dragone, 2005). To enhance the interaction of participants in a meeting, H. N. Charif proposed a vision system to analyze the activities of occupants in a smart meeting room. Occupants were tracked using a particle filter to monitor the trajectories of occupants' heads. Based on those trajectories, activities of occupants are extracted to enhance the interactions in the meeting room (Charif, 2006). Dai et al. (2009) put forward an event-based dynamic context model to analyze the group interaction in a smart meeting room. Based on the proposed dynamic model, the interactions such as lecturing, asking, talking etc. are extracted, which benefits the computer understanding of human actions and interactions and facilitates the construction of services for target users. Yu et al. proposed a solution for capture, recognition, and visualization of human interactions with the collaboration of multiple sensors, such as video cameras, microphones, and motion sensors. This collaborative approach was employed to analyze the group interactions in the smart meeting (Yu, 2010).

Although numerous studies on the social interactions have been conducted, they emphasized the interaction between the artificial system and humans. We aim to enhance the user experience in social interaction with the introduction of participatory sensing that enables users to gather, analyze, and share local knowledge. When users are involved in the process of sensing, they are more likely to obtain good user experience and may be positive to share more information in the system (Burke, 2006).

Mobile Social Networking

The advent of mobile social networking comes along with the popularity of smart phones and the maturity of wireless communication. The mobile social networking has been a hot topic in recent years, and many applications have been designed in this field. Some typical applications are as follows.

Several works were dedicated to inferring friendship network structure by using mobile phone data. The social contexts, such as call logs, text message, location, proximity and time etc. are taken into account. The Location-Based Services (LBS) are triggered based on the location of the user and its peers (Eagle, 2009; Mirisaee, 2010; Ankolekar, 2009). Currently, many positioning technologies such as the GPS and the locating in ad-hoc network are available. PeopleTones detects the buddy proximity and notifies the user with mobile phone (Li, 2008).

There have been already numerous kinds of applications deployed on the campus. However, how to manage lots of applications in an open environment is still a challenge. Eunhee et al. developed a smart campus test bed based on the extensible Service Orientated Architecture (SOA), which runs a suite of applications that link "People-to-People-to-Place," called P3-System. The P3-system includes several social applications such as CampusWiki, providing editable pages about information of the campus, and CampusMesh, a location-aware social reminding (Kim, 2007). The CenceMe System transparently gathers contexts through the sensors embedded in mobile phones, and adopts reasoning techniques to extract high-level contexts for mobile social applications. Although the CenceMe fills the gap left between manual status updates by automatically updating activity and location presence information, the semantic reasoning is time-consuming which may lead to poor user experience due to the delay of response. Furthermore, the reasoning process is performed on the smart phone, which leads to the sharp increase of energy consumption (Campbell, 2008; Miluzzo, 2008).

It is obvious that the research on mobile social networking is still at its early stage. The features of the resource-limited devices such as PDAs or mobile phones are not fully taken into consideration when designing the architecture. Furthermore, the architecture should provide a mechanism to maintain system performance in heterogeneous networks. We aim to support social interactions among students and the faulty members on campus with the mobile social networking technology. Social contexts are extracted from mobile devices apart from social networking sits. A lightweight architecture is presented to support mobile social networking applications.

SYSTEM REQUIREMENTS

In the design of the system architecture, requirements should be concerned from both the function perspective and system perspective. The function requirements embody the basic functionalities to implement; the system requirements depict some essential guidelines to be followed in the system architecture.

From the Function Perspective

There are many mobile devices, services, and users in a smart campus. The architecture should manage those entities and provide a mechanism to access the services in the mobile environment. Some basic modules, such as semantic extraction, pattern mining, ubiquitous search, and location management, are very important for the implementation of applications. The modules provide useful information for the high-level services or applications.

Semantic Extraction

There are lots of data sources such as sensors from mobile phones, call logs and interaction records from social networking sits etc. The diversity of data sources makes it necessary to introduce the semantic extraction component to extract and manage the large amount of contexts. Semantic extraction aggregates the contexts collected from users themselves and users' environment and represents them in a machine understandable way, which not only facilitates the integration of contexts from heterogeneous networks, but also benefits the representation of relationships among different entities. The semantic extraction also infers high-level contexts based on the semantic representation of basic contexts.

Pattern Mining

With both of the low-level and high-level contexts semantically extracted, pattern mining can be leveraged to discover higher-level knowledge about users. For example, the tri-axial accelerometer can be used to measure the acceleration of the movement (low-level context) as well as to recognize user activities (high-level context). Based on these contexts, higher-level knowledge, such as the energy expenditure in daily life, and the traffic modes can be mined. These new knowledge is significantly useful for the construction of proactive services and the design of the recommendation services.

Ubiquitous Search

Ubiquitous search serves as the interface between human being and entities in the mobile social networking systems. Different from the traditional search engine, the ubiquitous search provides more complex retrieve of interesting information. It supports the tempo-spatial query of entities in the real world, such as interesting events, user activities etc. Furthermore, the ubiquitous search utilizes relationships among objects to explore the related information. For example, "Who called me while I was jogging in the park?," "Where his classmates are when he is on the play yard?," etc.

Location Management

Location is one of the most important contexts, which not only drives the development of location-based services with the perfection of positioning technology, but also reveals the situation. The location management, on one hand, tracks the positions of users in the environment; on the other hand, it stores the trajectory of users and provides the data inputs for pattern mining. Furthermore, the location management should support both indoor positioning and outdoor positioning. The GPS offers the outdoor positioning and the Wi-Fi Access Points (AP) contribute to indoor positioning.

From the System Perspective

The heterogeneity and the mobility pose challenges to the design of the mobile social networking system in university campus. Thus, some basic rules should be followed to handle the upcoming problems in applications. Here, we emphasize the scalability and the compatibility for the server. Meanwhile, the lightweight is concerned for the client on the mobile devices.

Scalability

There are many mobile devices, services, and applications in the smart campus environment. The handheld mobile devices are different from each other in terms of their capabilities. Thus, the architecture should be scalable to support heterogeneous mobile devices and the deployment of the increasing number of applications. The scalability is important for the system due to the mobility and dynamics of the objects in the social environment.

Lightweight

Mobile devices not only collect social contexts such as call logs, text messages, sensory data, but also take charge of communication between peers and the server. Due to the diversity of contexts, a middleware is required to aggregate those social contexts. However, the mobile devices are limited in terms of battery and CPU speed. Considering the multiple functions and the constraint resource of a mobile device, the middleware deployed on the mobile phone should be lightweight, which means less power consumption.

Compatibility

The diversity of mobile devices and software brings challenges to the construction of the mobile social networking. The connectivity among heterogeneous devices needs the compatibility of the communication networking. It should be guaranteed that the information transformation among different network protocols. Thus, in the design of system architecture, the network layer plays the role of bridge, which transfers information among different protocols. Furthermore, the compatibility of software is very important for the success of applications. Currently, many communication services are available such as Wi-Fi, Bluetooth, GSM, and 3G. Services are situated in such a heterogeneous network. Those services should be capable of communicating with each other in this heterogeneous environment. Thus, the architecture should provide a mechanism to guarantee the connectivity among different communication services.

SYSTEM ARCHITECTURE

System Overview

The main target of a smart campus system is to collect and analyze data from the public to enhance human social experience in a campus. From a system's view, its basic functions consist of semantic extraction, pattern mining, ubiquitous search, location management, etc. To develop a scalable, weightless, and robust architecture, the first challenge for the system design is to support reliable and scalable communication in heterogeneous network. Since the data is produced by users' devices, and collected through access points or the Internet, multiple network protocols are necessitated to guarantee the robustness of the system.

The system architecture is illustrated in Figure 2. It consists of two main entities, Servers and Mobile Clients. The servers are Ethernet-connected, which are equipped with rich storage and computational power. A mobile client is a smart phone (linking to its custodian), which are equipped with various sensors, such as accelerometer, thermometer, GPS, etc. The mobile clients get original sensing data from the built-in sensors and submit to the server through wireless network. The server extracts the social contexts from the sensory data and returns relevant results to the appropriate client. Separating the tasks of sensing and processing can enhance the system scalability.

The system mainly works in the infrastructure mode. The mobile devices access to the Internet and interact with servers via the wireless AP. As the infrastructure network might be not available in some situations, the system also supports the

Figure 2. System architecture overview

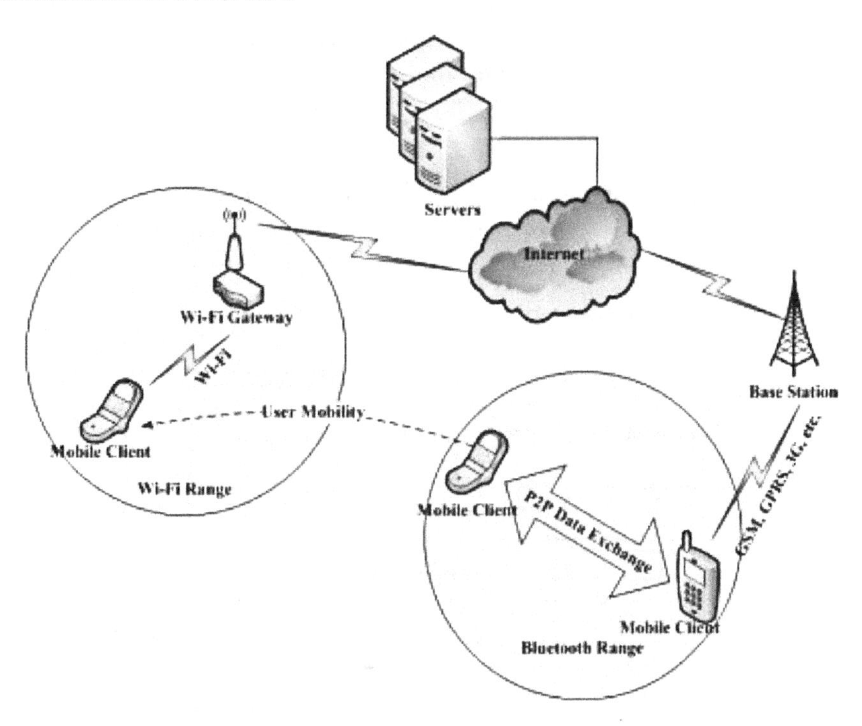

ad-hoc mode when the devices come within the range of each other using opportunistic network, e.g., Bluetooth. This guarantees the robustness of the system, and makes services available in most cases.

Server Components

The sever backend consists of several modules ranging from context aggregation, social network analysis, context storage, knowledge mining to peer communication and admission control. The server architecture is depicted in Figure 3, which consists of three layers: Network Layer, Service Layer, and Application Layer.

The Network Layer provides the communication mechanism for connecting server and clients. Due to the heterogeneity of network, the Network Layer supports multiple network connections including wired and wireless, and supports multiple network protocols such as GSM, 3G, Wi-Fi, and Bluetooth. All the mechanisms aim to guarantee the robustness of the architecture in heterogeneous network environments.

The Service Layer is the core of the server, which includes two main parts: OSGi (Open Services Gateway Initiative) framework (Marples, 2001) and Common Services. The OSGi, a service oriented framework, provides a public platform to manage services registered in the server. The Common Services is a set of public services, which are implemented based on the OSGi specification. These services compose the central functionalities of the server and provide high-level contexts to the Application Layer.

To achieve the scalability and lightweight, the OSGi framework is adopted in the server. OSGi is a service-oriented framework, which supports the module management and benefits the loose coupling of modules (Lee, 2003). On the other hand, it is hot swapping, which means the services could be installed or stopped without restarting the server. Advantages of this framework make it suitable for managing the mobile social ap-

Figure 3. Server architecture

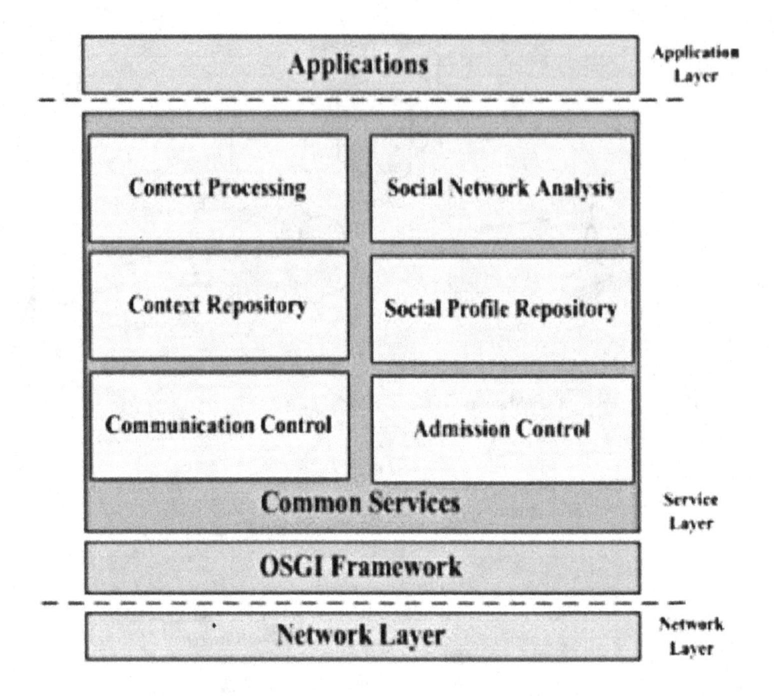

plications. Furthermore, OSGi is a lightweight framework for delivering and executing service-oriented applications. It offers benefits including platform independence, multiple service support, security, service collaboration, and multiple network support.

The Common Services include various basic services, which are described as follows. The communication control component is responsible for the reliable and secure communication, and supports multiple networks. It hides the details of the underlying network layer, and provides seamless and transparent communication with mobile clients. The system only allows legal users to register, and admission control service is introduced to manage users and provides a mechanism for efficient user validation. Context repository stores context instances such as location, time, and activity, etc. These contexts are collected from mobile clients and submitted to context repository. The context-processing module infers high-level contexts from low-level contexts; it also handles context conflict resolution. Social profile repository deals with user's social profile data storage, which includes personal details, social relations, preferences, etc. Social network analysis focuses on analyzing relationships among users, and determining the role and impact of individuals in a particular community.

Client Side on Smart Phones

Clients, on one hand, serve as the sensing node of social contexts. It sends the sensory data to the server through different networks, such as GSM and Wi-Fi. On the other hand, it serves as the terminal for users acting as interactive interface. Users could send and retrieve information from the server. As there are many sensors and functions deployed on the mobile phone, managing the sensors is quite important. We designed a middleware to collect social contexts and communicate with the server. The client architecture consists of three layers: Hardware Layer, Middleware Layer, and Application Layer (see Figure 4).

Figure 4. Client architecture on the smart phone

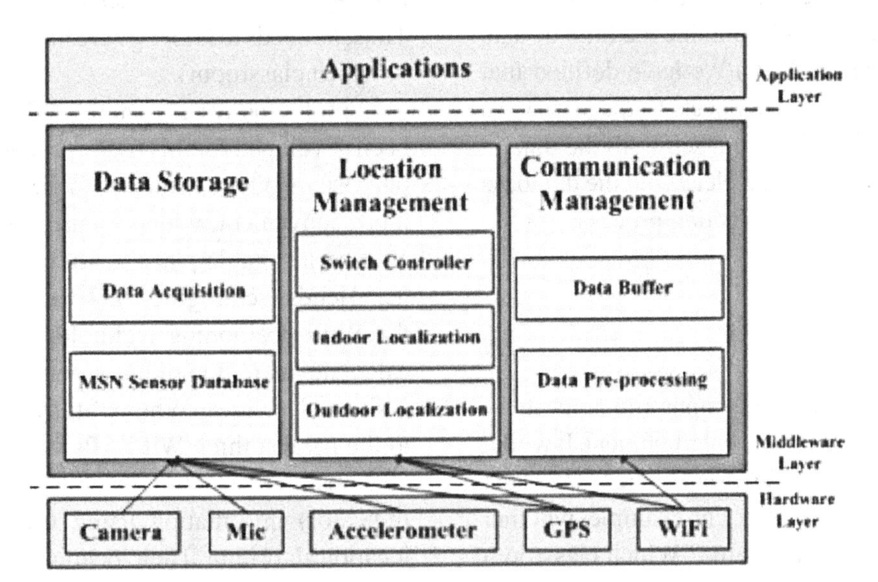

The Hardware Layer is composed of several sensors, such as camera, microphone, accelerometer, GPS, WiFi, Bluetooth, etc.

The Middleware Layer consists of three function modules: Data Storage, Location Management, and Communication Management. The Data Storage consists of two parts. One is Data Acquisition part, which offers interfaces for applications to access the sensory data stored in the mobile devices automatically. The other is MSN Sensor Database, which is used to store the sensor data, the user's moving trajectory and the user's contact history. As GPS works well in outdoor environment while WiFi localization is suitable in indoor environment, Location Management module is used to combine both. Its main function is to switch between the indoor localization mode and the outdoor localization mode. The switching operation is controlled by the Switch Controller. The third module of the Middleware Layer is the Communication Management module. The client communicates with the server through the nearby Access Points (AP). The Data Buffer part is used to cache data, which is sent to the server or from the server. The Data Pre-processing part is used to perform certain initial data processing before sending or upon receiving data. For the data to be sent, it will be enveloped with packet header, which contains the data type, user ID, and other useful information. For the data received, the Data Pre-processing will check the user related information contained in the packet header to see whether it matches the local user's information and if so, more useful information will be read from the packet header for further processing, otherwise the data received will be dropped.

On top of the Middleware Layer is the Application Layer. We have implemented several applications to help users enjoy the mobile social network and to make the campus life more convenient and interesting. The details of applications will be presented in the next section.

Mobile devices have limited memory and energy, thus memory saving and energy saving are important for the system. To achieve memory saving, the basic strategy is to reduce redundant data. When the Middleware Layer receives a new sensor reading, it will first compare the reading with the previous value to check whether there is an update. If they are the same, the new data will not be stored in the database. Another energy-consuming process is data transmission from the client to the server. For energy-saving, an effective strategy is to minimize the time of data transmission (Sun, 2010) We have defined that when a mobile device connects to the server the first time in a day, the client sends all the data in database to the server and cleans out the database and starts a new data-collecting process.

APPLICATIONS

Generally speaking, the campus life consists of study, communication and entertainment. It would be helpful for people to learn about the usage of the facilities before planning. For example, whether the study lounge is available? Which classroom does my classmate sit in? Whether the tennis-court is crowded? Three applications were implemented and deployed based on the proposed architecture, which are closely related to the daily campus life and aim to enhance the social interactions among college students.

The three applications are named Where2Study, I-Sensing, BlueShare, respectively. The two applications, Where2Study and I-Sensing, are developed on the Android 2.1 platform; the BlueShare is designed on the Symbian S60 system. The client software is installed on the mobile phones for data sensing and context gathering. The WiFi network with complete coverage of the campus is available. The server is implemented based on the OSGi specification, which provides a service oriented architecture to manage the service registered on the server.

Where2Study

Scenario 1: A student named John is going to look for a study lounge to prepare for the coming final exam. He wonders which classroom is available (i.e., having free seats). To solve John's problem, a retrieve request is sent to the server and information about all classrooms is presented on his cell phone. Later, he encounters a question in his study and wants to discuss it with his friends. Thus, he needs to know where his friends are (i.e., in which classroom).

A prototype system, namely Where2Study has been developed for this usage scenario. The main purpose of Where2Study is to make campus life more convenient, which focuses on helping users find a suitable place to study and locate his/her friends (see Figure 5). To achieve this goal, the WiFi positioning technology is adopted to calculate the region of classroom where the user is situated in. Every Where2Study client connects to the nearest three WiFi APs, and calculates its own position by RSSI (Received Signal Strength Indicator) information using triangle centroid location algorithm. Their location is collected and processed by the server. Thus, the system could determine whether a classroom has empty seats for individual study.

Where2Study not only presents the navigation map of a building to help students find classrooms (Figure 5a), but also shows the status of all classrooms as Figure 5b, such as which classrooms are full and which ones have free seats. To check the detailed information of a particular room, a user can click a button in Figure 5b and then the status of the seats in the room is displayed, as shown in Figure 5c. Furthermore, the application supports to query the location and activity of user friends, as shown in Figure 5d.

A key feature of this application is the capability to browse the status (e.g., name, location, and activity) of close friends. This allows users to reach out and be aware of their social network established by friendship, which will facilitate each other to study (e.g., all my friends are studying at the moment). In addition, when a user encounters a problem during study, he or she could turn to their friends for discussion according to their location shown on the mobile phone.

Figure 5. Where2Study user interface

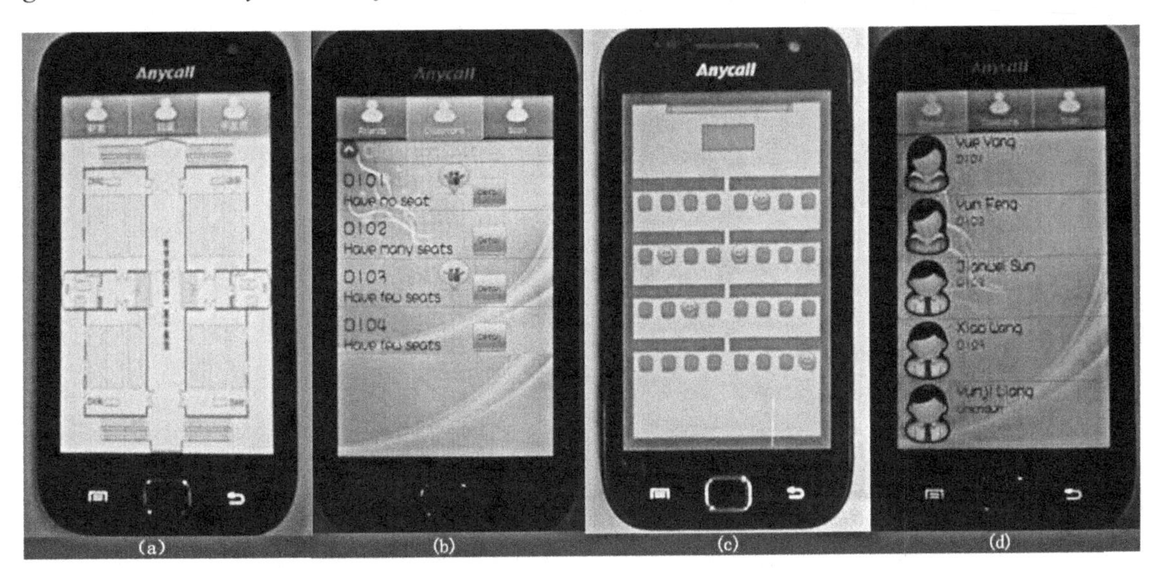

I-Sensing

Scenario 2: On Sunday, James wants to play basketball with his friends but he does not know whether the basketball court is available. So he edits a sensing task (e.g., is the court crowded?) with the I-Sensing client and sends it to the server. On receiving the task, the server transmits it to the users around the basketball court. When a user receives the task, he can choose different ways to accomplish the task, for example, writing a text, taking a photo and so on. Once receiving the replies, the server transmits them to James and James will know the information of the basketball court and then make a decision.

People are often interested in the information about a place while they are not there. There is no doubt that the people who are there can offer you the most accurate in-situ information. Therefore here comes the question that how to locate the mobile devices that are within the range of your interesting place. Analyzing the above mentioned scenario in detail, we can find that it has many things in common with an emerging sensing system called Participatory Sensing, where the mobile devices can form an interactive, participatory sensor network that enables users to gather, analyze and share local knowledge. I-Sensing is a campus information sharing system based on participatory sensing, which is deployed on mobile devices to gather information of place of interest. We developed real-name registration mechanism and reliability mechanism to intensify the data quality and introduce a friend-ranking mechanism to encourage user participation. We have implemented it on Android platform.

There are four parts in the application: Task Composing, Task Search, Personal Page, and Friends Ranking. As shown in Figure 6a, a user on campus can use the application to propose a sensing task, which indicates what the user wants to know and the prize that someone may get after completing the task. The user may get several replies from different sensing sources. He can choose the type of reply that he wants to read, for example, a reply with pictures, which is shown in Figure 6b. At Task Search part, a user can see the sensing tasks which are launched by other users and transmitted by the server according to their location information as Figure 6c shows. At last, a user can browse his or her points and ranking among his or her friends (see Figure 6d).

Figure 6. I-sensing user interface

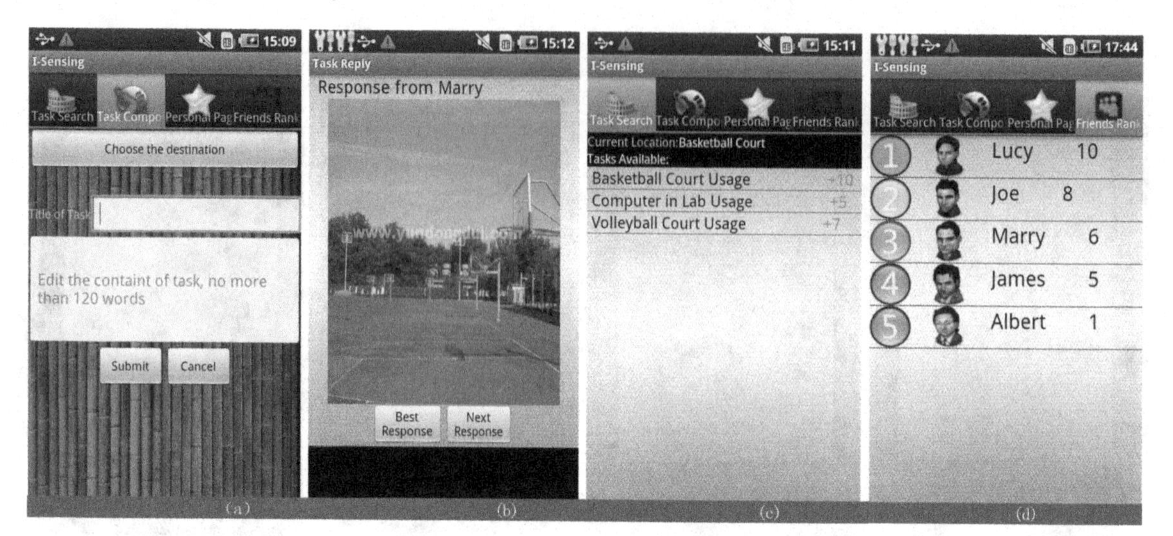

BlueShare

Scenario 3: A teacher named Jim found copying courseware is time consuming during the break time. When he wants to send some references or slides to his students, the students have to take turns to copy those files from his computer. To relieve Jim and his students' workload, an application uses Bluetooth protocol to deliver those assignments or courseware. Specifically, the nearby devices receive data directly from Jim's computer, and then forward the received data to their neighbors. Only a few minutes, all the students carrying Bluetooth devices receive the data.

We have implemented a prototype system named BlueShare on the Symbian S60 platform that runs on a variety of mobile phones. BlueShare is a media sharing application among Bluetooth devices based on the opportunistic network. The interesting media is sent to all users nearby the Bluetooth devices.

The mainstream of computing paradigms like Internet have many constraints on network communication and the connectivity is exceedingly required, which is not feasible for applications in certain situations under which emergencies might occur and the communication facilities are not always available. BlueShare is a resource-sharing system based on opportunistic network, which mainly use Bluetooth to transfer data. Opportunistic networks use the communication opportunities to forward data when any two nodes encounter. This new network mode is different from the traditional network communication mode, in which connectivity is not essential in opportunistic forwarding, and few infrastructures are needed.

With BlueShare, users share files with other participants in a store-and-forward mode. In opportunistic forwarding, we adopt a hybrid of redundancy-based and utilize-based algorithm. The scheme ensures a high success rate of opportunistic forwarding, and takes into account the network performance as well.

Two working modes are provided with the application, i.e., positive mode and negative mode. In the positive mode, the devices actively send files to proximate Bluetooth devices; On the other hand, negative mode indicates that the devices are passive and ready to response to the connection request from positive nodes. As shown in Figure 7a, the mode of devices can be configured. The operations on the positive nodes are presented in Figure 7b.

Figure 7. BlueShare user interface

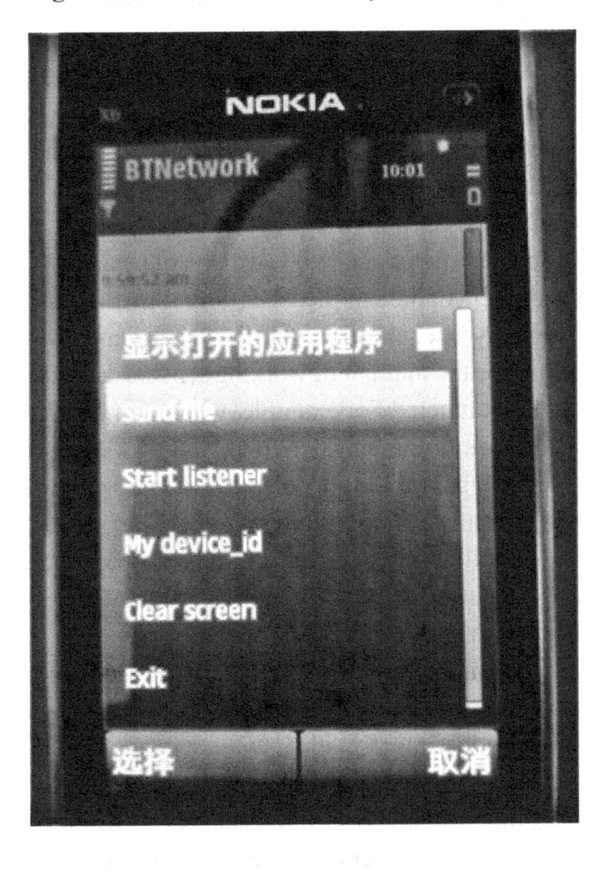

(a)

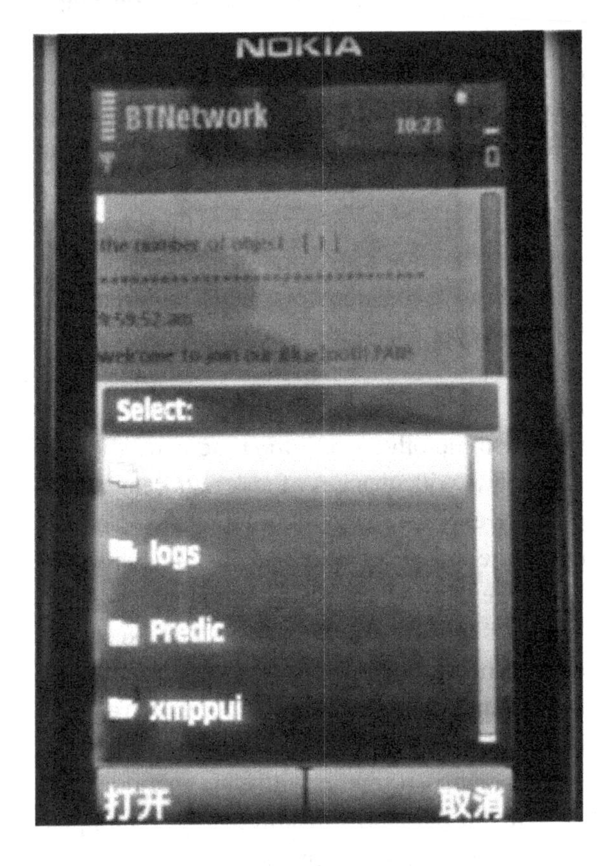

(b)

The aforementioned three applications are typical ones in the campus life, which aim to enhance social interactions among college students in their daily life including study, communication, and entertainment. The applications make full use of the social connections and mobile phones to promote the development of a smart campus.

EXPERIMENTAL RESULTS

To evaluate the proposed system, we measure the system performance in terms of the response time and the energy consumption. The response time is the duration when the request is generated from clients until a response is delivered from the server. The energy consumption indicates the power consumption of the middleware on the mobile phone. Besides system performance, a user study is performed to investigate to what extent the aforementioned applications support social interaction on a campus.

System Performance

Our system is composed of two parts—the central server and mobile clients. Both of them play significant roles in the system performance. The response time of sever is a crucial metric for the success of the prototype, as the latency is likely to decline user experience as commercial products. Likewise, the energy consumption is a vital bottleneck for the success of mobile applications. Therefore, we evaluate the system performance in terms of response time and energy consumption for the server and the client respectively. The experiments about the server were conducted on a workstation with 2.5 GHz i5-2520M CPU and 2 GB RAM and the experiments about the client were carried out on Samsung I909.

To measure the response time of the server, we analyzed the time consumption for the three applications. We construct lots of task requirements for the server and record the time consumption for each task. As shown in Figure 8, the response time is about 200 ms, which is acceptable for users. On the other hand, the time consumption is dynamic with changes of task requirements. A possible explanation to the jitter of the response time is the changes of instantaneous computational load. The instantaneous computational load may have a significant influence on the execution time. When numerous processes are running in the current situation, the target application cannot occupy the computational resources immediately and has to share resources with others periodically.

With the popularity of smart phones equipped with unprecedented sensing capabilities, the context-aware applications on mobile devices are promising. However, the long-term sensing with the full working load of sensors is energy consuming. The energy saving is crucial for the success of mobile applications. To reduce energy consumption, we reduce the sampling frequencies of sensors and minimize the working sensors in the current situation.

On one hand, lower sampling frequency means less work time for the heavy-duty sensor. Thus, we reduce the sampling frequencies of sensors such as accelerometer and WiFi, which contributes to the descent of the sampled data with the reduction of energy consumption. On the other hand, we minimize the number of working sensors according to situations. For example, when users go to outdoor environment, the WiFi scanner for indoor positioning is turned off and the GPS is utilized for outdoor positioning. Likewise, when user enters into a building, the sampling frequency of GPS is extended (e.g., once 10 minutes) and the WiFi scanner is trigged for indoor positioning.

To demonstrate the efficiency of the proposed strategies for energy consumption, we performed several experiments with or without the proposed solutions. First, we analyze the contribution of sampling frequencies to the energy consumption; then the function of the sensor selection is presented. Our experiments were carried out on Samsung I909. We measured the time spans with changes of sampling frequencies about the accelerometer when 90% of battery power is

Figure 8. Response time of the server

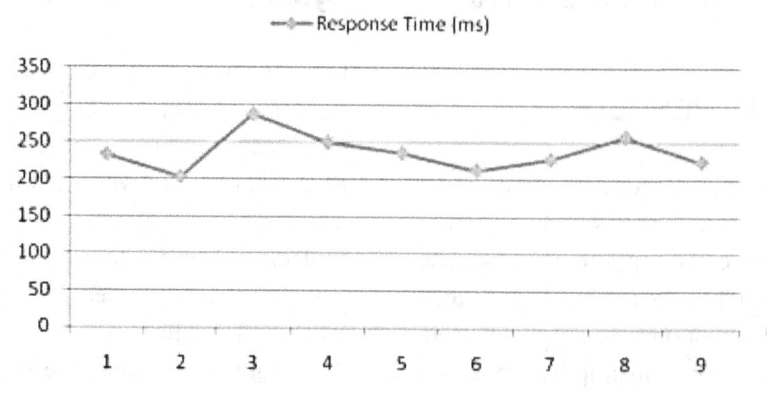

consumed. The tendencies of related metrics with the changes of sampling frequencies are shown in Figure 9. It is obvious that the battery life declines with the increase of sampling frequencies. For a resource-constraint device, the high sampling frequency leads to the rapid depletion of power. Our experimental results show that the sampling frequency should be concerned in the design of the prototype system and low sampling frequency contributes to the descent of energy consumption.

In addition, the selection of the working sensors should be concerned. The proposed system minimizes the size of the working sensors based on the situated scenarios. To evaluate the efficiency of the proposed strategies, we analyze the collaboration of accelerometer and GPS according to the changes of user locations. In our experiments, when the issue about the energy is not concerned, the time span is about 3 hours, which means continuous working of sensors including the GPS, WiFi scanner, accelerometer, etc. will deplete the cell phone battery within 3 hours. Intuitively, it is unacceptable in our daily life for the frequency battery charging. By contrast, the introduction of the proposed solution extends the battery life to more than 6 hours.

User Study

To evaluate the system from the user perspective, a user study was conducted. We mainly analyzed the user acceptance toward the knowledge sharing, content provisioned, and user interface. We randomly select 20 participants from our system and send a questionnaire to those target objects by our system automatically. When the questionnaire was completed, the target object sent it back to the server. Then we collected their questionnaires and analyzed those feedbacks. The questionnaire used for our user study is shown as follows:

1. Are you willing to share your location with others in the system?
2. Are you willing to respond to the task generated by the server?
3. Are you satisfied with the content provisioned by the system?

Figure 9. Tendencies of time span and recognition rate with changes of sampling frequencies

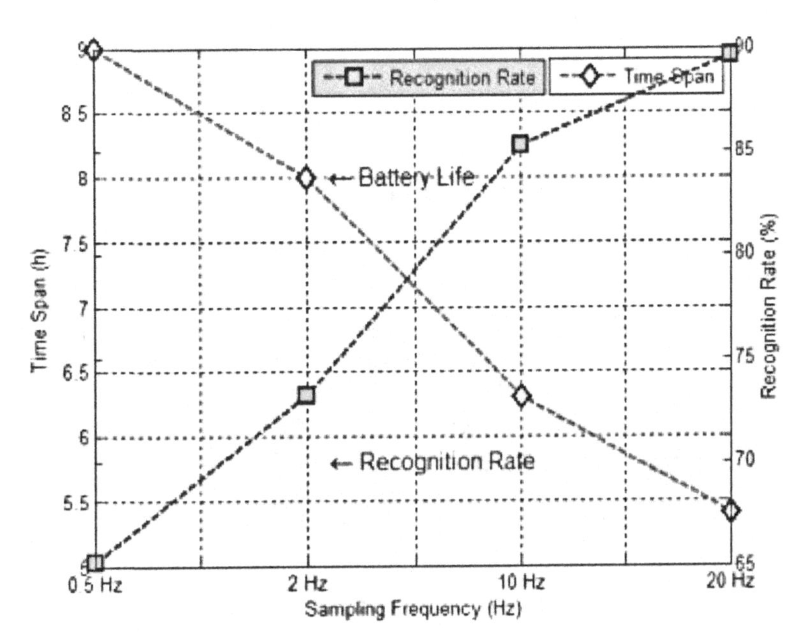

4. Is it convenient to operate the applications on mobile phones?
5. Are you willing to use the applications in your daily life?

All questions were answered using the following scale: 5=strongly agree, 4=agree, 3=neutral, 2=disagree, 1=strongly disagree. The results are illustrated in Figure 10. College students were generally willing to share their locations with their friends in campus (Q1). Fortunately, most subjects are willing to share their contexts with others, especially for their friends. The results show most subjects are willing to join in the system, which makes it possible to obtain responses from others in time (Q2). The positive responses from subjects contribute to the reduction of response time and enhance the user experience for the task launchers. Although there were some errors in content provisioning due to the update of sampled data occasionally, the user test shows that most of the subjects thought the content provisioned was helpful (Q3). The applications bring more convenience to campus life and enhance social interaction. An important observation on the user interface is that the participants had positive feelings about multimodal inputs (Q4). Currently, smart phones are very popular especially for college students. Thus, the multimodal inputs on the smart phones are more convenient to interact with the mobile devices. Especially, the handwriting recognition on the mobile phone makes it more efficient for inputs. Furthermore, most of the participants were interested in the applications and would use them to enhance social interaction in campus (Q5). This demonstrates that the proposed system is welcome among college students and is considered to be useful to enhance the social interaction among college students.

CONCLUSION AND FUTURE WORK

Campus is essentially a social environment where lots of social interaction occurs. To bring more convenience to campus life and enhance social interaction, we propose to build a smart campus based on mobile social networking. To efficiently manage services in the system, we proposed scalable system architecture based on the OSGi specification. Three applications were developed to demonstrate the proposed architecture. The experimental results show that the average response time and energy consumption are acceptable. Furthermore, a user study about the proposed system is conducted.

Figure 10. User study results

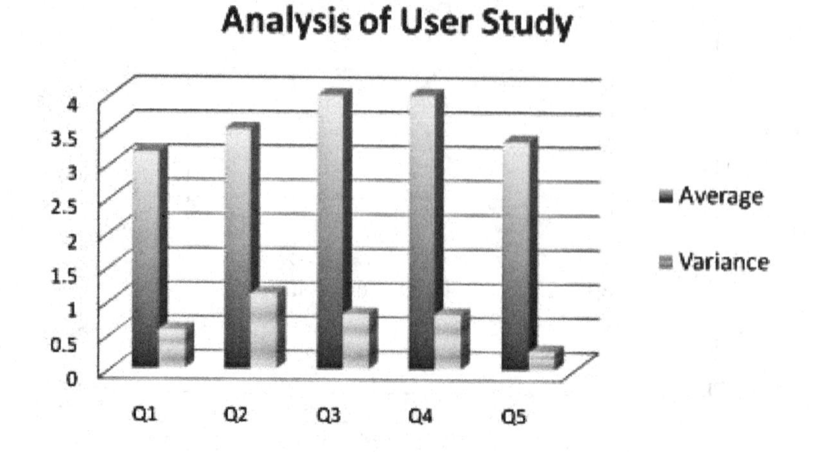

The construction of the smart campus is far-reaching. We plan to develop more applications based on the proposed architecture. Meanwhile, there are lots of hot issues about mobile social computing to enhance social interaction on campus. For the server, we plan to analyze user behaviors and patterns with the fusion of contexts from mobile social networking systems; on the client, we are going to conduct research on energy-aware sensing to extend the battery life.

ACKNOWLEDGMENT

This work was partially supported by the National Basic Research Program of China (973 Program) (No. 2012CB316400), the National Natural Science Foundation of China (No. 61103063, 60903125), the Program for New Century Excellent Talents in University (No. NCET-09-0079), and Microsoft Corp.

REFERENCES

Ankolekar, A., Szabo, G., Luon, Y., Huberman, B. A., Wilkinson, D., & Wu, F. (2009). Friendlee: A mobile application for your social life. In *Proceedings of the 11th International Conference on Human-Computer Interaction with Mobile Devices and Services (MobileHCI 2009)*. Bonn, Germany: MobileHCI.

Barkhuus, L. (2005). Bring your own laptop unless you want to follow the lecture: Alternative communication in the classroom. In *Proceedings of the Internal ACM SIGGROUP Conference on Supporting Group Work (GROUP 2005)*. Sanibel Island, FL: ACM Press.

Burke, J., Estrin, D., Hansen, M., & Parker, A. (2006). Participatory sensing. In *Proceedings of ACM Sensys World Sensor Web Workshop (WSW 2006)*. Boulder, CO: ACM Press.

Campbell, A. T., Eisenman, S. B., Fodor, K., & Lane, N. D. (2008). Transforming the social networking experience with sensing presence from mobile phones. In *Proceedings of the 6th ACM Conference on Embedded Network Sensor Systems (SenSys 2008)*. New York, NY: ACM Press.

Charif, H. N., & McKenna, S. J. (2006). Tracking the activity of participants in a meeting. *Machine Vision and Applications*, *17*(2), 83–93. doi:10.1007/s00138-006-0015-5

Dai, P., Di, H., Dong, L., Tao, L., & Xu, G. (2009). Group interaction analysis in dynamic context. *IEEE Transactions on Systems, Man, and Cybernetics. Part B, Cybernetics*, *39*(1), 34–42. doi:10.1109/TSMCB.2008.2009559

Dong, F., & Gou, X. (2009). An advanced location based service on mobile social network. In *Proceedings of the 2nd IEEE International Conference on Broadband Network and Multimedia Technology (IC-BNMT 2009)*. Beijing, China: IEEE Press.

Dragone, M., Duffy, B. R., & O'Hare, G. M. P. (2005). Social interaction between robots, avatars & humans. In *Proceedings of 2005 IEEE International Workshop on Robots and Human Interactive Communication (RO-MAN 2005)*. Nashville, TN: IEEE.

Eagle, N., Pentland, A., & Lazer, D. (2009). Inferring friendship network structure by using mobile phone data. *Proceedings of the National Academy of Sciences of the United States of America*, *106*(36), 15274–15278. doi:10.1073/pnas.0900282106

Gaonkar, S., Li, J., & Choudhury, R. R. (2008). Micro-blog: Sharing and querying content through mobile phones and social participation. In *Proceedings of MobiSys 2008*. Breckenridge, CO: MobiSys.

Gou, L., Kim, J. H., & Chen, H. H. (2009). MobiSNA: A mobile video social network application. In *Proceedings of MobiDE 2009*. Providence, RI: MobiDE.

Grisword, W. G., Shanahan, P., & Brown, S. W. (2004). ActivieCampus: Experiments in community-oriented ubiquitous computing. *IEEE Computer, 37*(10), 73–81. doi:10.1109/MC.2004.149

Halawani, T., & Mohands, M. (2003). Smart card for smart campus: KFUPM case study. In *Proceedings of 10th IEEE Internal Conference on Electronics, Circuits and Systems (ICECS 2003)*. Sharjah, United Arab Emirates: ICECS.

Hodgson, M. (2003). Classtalk system for predicing and visualizing speech in noise in classrooms. *Canadian Acoustics, 31*(3), 62–63.

Kim, E., Plummer, M., Hiltz, S. R., & Jones, Q. (2007). Perceived benefits and concerns of prospective users of the SmartCampus location-aware community system test-bed. In *Proceedings of the 40th Annual Hawaii International Conference on System Sciences*. Big Island, HI: IEEE.

Lee, C., Nordstedt, D., & Helal, S. (2003). Enabling smart spaces with OSGi. *IEEE Pervasive Computing / IEEE Computer Society [and] IEEE Communications Society, 2*(3), 89–95. doi:10.1109/MPRV.2003.1228530

Li, K. A., Sohn, T. Y., Huang, S., & Griswold, G. (2008). PeopleTones: A system for the detection and notification of buddy proximity on mobile phones. In *Proceedings of the 6th International Conference on Mobile System, Application and Services, (MobiSys 2008)*. New York, NY: MobiSys.

Marples, D., & Kriens, P. (2001). The open services gateway initiative: An introductory review. *IEEE Communications Magazine, 39*(12), 110–114. doi:10.1109/35.968820

Miluzzo, E. (2008). Supporting personal sensing presence in social networks: Implementation, evaluation and user experience of the CenceMe application using mobile phones. In *Proceedings of the 6th ACM Conference on Embedded Network Sensor Systems (SenSys 2008)*. New York, NY: ACM Press.

Mirisaee, S. H., Noorzaden, S., & Sami, A. (2010). Mining friendship from cell phone switch data. In *Proceedings of the 3rd International Conference on Human-Centric Computing (HumanCom 2010)*. HumanCom.

Nakanishi, H. (2004). FreeWalk: A social interaction platform for group behaviour in a virtual space. *International Journal of Human-Computer Studies, 60*(4), 421–454. doi:10.1016/j.ijhcs.2003.11.003

Pietilainen, A. K., Oliver, E., & LeBrun, J. (2009). MobiClique: Middleware for mobile social networking. In *Proceedings of WOSN 2009*. Barcelona, Spain: WOSN.

Rohs, M., & Bohn, R. (2003). Entry points into a smart campus environment – Overview of the ETHOC system. In *Proceedings of the 23rd Internal Conference on Distributed Computing Systems (ICDCSW 2003)*. ISDCSW.

Shi, Y., Xie, W., & Xu, G. (2003). The smart classroom: merging technologies for seamless tele-education. *IEEE Pervasive Computing/ IEEE Computer Society [and] IEEE Communications Society, 2*(2), 47–55. doi:10.1109/MPRV.2003.1203753

Sun, T., Shi, X. Y., & Shen, Y. M. (2010). iLife: A novel mobile social network services on mobile phones. In *Proceedings of the 10th International Conference on Computer and Information Technology (CIT)*. Bradford, UK: CIT.

Suo, Y., Miyata, N., & Morikawa, H. (2009). Open smart classroom: extensible and scalable learning system in smart space using web service technology. *IEEE Transactions on Knowledge and Data Engineering, 21*(6), 814–828. doi:10.1109/TKDE.2008.117

Yau, S. S., Gupta, S. K. S., & Fariaz, K. (2003). Smart classroom: Enhancing collaborative learning using pervasive computing technology. In *Proceedings of Annual Conference and Exposition: Staying in Tune with Engineering Education (ASEE)*. Nashville, TN: ASEE.

Yu, Z. W., Yu, Z. Y., Aoyama, H., Ozeki, M., & Nakamura, Y. (2010). Capture, recognition, and visualization of human semantic interactions in meetings. In *Proceedings of the 8th IEEE International Conference on Pervasive Computing and Communications (PerCom 2010)*. Mannheim, Germany: IEEE Press.

KEY TERMS AND DEFINITIONS

Middleware: The software that provides common functions for applications.

Mobile Social Network: The human social network collected with mobile devices, such as smart phones.

OSGi (Open Service Gateway Initiative): A module system and service platform for the Java programming language that implements a complete and dynamic component model.

Participatory Sensing: An approach to data collection and interpretation in which individuals, acting alone or in groups, use their personal mobile devices and Web services to systematically explore interesting aspects of their worlds ranging from health to culture.

Smart Campus: The campus deployed with various smart devices and sensors, through which persons are sensed and served with adaptive services.

Smart Phone: The cell phones integrated with various sensors and programmable.

Social Interaction: The interaction and communication among human being.

Chapter 13
Recommending Related Microblogs

Lin Li
Wuhan University of Technology, China

Huifan Xiao
Wuhan University of Technology, China

Guandong Xu
University of Technology Sydney, Australia

ABSTRACT

Computing similarity between short microblogs is an important step in microblog recommendation. In this chapter, the authors utilize three kinds of approaches—traditional term-based approach, WordNet-based semantic approach, and topic-based approach—to compute similarities between micro-blogs and recommend top related ones to users. They conduct experimental study on the effectiveness of the three approaches in terms of precision. The results show that WordNet-based semantic similarity approach has a relatively higher precision than that of the traditional term-based approach, and the topic-based approach works poorest with 548 tweets as the dataset. In addition, the authors calculated the Kendall tau distance between two lists generated by any two approaches from WordNet, term, and topic approaches. Its average of all the 548 pair lists tells us the WordNet-based and term-based approach have generally high agreement in the ranking of related tweets, while the topic-based approach has a relatively high disaccord in the ranking of related tweets with the WordNet-based approach.

INTRODUCTION

Measuring the similarity between documents and queries has been extensively studied in information retrieval. However, there are a growing number of tasks that require computing the similarity between two very short texts. One of microblog recommendation methods is to suggest micro-blogs related to what a user has issued or trending topics. One indispensable step in realizing effective recommendation is to compute short text similarities between micro-blogs.

Traditional term-based similarity computing measure perform poorly on such tasks because

DOI: 10.4018/978-1-4666-2806-9.ch013

of data sparseness and the lack of context, it rely heavily on terms occurring in both two documents. If two texts do not have any terms in common, then they receive a very low similarity score, regardless of how topically related they actually are. This is well known as the vocabulary mismatch problem. For example, "UAE" and "United Arab Emirates" are semantically equivalent, yet they share no terms in common. This problem is only exacerbated if we attempt to use traditional measures to compute the similarity of two short segments of text (Metzler, Dumais, & Meek, 2007). According to conventional measures, the more overlaps of words two texts have, the higher similarity score they will receive, which is unreasonable and inaccurate. For example, "apple pie" and "apple phone" share one word apple yet have low semantic relation.

To overcome the above difficulty, in this work, we investigate a topic-based approach and a WordNet-based approach to estimate similarity scores between microblogs and recommend top related ones to users. The WordNet based approach utilizes dictionary-based algorithms to capture the semantic similarity between two texts based on the WordNet taxonomy dictionary. Latent Dirichlet Allocation (LDA) (Blei, Ng, & Jordan, 2003), a topic model for text or other discrete data, allows us to analyze of corpus, and extracts the topics that combined to form its documents. Empirical study is conducted to compare their recommendation effectiveness in terms of precision measure.

BACKGROUND

Many twitter related work and problems have been investigated in the literature. Kwak et al. (2010) have made the first quantitative study on the entire Twitter sphere and information diffusion on it. They studied the topological characteristics of Twitter and its power as a new medium of information sharing and have found a non-power-law follower distribution, a short effective diameter, and low reciprocity, which all mark a deviation from known characteristics of human social networks. Yin et al. (2011) analyze link formation in micro-blogs. They found that 90 percent of new links are to people just two hops away and the dynamics of new link creation are affected by the users account age. Their experimental results showed that in the very beginning (within 100 days), the users add many friends and then for the older users (100-400 days), their friends seem more stable, while for much older users (more than 500 days), their number of new friends is larger and larger. Results also showed that the older the user, the larger the increase in followers.

On the other hand, computation of short text similarity has also been studied in many researches through various points of view. Many techniques have been proposed to overcome the vocabulary mismatch problem, including stemming (Krovetz, 1993; Porter, 1980), LSI (Deerwester, Dumais, Landauer, Furnas, & Harshman, 1990), translation models (Berger & Lafferty, 1999), and query expansion (Zhai & Lafferty, 2001; Lavrenko & Croft, 2001). Query expansion is a common technique that used to convert an initial, typically short, query into a richer representation of the information need (Zhai & Lafferty, 2001; Lavrenko & Croft, 2001; Rocchoi, 1971). This is accomplished by adding terms that are likely to appear in relevant or pseudo-relevant documents to the original query representation. Sahami and Heilman proposed a method of enriching short text representations that can be construed as a form of query expansion (Sahami & Heilman, 2006). Their proposed method expands short segments of text using Web search results. The similarity between two short segments of texts can then computed in the expanded representation space.

TERM-BASED APPROACH

In this section, we introduce traditional term based approach. A weight is assigned to each term in a tweet depending on the number of occurrences of the term in the tweet. The way we used in this work is to assign the weight to be equal to the number of occurrences of term ***t*** in a tweet, which is referred to as term frequency and is denoted ***tf***. At this point, we may view each tweet as a vector with one component corresponding to each term in the dictionary, together with a weight for each component that is given by ***tf***. For dictionary terms that do not occur in tweet, this weight is zero. The standard way of quantifying the similarity between two tweets T_1 and T_2 is to compute the cosine similarity of their vector representations, $\vec{V}(d_1)$ and $\vec{V}(d_2)$.

$$sim(d_1, d_2) = \frac{\vec{V}(d_1) \cdot \vec{V}(d_2)}{\left|\vec{V}(d_1)\right|\left|\vec{V}(d_2)\right|} \quad (1)$$

$$sim(d_1, d_2) = \vec{V}(d_1) \cdot \vec{V}(d_2) \quad (2)$$

The effect of the denominator of equation 1 is thus to length-normalize the vectors $\vec{V}(d_1)$ and $\vec{V}(d_2)$ to unit vectors $\vec{V}(d_1) = \vec{V}(d_1) / \left|\vec{V}(d_1)\right|$ and thus, equation 2 can be viewed as the dot product of the normalized versions of the two document vectors. This measure is the cosine of the angle θ between the two vectors.

Viewing a collection of N texts (e.g., tweets) as a collection of vectors leads to a natural view of a collection as a term-text matrix. It is an M × N matrix whose rows represent the M terms (dimensions) of the N columns, each of which corresponds to a text. As always, the terms being indexed could be stemmed before indexing. For instance, "jealous" and "jealousy" would under stemming be considered as a single dimension.

The following is a basic introduction to the overall process of term-based approach that used to measure tweets similarity:

1. Split one tweet (short text message) into several terms.
2. Normalize ill-formed terms and pick out all the stop words, punctuation marks and signs, say " ", "#", "@", and numbers.
3. Find distinct terms and establish continuous and unique index for them. After indexing, the term-text matrix can be created using distinct term as horizontal dimension, the tweets as vertical dimension, and each matrix unit represents how many times the specific row term appeared in the correspondent column tweet.
4. Finally, compute Euclidean length for each tweet vector (the column of term-text matrix) and utilize equation 2 to get similarity score between two tweets.

WORDNET-BASED APPROACH

WordNet is a lexical database that is available online and provides a large repository of English lexical items. The whole dictionary can be treated as a large graph with each node being a synset and the edges representing the semantic relations. A shared parent of two synsets is known as a sub-sumer. The Least Common Subsumer (LCS) of two synsets is the sumer that does not have any children that are also the sub-sumer of two synsets. In other words, the LCS of two synsets is the most specific sub-sumer of the two synsets. There are many proposals for measuring semantic similarity between two synsets. In this work, we take into account both path length and depth of the least common sub-summer (Dao & Simpson, 2005). The similarity definition is as follow:

Sim(s, t) = 2 * depth (LCS)/ [depth(s) + depth(t)] (3)

In Equation 3, t denotes the source and target words being compared; depth(s) is the shortest distance from root node to a node S on the taxonomy where the synset of S lies; LCS denotes the least common sub-submer of s and t.

In one taxonomy, the shorter the path from one node to another, the more similar they are. The path length is measured in nodes/vertices rather than in links/edges and the length of the path between two members (words) of the same synsets is 1 (synonym relations). For example, we see that "book" and "volume" in Figure 1 are within the same synset indicating that they have the same meaning exactly. Thus, the distance is zero.

Here, we treat each microblog as a sentence. Similarity computation between microblogs is converted to sentence similarity calculation. Steps for computing semantic similarity between two sentences are as follows (Dao & Simpson, 2005).

1. First, each sentence is partitioned into a list of words and we remove the stop words. Stop words are frequently occurring, insignificant words that appear in a database record, article, or a Web page, etc. As a result, each sentence is turned to be a list of tokens.
2. Second, the task is part-of-speech disambiguation (or tagging) to identify the correct Part of Speech (POS – like noun, verb, pronoun, adverb. . .) of each word in the sentence.
3. Third, stemming words are done. We use the Porter stemming algorithm. Porter stemming is a process of removing the common morphological and inflexional endings of words. It can be thought of as a lexicon finite state transducer with the following steps: Surface form -> split word into possible morphemes -> getting intermediate form -> map stems to categories and affixes to meaning -> underlying form, i.e., foxes -> fox + s -> fox.
4. Fourth, we find the most appropriate sense for every word in a sentence (Word Sense Disambiguation). The Lesk algorithm (Lesk, 1986) uses dictionary definitions (gloss) to disambiguate a polysemous word in a sentence context. The major objective of his idea is to count the number of words that are shared between two glosses. The more overlapping the words, the more related the senses are. To disambiguate a word, the gloss of each of its senses is compared to the glosses of every other word in a sentence. A word is assigned to the sense whose gloss

Figure 1. WordNet concept tree (Zhao, Du, & Nauerz, 2008)

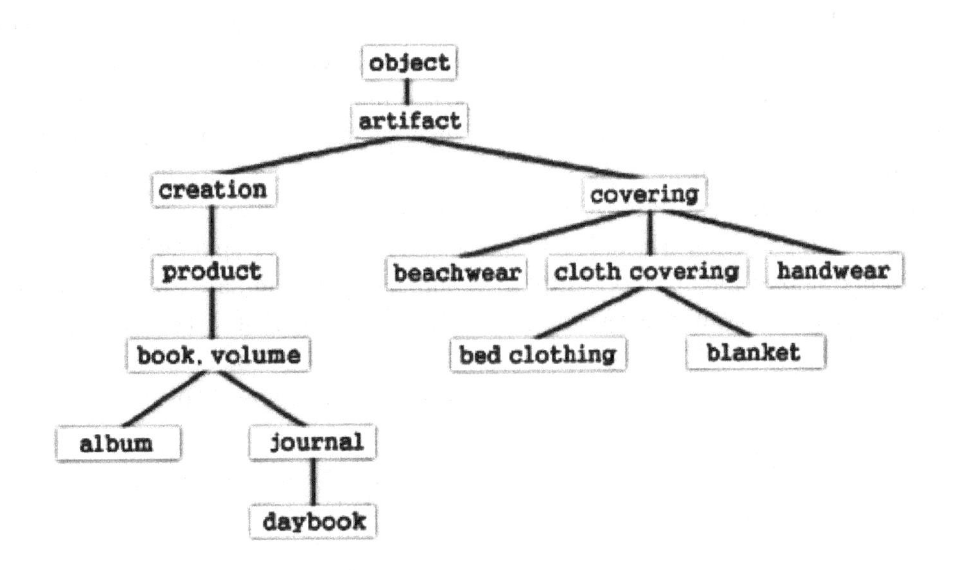

shares the largest number of words in common with the glosses of the other words.

5. Finally, the similarity between the sentences is computed based on the similarity of the pairs of words. We capture semantic similarity between two word senses based on the path length similarity, in which we treat taxonomy as an undirected graph and measure the distance between them in WordNet.

TOPIC-BASED APPROACH

We chose LDA to do topic analysis on microblog texts. Similarity computation between microblogs is equivalent to the dot product of two normalized topic vectors. Steps we take are as follow.

1. We need to pre-process the original data into required data format. During the process of analyzing our dataset (548 tweets), the term index and term-text matrix would be created, which provide great convenience for the transformation of all the original 548 tweets into the data format that LDA-c[1] implementation required.
2. We employ the variational EM algorithm (Blei, Ng, & Jordan, 2003) to find the variational parameters that maximize the total likelihood of the corpus with respect to model parameters. Meanwhile, the calculated estimation parameters can be used to infer topic distribution of a new document by performing the variational inference.
3. After topic inference, each tweet can be represented by a topic vector. The dot product of two normalized topic vectors is the similarity score of corresponding tweets.

WORDNET AND TOPIC COMBINED APPROACH

We utilize the Borda count (Borda, 1781) method to combine the experimental results of Wordnet-based approach and topic-based approach. The reason that we fuse two approaches is given in Section Experimental Results.

The Borda count is a single-winner election method in which voters rank candidates in order of preference. It determines the winner of an election by giving each candidate a certain number of points corresponding to the position in which he or she is ranked by each voter. Once all votes have been counted, the candidate with the most points is the winner. Because it sometimes elects broadly acceptable candidates, rather than those preferred by the majority, the Borda count is often described as a consensus-based electoral system, rather than a majoritarian one. The number of points given to candidates for each ranking (list) is determined by the position of candidates standing in the list. Under the simplest form of the Borda count (the exact form we applied), if there are five candidates in an list then a candidate will receive five points each time they are ranked first, four for being ranked second, and so on, with a candidate receiving 1 point for being ranked last (or left unranked). In other words, where there are n candidates a candidate will receive n points for a first preference, $n - 1$ points for a second preference, $n - 2$ for a third, and so on. Since lists are all in descending order, therefore, the first elements will receive 548 points, the second 547, and the last receiving 1 point.

Both of the two approaches would get 548 lists indicating the similarity of each tweet to all the 548 tweets (including itself) in descending order. Borda count method combines two experimental results in following steps:

1. Points will be assigned to each candidate in one list in the way described above.

2. Each borda count list produced by WordNet-based approach is combined with its corresponding list produced by topic-based approach, that is, to add the points assigned for the same tweet in two lists together.
3. The 548 combined lists are sorted by points in descending order and the top 10 candidates are picked up for each ranking, which are exactly the top 10 most related tweets under the combined method.

The above three steps will be conducted for three times with varied topic number k in LDA at 10, 15 and 20.

DATASET AND EVALUATION METHODOLOGY

We demonstrate the working of the two approaches on the dataset extracted from (Han & Baldwin, 2011). It contains 548 English messages sampled from Twitter API (from August to October 2010) and contains 1184 normalized tokens. All ill-formed words had been detected, and recommended candidates are generated based on morphophonemic similarity. Both word similarity and context are then exploited to select the most probably correct candidate for the word. Popular precision and Kendall tau distance are used as evaluation metrics. Given the recommended list for each tweet in dataset, we manually judge how many of these relatively highly related tweets are really related. The precision metric is the ratio of the number of correctly selected relative tweets to the number of pairs of manually judged related tweets, defined as in Equation 4.

$$precision = \frac{correctly\ selected\ related\ tweets}{manually\ judged\ related\ tweets} \quad (4)$$

The Kendall tau distance is a metric of comparing the disagreement between two lists by counting the number of pairwise disagreements between two lists. The larger the distance, the more dissimilar the two lists are. The Kendall tau distance between two lists τ_1 and τ_2 is shown in Box 1.

$K(\tau_1, \tau_2)$ will be equal to 0 if the two lists are identical and 1 if one list is the reverse of the other. Often Kendall tau distance is normalized by dividing by n(n−1)/2 so a value of 1 indicates a maximum disagreement. Here it is used to measure the agreement of the two recommendation lists produced by our WordNet- and topic-based approaches.

EXPERIMENTAL RESULTS

Before showing our experimental results, what should be mentioned here is that, because of lacking of clear and specific criteria, and because of that tweet is short English messages without context, it is difficult to judge accurately whether two tweets are within one topic. Furthermore, since the quantity of our dataset is small, the total number of related tweets is also scant, which means the denominator of metric precision is in small number.

The precision metric is defined as the ratio of the number of correctly selected relative tweets

Box 1.

$$K(\tau_1, \tau_2) = \left|(i,j) : i < j, (\tau_1(i) < \tau_1(j) \wedge \tau_2(i) > \tau_2(j)) \vee (\tau_1(i) > \tau_1(j) \wedge \tau_2(i) < \tau_2(j))\right|$$

to the number of pairs of manually judged related tweets. The accuracy of precision may be affected by personal understanding toward tweets, which contain large amounts of informal abbreviations and expressions. By varying the number of latent topics (K) in LDA, our results say that the highest precision score is obtained when K=15 as shown in Table 1. The comparison between the three approaches is given in Table 2.

We can see the results from Table 2 that the precision scores for the three approaches are similar, while the WordNet-based approach is slightly higher than the cosine-based one, even the total amounts of related tweets(119 pairs of related tweets by one human judger) is small. In addition, topic-based one has the least precision value. According to previous experience, we believe when the experimental dataset become much larger, WordNet-based approach would get higher precision as studied in Zhao and Du (2008).

Kendull tau distance is used to measure the agreement of the two recommendation lists produced by every two of the three approaches. The average Kendall tau distance between 548 lists generated by the WordNet based approach and their corresponding lists generated by the term based approach is approximately 0.0397922914, and the average Kendall tau distance between 548 lists generated by the term based approach and their corresponding lists generated by the topic based approach is approximately 0.073951436706234, which shows general high agreement in the order of lists. This is due to the fact that lists generated by the term based approach is sparse, and most of its elements are zero. The average Kendall tau distance between 548 lists generated by the WordNet based approach and their corresponding lists generated by the topic-based approach is 0.570537528674173, which indicates the relative high disaccord in the ranking of lists by the two approaches. In other words, the recommendation made by the WordNet based approach is different from that made by the topic based approach. We reckon the rational for this observation is that these two approaches tackle the recommendation from different aspects. The essence of topic based method lies on the assumption that there exists an unseen structure of "topics" or "themes" in the text corpus, which governs the co-occurrence observations, while WordNet-based method is more concerned with the semantics of words.

The precision scores of combined method at varied number of topics (k) are shown in Table 3. We can see from Table 3 that the precision scores for top 5 and 10 when k equals to 10 are all slightly lower than the results of the WordNet approach, while all slightly higher than it when k equals to 15 and 20. Moreover, precision scores for both top 5 and 10 are best when topic number (k) is equal to 20.

Table 1. Precision scores at different number of topics (K)

	K=10	K=15	K=20
Top 5	0.8246	0.8246	0.8231
Top 10	0.8246	0.8261	0.8231

Table 2. Precision scores of the three approaches at top 5 and top 10 recommendations

	Cosine-based	WordNet-based	Topic-based
Top 5	0.8276	0.8306	0.8246
Top 10	0.8350	0.8366	0.8261

Table 3. Precision scores of combined method at different number of topics (K)

	K=10	K=15	K=20
Top 5	0.8261	0.8306	0.8351
Top 10	0.8351	0.8381	0.8411

CONCLUSION

In this chapter, we studied the problem of measuring related micro-blogs, whose first step is to compute similarity between short texts. We compared three types of similarity computing approaches, i.e., traditional term-based approach, WordNet-based semantic approach, and topic-based approach. We showed that WordNet-based approach works best in finding related short microblogs under a small number dataset containing only 548 tweets. The combination of the WordNet- and topic-based approach can further improve recommendation precision. We will conduct experiment on larger dataset in the future to see whether WordNet-based approach would get much better result as we thought. We think that topic models are applicable for long and rich training texts, but not effective for short and sparse text. WordNet shows stable and acceptable performance in both long and short texts. Furthermore, the combined method of WordNet and LDA shows best effect in terms of precision (highest precision score when k is equal to 20).

REFERENCES

Berger, A., & Lafferty, J. (1999). Information retrieval as statistical translation. [ACM Press.]. *Proceedings of SIGIR, 1999*, 222–229.

Blei, D. M., Ng, A. Y., & Jordan, M. I. (2003). Latent dirichlet allocation. *Journal of Machine Learning Research, 3*, 993–1022.

Borda, J. (1781). Mémoire sur les élections au scrutin. *Comptes rendus de l'Académie des Sciences, 44*, 42–51.

Dao, T., & Simpson, T. (2005). Measuring similarity between sentences. Technical Report. *WordNet*. Retrieved from http://wordnetdotnet.googlecode.com/svn/trunk/Projects/Thanh/Paper/WordNet-DotNet_Semantic_Similarity.pdf

Deerwester, S., Dumais, S., Landauer, T., Furnas, G., & Harshman, R. (1990). Indexing by latent semantic analysis. *Journal of the American Society for Information Science and Technology, 41*(6), 391–407. doi:10.1002/(SICI)1097-4571(199009)41:6<391::AID-ASI1>3.0.CO;2-9

Han, B., & Baldwin, T. (2011). Lexical normalisation of short text messages: Maknsens a #twitter. In *Proceedings of ACL*, (pp. 368–378). ACL.

Krovetz, R. (1993). Viewing morphology as an inference process. [ACM Press.]. *Proceedings of SIGIR, 1993*, 191–202.

Kwak, H., Lee, C., Park, H., & Moon, S. B. (2010). What is twitter, a social network or a news media? In *Proceedings of WWW*, (pp. 591–600). IEEE.

Lavrenko, V., & Croft, W. B. (2001). Relevance based language models. [ACM Press.]. *Proceedings of SIGIR, 2001*, 120–127.

Lesk, M. (1986). Automatic sense disambiguation using machine readable dictionaries: How to tell a pine code from an ice cream cone. In *Proceedings of the 5th Annual International Conference on Systems Documentation*, (pp. 24-26). New York, NY: ACM Press.

Metzler, D., Dumais, S. T., & Meek, C. (2007). Similarity measures for short segments of text. In *Proceedings of ECIR*, (pp. 16–27). ECIR.

Porter, M. F. (1980). An algorithm for suffix stripping. *Program, 14*(3), 130–137. doi:10.1108/eb046814

Rocchio, J. J. (1971). *Relevance feedback in information retrieval*. Upper Saddle River, NJ: Prentice-Hall.

Yin, D., Hong, L., Xiong, X., & Davison, B. D. (2011). Link formation analysis in microblogs. [Beijing, China: ACM Press.]. *Proceedings of SIGIR*, *2001*, 24–28.

Zhai, C., & Lafferty, J. (2001). Model-based feedback in the language modeling approach to information retrieval. [CIKM.]. *Proceedings of CIKM*, *2001*, 403–410.

Zhao, S., Du, N., & Nauerz, A. (2008). Improved recommendation based on collaborative tagging behaviors. In *Proceedings of the 13th International Conference on Intelligent User Interfaces*. New York, NY: IEEE.

ENDNOTES

1 http://www.cs.princeton.edu/blei/lda-c/

Chapter 14
On Group Extraction and Fusion for Tag-Based Social Recommendation

Guandong Xu
University of Technology Sydney, Australia

Yanhui Gu
University of Tokyo, Japan

Xun Yi
Victoria University, Australia

ABSTRACT

With the recent information explosion, social websites have become popular in many Web 2.0 applications where social annotation services allow users to annotate various resources with freely chosen words, i.e., tags, which can facilitate users' finding preferred resources. However, obtaining the proper relationship among user, resource, and tag is still a challenge in social annotation-based recommendation researches. In this chapter, the authors aim to utilize the affinity relationship between tags and resources and between tags and users to extract group information. The key idea is to obtain the implicit relationship groups among users, resources, and tags and then fuse them to generate recommendation. The authors experimentally demonstrate that their strategy outperforms the state-of-the-art algorithms that fail to consider the latent relationships among tagging data.

1. INTRODUCTION

Tag-based services, e.g., Del.icio.us[1], Last.fm[2], and Flickr[3] have undergone tremendous growth in the past several years. All of the above services allow users to express their own opinions on resources with arbitrary words. Making use of social tagging data for recommendation is emerging as an active research topic in the field of recommender systems recently. Traditional recommender systems focus on the explicit rating data of users, e.g., movie ratings, to gain the user preference and make predictions for new items. Different from rating data, social tagging data does

DOI: 10.4018/978-1-4666-2806-9.ch014

not contain user's explicit preference information on resources, instead, reflecting the personalized perceptions on resources by users. In particular, such data involves three types of objects, i.e., user, resource, and tag. These differences bring in new challenges as well as opportunities to deal with recommendation problems in the context of social tagging systems. A primary concern of recommender systems in tag-based recommender systems is to present users with avenues for navigation that are most relevant to their information needs. Tags serve as intermediaries between users and resources; therefore, the key challenge in social annotation recommender systems is how to accurately capture user preferences through tags.

Figure 1(a) is a typical social annotation recommender system. It has three types of entities that are considered by the recommender system: <user, resource, tag>. One user prefers some resources, which he is interested in and annotates them with some words. In this case, one resource can be tagged by several tags or one tag can be annotated on several resources. Therefore, tags just serve as intermediaries between users and resources. So far, we can see that one user may be interested in some resources and annotate tags on them. Here, the tag which has been annotated on resource describes user's own opinion and indicate his interests. Likewise, for the same resource, different users may use different tags to annotate. If we want to retrieve resources via these ambiguous tags, it is very common that we cannot find the desired results through just browsing the returned resources. However, there are also some tags the users have common view, i.e., these tags can also represent resources properly. Therefore, a user may annotate some tags on various resources, we can illustrate these activities as a user preferring resources based on their interests. Likewise, some common view tags reflect the topic information of resources. We illustrate these two scenarios in Figure 1(b) and (c), respectively, after extracting this information. From the figures, we can see that, tags form different aggregate based on user or resource views. Here, we call these aggregates as "Groups." Moreover, different users may reach via resources different groups. Let's take the following example, consider one user Tom, who annotated the Website of Chicago City with the ambiguous tags "Jordan," "NBA," "basketball," or "Bulls," etc., while another user John annotated with "tour," "relax," "O'Hare," etc. From these tags, we can see that, Tom is a basketball fan while John is a tour pal. The tags reflect the interest of users. If John intends to retrieve resources based on the tags, which Tom annotated, he may not obtain his desired results because they have different interests. However, there are also tags, such

Figure 1. Tag-based recommender system

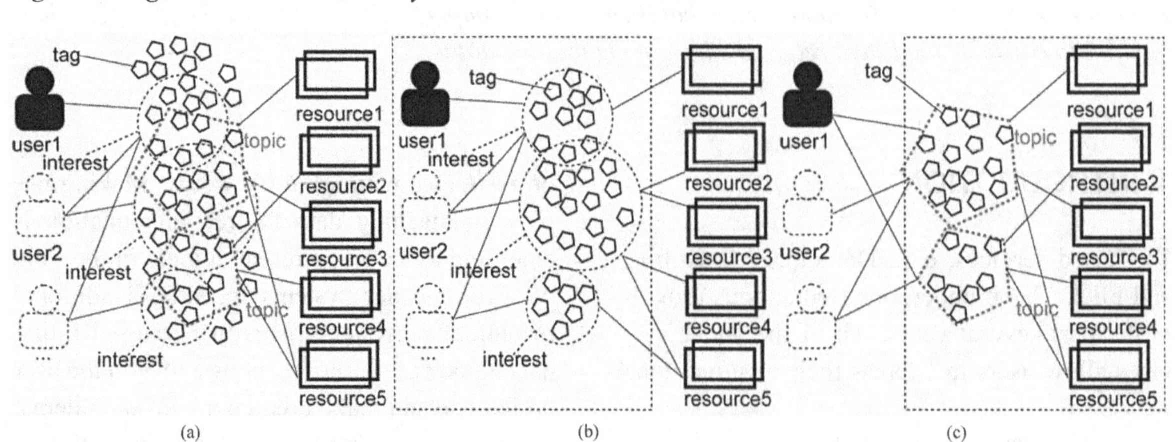

as "Bulls," "O'Hare" which have the common view since "Chicago Bulls" is the Urban Business Card of the city and "O'Hare" is the name of the international airport. Therefore, these two tags can represent what the resource is, i.e., they reflect the real topic of the resource. How can we utilize the user annotation to derive user interest and resource topic is becoming the main challenge.

In this chapter, we aim to make use of the information between tags and resources and between tags and users and to form aggregates. We call these aggregates as Groups. Through these groups, users can obtain their desired information more accurately. We extract two kinds of groups based on different affinity relationships, Topic-Groups based on the bipartite relationship between tags and resources; and Interest Groups based on the bipartite relationship between tags and users. To obtain these latent relationships, it is better to organize the different social relational groups. We regard that, by fusing these two kinds of groups, the more accurate recommendation can be obtained. Some researchers have worked on integrating the social relationship into the recommender system (Groh., 2007; Hummel, 2007), yet they just considered the single relationship, such as friendship data (Acquisti, 2005). Prior approaches lack the exploration of other possible ways of revealing the latent relationship among users, tags and resources more effectively. As the social relationship is inherent in the bipartite (Sun, 2005; Li, 2008) graph of tags and resources and tags and users, we are able to extract such relationships through this graph.

We believe that tags contain more representative information about the users' preferences, such as their interests in some specific resources, but with huge redundancy information due to the freedom of chosen words. We can cluster the tags to obtain such groups. Through such groups, one can get more accurate recommendation. We are interested in understanding whether this group information could produce any practical benefits. In this chapter, we cluster the tags based on the bipartite graph between tags and resources, or between tags and users. Our objective is to study whether and how to better partition these tags to organize the groups and adjust their weights. We believe that our approach can shed light on the applicability of latent social relationship analysis in boosting the intelligence of the current tag-based recommender systems.

The contributions of our chapter are as follows:

- We address the problem of extracting the group information from the tags based on the bipartite graph between tags and resources or between tags and users to obtain the latent social relationships. We call these groups as Topic Groups and Interest Groups.
- We propose the group formulation approach by clustering tags based on such group information. Through this, we can deal with the redundancy of the tags.
- We empirically investigate how to tune the weight of each group and make it adaptive to manage the impact of Topic Groups and Interest Groups on recommendations.
- We conduct comprehensive experiments on the real dataset. The evaluation results demonstrate the effectiveness of the proposed solutions.

The remainder of this chapter is organized as follows. We introduce the preliminaries in Section 2. The group extraction and organization solutions are presented in Sections 3. Experimental evaluation results are reported in section 4. Section 5 introduces the related work and Section 6 concludes the chapter.

2. PRELIMINARIES

2.1. Social Tagging System Model

In this chapter, our work is to generate top-N recommendation with tagging data. A typical social tagging system has three types of objects, users, tags and resources which are interrelated with one another. Social tagging data can be viewed as a set of triples (Guan, et al., 2010). Each triple (*u, t, r*) represents user *u* annotate a tag *t* to resource *r*. A social tagging system can be described as a four-tuple, there exists a set of users, *U*; a set of tags, *T* ; a set of resources, *R*; and a set of annotations, *A*. We denote the data in the social tagging system as *D* and define it as: $D =< U, T, R, A >$. The annotations, *A*, are represented as a set of triples containing a user, tag and resource defined as: $A \subseteq < u, t, r >: u \in U, t \in T, r \in R$. Therefore a social tagging system can be viewed as a tripartite hypergraph (Mika, 2005) with uscrs, tags, and resources represented as nodes and the annotations are represented as hyper-edges connecting users, tags and resources.

2.2. Standard Recommendation Model in Social Tagging System

Standard social tagging systems may vary in the ways of their ability of handling recommendation. In previous studies, the possible approaches include recency, authority, linkage, popularity, or vector space models. In this chapter, we conduct our work on the vector space model, which is derived from the information retrieval principle. Under such scheme, each user, u, is modeled as a vector over the set of tags, where $w(t_i)$, in each dimension corresponds to the relationship of a tag t_i with this user, u, $u = \left\langle w(t_1), w(t_2), \cdots, w(t_{|T|}) \right\rangle$. Likewise each resource, r, can be modeled as a vector over the set of tags, $r = \left\langle v(t_1), v(t_2), \cdots, v(t_{|T|}) \right\rangle$, where $v(t_i)$ indicates the relationship of a tag t_i with this resource. Some work (Shepitsen, Gemmell, Mobasher, & Burke, 2008; Hotho, Jschke, Schmitz, & Stumme, 2006) use the tag frequency, $Ftx = \left| a = \langle u, r, t \rangle \in A : u \in U \right|$, to calculate the weight of the vector. After that, similarity measures such as the Jaccard similarity coefficient or Cosine similarity is applied to obtain the similarity scores between users or resources via various recommendation strategies, such as user-based or resource-based approaches. In this chapter, we mainly adopt the binary expression of tagging, i.e. the user annotation activity. We take the happening of tagging into consideration only. That means if a user, u, annotate a tag, t, on a resource, r, w will be "1" in this 3D matrix; otherwise "0."

3. GROUP EXTRACTION AND FUSION STRATEGY ON SOCIAL RECOMMENDER SYSTEM

Definition 1 (Group): A group is defined to be a virtual expression, in which tags have similar co-occurrence. Such group information comes from the bipartite graph between tags and resources (Topic Group) and between tags or users (Interest Group).

The framework of our Group Extraction and Group Fusion Social Recommender System is illustrated in Figure 2. It mainly includes three steps to obtain the final recommendation.

3.1. Affinity Graph-Based Group Extraction Based on Bipartite Graph

In this section, we illustrate how to extract the group information between resources and tags or between users and tags. Since the data in social annotation recommender system is a triple model as <user, tag, resource>, we decompose this to obtain two different pairs: <resource, tag> and

Figure 2. The framework of group extraction and group fusion social recommender system (three steps: [1] tag clustering for group extraction; [2] group organization; [3] group fusion for the recommendation generation)

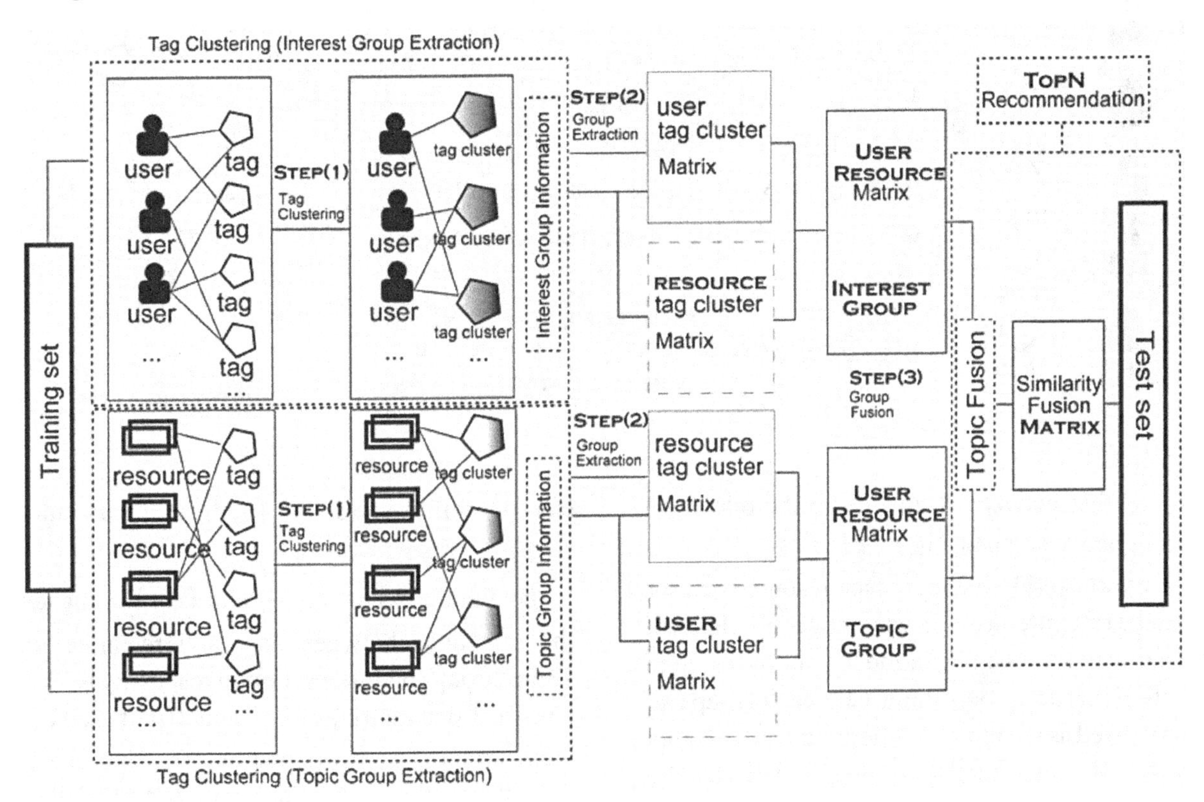

<user, tag>, respectively, which can be represented by a bipartite graph as shown in Figure 3. Based on the bipartite graph between resources and tags and between users and tags, we can obtain the affinity graphs of tags based on these two different relationships. We explain the process based on the <user, tag> pair. From Figure 3, we find that each tag vector over users denotes that whether this tag has been annotated on a resource by this user. However, because of the ambiguity of the tags, we cannot obtain the group information easily from individual tag. Therefore, we employ the Hierarchical Agglomerative Clustering algorithm (Shepitsen, Gemmell, Mobasher, & Burke, 2008) to cluster these tags and to extract group information. There are many clustering algorithms available in data mining research domain, here we mainly adopt the Hierarchical Agglomerative Clustering algorithm due to its visible clustering outcomes and competitive clustering capability. From Figure 3, we can see that, we obtain two clusters of tags, i.e., (tag1, tag4, tag5) and (tag2, tag3) after clustering. By remapping from the original co-occurrence between tags and users, we obtain the new co-occurrence data between tag clusters and users. Here, we regard that the clusters of tags as the user's interest group information. By this step, we obtain interest topics based on the relationship between tags and users. In a similar way, we can also obtain the topic groups from the resource-tag relationships.

3.2. Group Formation and Fusion

Lemma 1 (Relevance Propagation): We denote r_{ab} as the relevance between node a and b and

Figure 3. Affinity graph-based group extraction (an example on tag-user model)

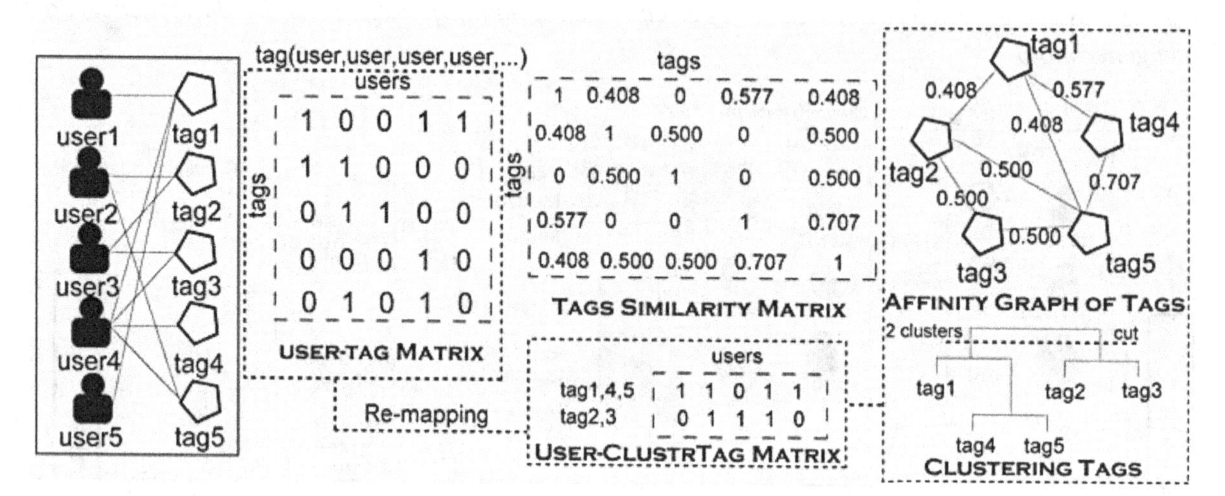

r_{bc} as that between b and c. So the relevance score between a and c is $r_{ab} * r_{bc}$.

By clustering the tags, each resource or user can be re-expressed by a vector over tag clusters. That is to say, we obtain the weight of resource or user on each group. Intuitively, each group can be viewed as the representation of an interest area or a topic area. The higher weight indicates the user is more interested in such group or the resource is more closed to such group. By far, we obtain two types of vectors $u_i = \left\langle w\left(R_{itc_1}\right), w\left(R_{itc_2}\right), \cdots, w\left(R_{itq_{|TC(z)|}}\right)\right\rangle$ and $r_t = \left\langle v\left(R_{ttc_1}\right), v\left(R_{ttc_2}\right), \cdots, v\left(R_{ttq_{|TC(z)|}}\right)\right\rangle$ which is illustrated in solid rectangle in step 2 of Figure 2 and we also can obtain $u_t' = \left\langle w'\left(R_{itc_1}\right), w'\left(R_{itc_2}\right), \cdots, w'\left(R_{itq_{|TC|}}\right)\right\rangle$ $r_i' = \left\langle v\left(R_{itc_1}\right), v\left(R_{itc_2}\right), \cdots, v\left(R_{itq_{|TC|}}\right)\right\rangle$ by re-mapping from original <resource, tag> and <user, tag> pairs which are presented in dotted rectangle. Here, u_i and u_t' denote the user vector based on Interest Group and Topic Group respectively. Likewise, r_t and r_i' are the resource based on Interest Group and Topic Group. According the lemma 1 the relevance between user and resource can be represented as: $(u,r)_i = u_i * r_i'$ and $(u,r)_t = u_t' * r_t$.

By far, the vector $(u,r)_t$ and $(u,r)_i$ indicate the relevance between user and resource in Topic Group and Interest Group respectively. We construct two matrixes UR_T and UR_I based on these vectors. As mentioned before, for a given user, the relevance between him and resources can be computed through different groups. Interest groups are the interests exhibited by different user groups while topic groups can properly describe what the resources are. Here, topic information plays an important role as adjusting user to obtain the resources which they are really interested in. We fuse these two group information by a tunable factor $\lambda \in [0,1]$, that is:

$$UR = \lambda UR_T + (1-\lambda) UR_I$$

If $\lambda = 0$, which means such fusion is weighted on interest group only while $\lambda = 1$ indicates on topic group only. The value between 0 and 1 balance the weights between these two groups. In this work, we will empirically investigate the impact of choosing different λ on the recommendation performance.

3.3. The Recommender Generation

The final step is to generate the recommendation. The matrix UR denotes the user-resource relevance information. Each row (user) vector over resource indicates the relevance between this user and the resources. For example, $u = \left\langle p\left(R_1\right), p\left(R_2\right), \cdots, p\left(R_{|R|}\right)\right\rangle$ denotes the weight how a user prefer a resource. For a giving user, we sort the value in each vector and push the top-N new resources as the recommendations.

4. EXPERIMENTAL EVALUATIONS

To evaluate our approach, we conducted extensive experiments. We performed the experiments using an Intel Core 2 Duo CPU (2.0GHz) workstation with 1G memory, running Red Hat Linux 4.1.2-33. All the algorithms were written in C and compiled by GNU gcc. We conducted experiments on two real datasets, i.e., MovieLens and Delicious.

4.1. Experimental Datasets

The first dataset MovieLens is provided by GroupLens[4]. It is a movie rating dataset. Users were selected at random for inclusion. All users selected had rated at least 20 movies. Unlike previous MovieLens data sets, no demographic information is included. Each user is represented by an ID, and no other information is provided. The data are contained in three files, movies. dat, ratings.dat, and tags.dat. Also included are scripts for generating subsets of the data to support five-fold cross validation of rating predictions. We use tags.dat for it has the social annotation activities. The statistical result of the dataset is listed in Table 1. The second dataset delicious is downloaded from the social tagging system, delicious[5], which is combination of users, resources and tags. The statistical information regarding this dataset is tabulated in Table 2. Following the same division strategy, this dataset is divided into training and test set.

4.2. Evaluation Methodology and Metrics

We utilized the standard metrics in information retrieval domain to evaluate our approach. For each dataset, we judiciously divided the whole dataset into two parts by 80% (Training set) and 20% (Test set). We used this 80% of the data to train our groups upon which we generated the user-resource matrix for recommendation. That is to say, each element of the matrix represents the preference degree of one user on a specific resource. In evaluation, for a given user, we select the Top-N resources as recommendation based on the generated user resource matrix. Meanwhile, we assumed the predicted resource list occurred in the test dataset as well. Thus, we looked at the 20% of

Table 1. Statistics of experimental dataset of MovieLens

Property	MovieLens	Training set	Test set
Number of users	4,009	3,268	1,002
Number of resources	7,601	6,731	4,340
Number of tags	16,529	13,727	5,968
Total entries	95,580	76,464	19,116

Table 2. Statistics of experimental dataset of Del.icio.us

Property	Delicious	Training set	Test set
Number of users	3,713	2,361	1,021
Number of resources	7,320	6,012	4,230
Number of tags	1,3273	1,1940	8,948
Total entries	100,000	80,000	20,000

the dataset (i.e. the real preferences on resources in the test set) and compared the top-N predicted recommendation by our recommender system with the true preferences on new resources in the test set to calculate the precision, which measures the proportion of recommended resources that are ground truth resources. In particular, we selected the top-N predictions Recsys(N) and made these compared with the test set Rec(test). The metric of top-N precision is defined as follows: The metric of top-N precision is defined as follows:

$$\Pr e@N = \frac{\operatorname{Re} cSys(N) \cap \operatorname{Re} c(test)}{\operatorname{Re} cSys(N)}$$

In the following, we reported the results of recommending top-5, 10, 15, and 20 resources and based on the different group strategies and adjusting factors.

4.3. Evaluation Results and Discussions

At first, we conducted experiments on the distribution of Hierarchical Agglomerative Clustering to determine which distance strategy we should take on MovieLens dataset. We plotted two graphs on the relationship between HAC steps and the agglomerative similarity, here N(step)=N(tag)-N(cluster) (N[step], N[item] and N[cluster] denote the number of steps, tags and clusters). In Figure 4, we can see that if we apply AverageLinkage as the similarity strategy, the similarity and N(step) follows power-law, i.e., under such clustering, we can differentiate the tags into various clusters. While in SingleLinkage, there were several plain lines, which meant it was hard to distinguish tags into different groups. We drew same conclusions on similarity measure with delicious. In the following experiments, thus we selected AverageLinkage as our similarity strategy on clustering. We also investigated the impact of division coefficients, i.e., the value that we used to divide the HAC tree into clusters.

To evaluate the effectiveness of our strategy, we evaluated our approach against other compared strategies, such as no clustering and/or no fusion. Table 3 presents the evaluation results of four strategies employed on the real dataset. In this experiment, we fixed the division coefficient as 0.2. We termed no C, no F as no Clustering and no Fusion. Likewise, C(T), no F and C(I), no F are Clustering based on topic information and

Figure 4. Similarity distribution on MovieLens dataset

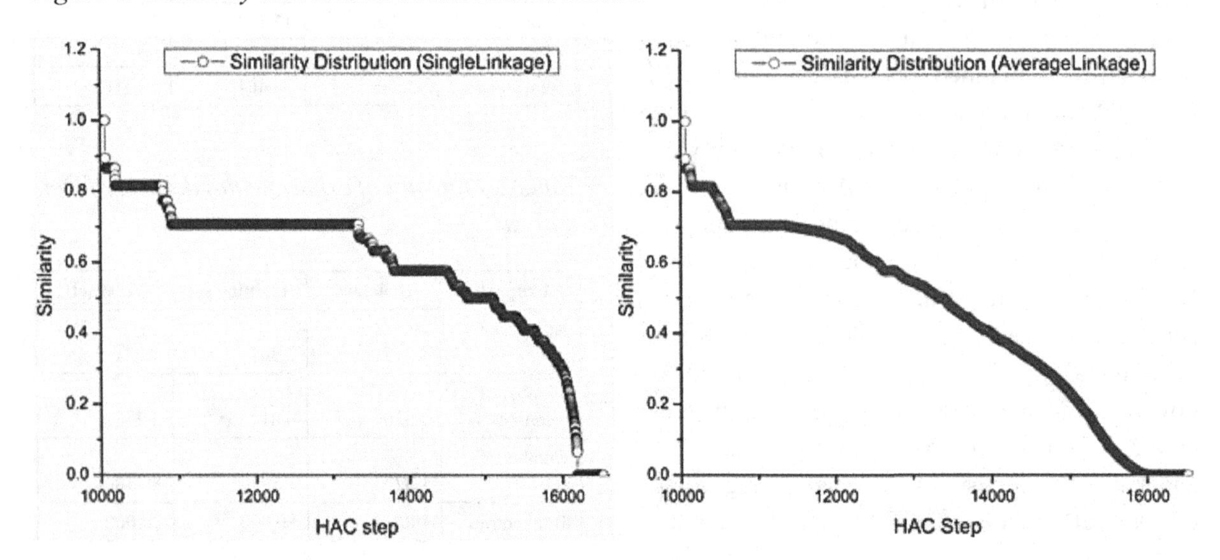

Table 3. Experimental evaluation on four approaches (no C=no clustering; C(T), no F=clustering but no fusion[λ=0]; C(I), no F=clustering but no fusion[λ=1]; C,F=clustering and fusion; division coefficient=0.2)

Strategy	Top 5	Top 10	Top 15	Top 20
no C, no F	0.082609	0.114815	0.163636	0.225000
C(T), no F	0.452345	0.485578	0.472935	0.532201
C(I), no F	0.457830	0.467482	0.478492	0.511359
C, F	**0.489371**	**0.490819**	**0.478638**	**0.574872**

based on interest information respectively. C, F denotes Clustering and Fusion. From the histograms, we can see that after clustering, the precision of "C(T), no F," "C(I), no F" and "C, F" increases significantly in comparison with other methods. This might be due to that tag redundancy was well tackled by clustering the tags. Both two strategies "C(T), no F" and "C(I), no F" have shown lower precision because these two strategies only focus on a single group information, however perform much better than the strategy of "no C, no F." Naturally, the Pre@N improves with the increase of the test window size (i.e. the N value).

As we discussed in previous sections, tag clustering can organized the tags into groups over clusters. That is to say, by clustering, tags can be aligned into groups, such as topic groups or interest groups. In the following, we will conduct the experiments to evaluate the effect on the division coefficient of our clustering. Figure 5(a) represents the precision changes under five different division coefficients. From the figure, we can see that the selection of the division coefficient is crucial to the success of our approach. If the division coefficient is too high, it may result in too many small clusters though tags in such cluster may be very close, but not well handling tag redundancies. That is to say, we cannot accurately extract the proper groups and cannot effectively tackle the redundancy of tags. In contrast, if the coefficient is set too small, it results in few clusters but with low cohesiveness, providing insufficient disambiguation. The topic cannot be extracted from the clusters distinctively. Seen from the peaks of both datasets, the appropriate division coefficient is set to be 0.2.

By far, we discussed different strategies of our approach against existing approaches in terms of precision, and also the effect of division coefficient when group extraction. Now, we discuss the effect on λ, the fusion factor. Figure 5(b) shows the effect by adjusting the factor λ. Seen from the figure both two figures achieve the higher precision using the fusion of two groups, i.e. topic groups or interest groups. Back to the figure, we can observe that when λ=0.8, the recommender system get the highest precision compared to other two values. The larger λ indicates the high weight of user's interest information fused into the recommender system. Therefore, the effect of interest group accounts for the higher importance than that of topic interest group in recommendation.

We reported the experimental results derived from delicious dataset as below. First, we presented the precision comparison between our approach and counterparts in Table 4. Likewise, we saw the strategy of "C, F" always outperforms other three strategies, which again justified our

Table 4. Experimental evaluation on four approaches (no C=no clustering; C(T), no F=clustering but no fusion[λ=0]; C(I), no F=clustering but no fusion[λ=1]; C,F=clustering and fusion; division coefficient=0.2)

Strategy	Top 5	Top 10	Top 15	Top 20
no C, no F	0.110941	0.123947	0.158473	0.169403
C(T), no F	0.5193485	0.537928	0.538959	0.540964
C(I), no F	0.498576	0.504896	0.517584	0.525875
C, F	**0.523506**	**0.539485**	**0.539981**	**0.605943**

Figure 5. The effect of two factors on precision (Movielens)

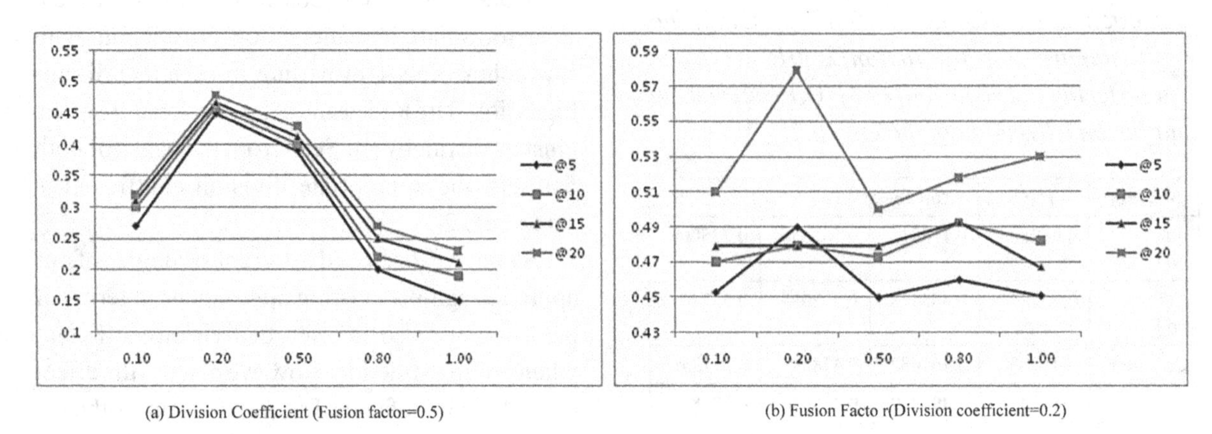

(a) Division Coefficient (Fusion factor=0.5)

(b) Fusion Facto r(Division coefficient=0.2)

claim that the group approach was able to well handle the inherent challenges in social tagging systems. Either group extraction approach based on interest group or topic group performs better than no clustering and no fusion significantly. Interestingly, the precision results from topic group alone approach were higher than those of interest group only approach, implying that the content information had somehow more important impact on extracting group information than user interest. This finding is consistent with the experimental results of fusion impact in the following.

The impact investigation on division coefficient and fusion factor is depicted in Figure 6. Here we gave two sets of plots in terms of N=5, 10,15 and 20. As for the impact of division coefficient, the finding was same as that on MovieLens dataset, i.e., selecting division coefficient as 0.2 to achieve the best precision results. This could be interpreted that we have to control the size of groups to achieve the best results (see Figure 6(a)). Figure 6(b) illustrates the changes of precision in response to the increase of fusion factor, λ. Different from the observation from MovieLens, the fusion weight of user interest aspect plays less significant role in recommendation, for either top-N settings, the best precision occurs at the point of λ=0.2. this suggests that we should take more contribution from topic group aspect into account to achieve better recommendation, which again justifies the comparison results seen from Table 4.

5. RELATED WORK

We reviewed the related literature of group extraction and fusion to improve the standard recommender system from three perspectives:

5.1. Tag Clustering

The efficiency of tag clustering is the ability of aggregate tags into topic domains. Gemmell, Shepitsen, Mobasher, and Burke (2008) and Shepitsen, Gemmell, Mobasher, and Burke (2008) demonstrated how tag clusters serving as coherent topics can aid in the social recommendation of search and navigation. In Hayes and Avesani (2007), topic relevant partitions are created by clustering resources rather than tags. By clustering resources, it improves recommendation by distinguishing between alternative meanings of a query. While in Chen and Dumais (2000), clusters of resources are shown to improve recommendation by categorizing the resources into topic domains. Our work is orthogonal to such works but different that our approach is to cluster

Figure 6. The effect of two factors on precision (Delicious)

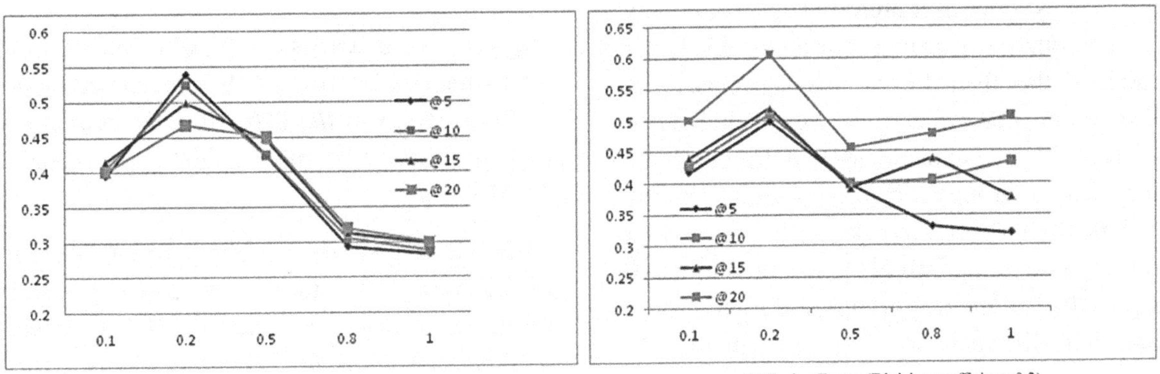

(a) Division Coefficient (Fusion factor=0.5)

(b) Fusion Factor (Division coefficient=0.2)

tags. We believe that tags can hold richer social information rather than resources can.

5.2. Information Extraction from Bipartite Graph

We studied several works that are related to our proposal. Sun, Qu, Chakrabarti, and Faloutsos (2005) used bipartite graph to do neighborhood formation and anomaly detection. Li, Yang, Liu, and Kitsuregawa (2008) also conduct the personalized query recommendation based on the Query-URL bipartite graph. In social tagging system, because of the triple relationships, it is difficult to extract the social information directly. i.e., we cannot apply their strategies to our recommender systems. Therefore, in this chapter, we extract our group information based on the bipartite graph which are tags-resources and tags-users respectively.

5.3. Fusion of Social Relationship

Recently, with the increasing development of social websites and appearance of social data, researchers have begun to pay attention to the social data and explored its usage in recommender systems. Groh (2007) used social network data for neighborhood generation. Konstas, Stathopoulos, and Jose (2009) adopted Random Walk with Restart to model the social tagging in a music track recommendation system. In addition, Hummel et al. (2007) proposed an online social recommender system attempting to use more social information for recommendation generation. All the work shows that, their fusion social information can benefit the recommender system. However, their work mainly focuses on friendship, i.e. the similarity between users. Compared to friendship, other community relationships, i.e., groups in this chapter, contain more information about users' activities (Spertus, Sahami, & Buyukkokten, 2005; Chen, Geyer, Dugan, Muller, & Guy, 2009). Our proposal is orthogonal but fuses the two different groups to obtain better recommendation.

6. CONCLUSION AND FUTURE WORK

In this chapter, we presented an approach on Group extraction and Group fusion in the social tag-based recommender system. In traditional tagging systems, the recommendation generated only based the similarities between the pairs of every two objects. However, in this chapter, we find that, by clustering the tags based on the bipartite graph of tags and resources, tags and users, topic and interest groups can be detected. After the groups were extracted from the tag information, we conducted the preferences which user preferred on resources by going through such groups. The adjustable factor λ controls the effect of two kinds of groups. From the experimental evaluations, we concluded that in the social recommender system, the combination of topic groups and interest groups is able to make better recommendation in comparison to using one group info alone or pure tag-based approach. The interest groups and topic groups contribute differently to the recommendation. In some cases, topic group information dominates in the fusion of two kinds of group information.

There are some further works to conduct. Firstly, we only demonstrated the λ on 5-fixed values. The intensive cross-validation should be taken to train the best λ value. Secondly, our tag clustering was yet based on the bipartite graph only. The weight is only based the occurrence between the tags and resources, or users pairs. An efficient tag clustering strategy can also promote the accuracy of the group extraction for the better precision of our approach. For the next step, we are interested in future exploring the effect of latent social relationship on recommender systems. One aspect is the time series. We can see that the triples of users, tags and resources and their annotation relationships combined by users' activities form a big graph. Regarding such consideration, we believe that the graph will evolve (Yang, Yu, Liu, & Kitsuregawa, 2010) as time goes by. Therefore, time information fusion can be considered to achieve better precision in social recommender systems.

REFERENCES

Acquisti, A., & Gross, R. (2005). Information revelation and privacy in online social networks. In *Proceedings of the 2005 ACM Workshop on Privacy in the Electronic Society*, (pp. 71-80). ACM Press.

Chen, H., & Dumais, S. (2000). Bringing order to the web: Automatically categorizing search results. In *Proceedings of the SIGCHI Conference on Human Factors in Computing Systems*, (pp. 145-152). ACM Press.

Chen, J., Geyer, W., Dugan, C., Muller, M. J., & Guy, I. (2009). Make new friends, but keep the old: Recommending people on social networking sites. In *Proceedings of CHI*, (pp. 201–210). CHI.

Gemmell, J., Shepitsen, A., Mobasher, M., & Burke, R. (2008). Personalization in folksonomies based on tag clustering. In *Proceedings of the 6th Workshop on Intelligent Techniques for Web Personalization and Recommender Systems*. ACM.

Groh, G. (2007). Recommendations in taste related domains: Collaborative filtering vs. social filtering. In *Proceedings of ACM Group 2007*, (pp. 127-136). ACM Press.

Guan, Z., Bu, J., Mei, Q., Chen, C., & Wang, C. (2009). Personalized tag recommendation using graph-based ranking on multi-type interrelated objects. In *Proceedings of the 32nd International ACM SIGIR Conference on Research and Development in Information Retrieval (SIGIR 2009)*, (pp. 540-547). ACM Press.

Guan, Z., Wang, C., Bu, J., Chen, C., Yang, K., Cai, D., et al. (2010). Document recommendation in social tagging services. In *Proceedings of WWW*, (pp. 391-400). IEEE.

Hayes, C., & Avesani, P. (2007). Using tags and clustering to identify topic-relevant blogs. In *Proceedings of the International Conference on Weblogs and Social Media*. IEEE.

Hotho, A., Jschke, R., Schmitz, C., & Stumme, G. (2006). Folkrank: A ranking algorithm for folksonomies. In *Proceedings of FGIR 2006*. FGIR.

Hummel, H. G., Berg, B. V., Berlanga, A. J., Drachsler, H., Janssen, J., & Nadolski, R. (2007). Combining social-based and information-based approaches for personalized recommendation on sequencing learning activities. *International Journal of Learning Technology*, *3*(2), 152–168. doi:10.1504/IJLT.2007.014842

Konstas, I., Stathopoulos, V., & Jose, J. M. (2009). On social networks and collaborative recommendation. In *Proceedings of the 32nd International ACM SIGIR Conference on Research and Development in Information Retrieval (SIGIR 2009)*, (pp. 195-202). ACM Press.

Li, L., Yang, Z., Liu, L., & Kitsuregawa, M. (2008). Query-url bipartite based approach to personalized query recommendation. In *Proceedings of AAAI*, (pp. 1189-1194). AAAI.

Mika, P. (2005). Ontologies are us: A unified model of social networks and semantics. *Lecture Notes in Computer Science*, *3729*, 522–536. doi:10.1007/11574620_38

Shepitsen, A., Gemmell, J., Mobasher, B., & Burke, R. (2008). Personalized recommendation in social tagging systems using hierarchical clustering. In *Proceedings of the 2008 ACM Conference on Recommender Systems*, (pp. 259-266). ACM Press.

Spertus, E., Sahami, M., & Buyukkokten, O. (2005). Evaluating similarity measures: A large-scale study in the orkut social network. In *Proceedings of KDD*, (pp. 678-684). KDD.

Sun, J., Qu, H., Chakrabarti, D., & Faloutsos, C. (2005). Neighborhood formation and anomaly detection in bipartite graphs. In *Proceedings of ICDM*, (pp. 418-425). ICDM.

Yang, Z., Yu, J. X., Liu, Z., & Kitsuregawa, M. (2010). Fires on the web: Towards efficient exploring historical web graphs. *Lecture Notes in Computer Science*, *5981*, 612–626. doi:10.1007/978-3-642-12026-8_46

ENDNOTES

1. http://del.icio.us/
2. http://www.last.fm/
3. http://flickr.com/
4. http://www.grouplens.org/
5. http://del.icio.us

Compilation of References

Abello, J., Resende, M., & Sudarsky, S. (2002). Massive quasi-clique detection. In *Proceedings of the 5th Latin American Symposium on Theoretical Informatics*, (pp. 598-612). Springer-Verlag.

Abrol, S., & Khan, L. (2010). TWinner: Understanding news queries with geo-content using Twitter. In *Proceedings of the 6th Workshop on Geographic Information Retrieval,* (pp. 1-8). IEEE.

Acquisti, A., & Gross, R. (2005). Information revelation and privacy in online social networks. In *Proceedings of the 2005 ACM Workshop on Privacy in the Electronic Society*, (pp. 71-80). ACM Press.

Adamic, L. A., & Glance, N. (2005). The political blogosphere and the 2004 US election: Divided they blog. In *Proceedings of the 3rd International Workshop on Link Discovery,* (pp. 36–43). IEEE Press.

Adams, B., Phung, D., & Venkatesh, S. (2006). Extraction of a social context and application to personal multimedia exploration. In *Proceedings of ACM Multimedia 2006*, (pp 987-996). ACM Press.

Adar, E., & Adamic, L. A. (2005). Tracking information epidemics in blogspace. In *Proceedings of the 2005 IEEE/WIC/ACM International Conference on Web Intelligence*, (pp. 207–214). IEEE Press.

Adomavicius, G., & Tuzhilin, A. (2011). Context-aware recommender systems. *Recommender Systems Handbook.* Retrieved from http://ids.csom.umn.edu/faculty/gedas/NSFCareer/CARS-chapter-2010.pdf

Agarwal, N., Liu, H., Murthy, S., Sen, A., & Wang, X. (2009). A social identity approach to identify familiar strangers in a social network. In *Proceedings of the 3rd International AAAI Conference of Weblogs and Social Media*. AAAI.

Agarwal, N., Liu, H., Salerno, J. J., & Sundarajan, S. (2008). Understanding group interaction in blogosphere: A case study. In *Proceedings of the 2nd International Conference on Computational Cultural Dynamics (ICCCD 2008)*. Washington, DC: ICCCD.

Agarwal, N., Liu, H., Salerno, J., & Yu, P. S. (2007). Searching for familiar strangers on blogosphere: Problems and challenges. In *Proceedings of the NSF Symposium on Next-Generation Data Mining and Cyber-enabled Discovery and Innovation (NGDM)*. NGDM.

Agarwal, N., Lim, M., & Wigand, R. T. (2011). Collective action theory meets the blogosphere: A new methodology. *Networked Digital Technologies*, *136*(3), 224–239. doi:10.1007/978-3-642-22185-9_20

Agarwal, N., & Liu, H. (2008). Blogosphere: Research issues, tools, and applications. *SIGKDD Explorations*, *10*(1), 18–31. doi:10.1145/1412734.1412737

Agrawal, R., & Srikant, R. (1994). Fast algorithms for mining association rules in large databases. In J. B. Bocca, M. Jarke, & C. Zaniolo (Eds.), *Proceedings of the 20th International Conference on Very Large Data Bases* (*VLDB*), (pp. 487-499). San Francisco, CA: Morgan Kaufmann.

Agrawal, R., & Srikant, R. (1994). Fast algorithms for mining association rules in large databases. In *Proceedings of the 20th International Conference on Very Large Data Bases*. IEEE.

Agrawal, R., & Srikant, R. (1994). Fast algorithms for mining association rules in large databases. In *Proceedings of the 20th VLDB Conference*, (pp. 487-499). VLDB.

Agrawal, R., & Srikant, R. (1994). Fast algorithms for mining association rules. In *Proceedings of VLDB*. Citeseer.

Agrawal, R., Imielinski, T., & Swami, A. (1993). Mining association rules between sets of items in large databases. *SIGMOD Record*, *22*(2), 207–216. doi:10.1145/170036.170072

Aït-Mokhtar, S., Chanod, J.-P., & Roux, C. (2002). Robustness beyond shallowness: Incremental deep parsing. *Natural Language Engineering*, *8*, 121–144. doi:10.1017/S1351324902002887

Albert, R., Jeong, H., & Barabasi, A. (1999). The diameter of the world wide web. *Nature*, *401*(130).

Ali-Hasan, N., & Adamic, L. A. (2007). Expressing social relationships on the blog through links and comments. In *Proceedings of International Conference on Weblogs and Social Media*. IEEE.

Andersen, R. (2008b). A local algorithm for finding dense sub-graphs. In *Proceedings of the Nineteenth Annual ACM-SIAM Symposium on Discrete Algorithms*, (pp. 1003-1009). Society for Industrial and Applied Mathematics.

Andersen, R., & Lang, K. J. (2008). An algorithm for improving graph partitions. In *Proceedings of the Nineteenth Annual ACM-SIAM Symposium on Discrete Algorithms*, (pp 651–660). Society for Industrial and Applied Mathematics.

Anderson, R., & Lang, K. (2006). Communities from seed sets. In *Proceedings of the 15th International Conference on World Wide Web (WWW 2006)*. IEEE.

Anderson, C. (2006). *The long tail: Why the future of business is selling more for less*. New York, NY: Hyperion.

Ankerst, M., Breunig, M. M., Kriegel, H. P., & Sander, R. (1999). OPTICS: Ordering points to identify the clustering structure. In *Proceedings of the 1999 ACM SIGMOD International Conference on Management of Data*. Philadelphia, PA: ACM Press.

Ankolekar, A., Szabo, G., Luon, Y., Huberman, B. A., Wilkinson, D., & Wu, F. (2009). Friendlee: A mobile application for your social life. In *Proceedings of the 11th International Conference on Human-Computer Interaction with Mobile Devices and Services (MobileHCI 2009)*. Bonn, Germany: MobileHCI.

Arora, S., Mayfield, E., Penstein-Rosé, C., & Nyberg, E. (2010). Sentiment classification using automatically extracted subgraph features. In *Proceedings of the NAACL HLT 2010 Workshop on Computational Approaches to Analysis and Generation of Emotion in Text, CAAGET 2010*, (pp. 131-139). Morristown, NJ: Association for Computational Linguistics.

Ashri, R., Denker, G., Marvin, D., Surridge, M., & Payne, T. (2004). Semantic web service interaction protocols: An ontological approach. *Lecture Notes in Computer Science*, *3298*, 304–319. doi:10.1007/978-3-540-30475-3_22

Au, T., Kuter, U., & Nau, D. (2005). Web service composition with volatile information. *Lecture Notes in Computer Science*, *3729*, 52–66. doi:10.1007/11574620_7

Baeza-Yates, R. (2009). *User generated content: How good is it?* Paper presented at the the 3rd Workshop on Information Credibility on the Web. Madrid, Spain.

Balahur, A., Steinberger, R., Kabadjov, M., Zavarella, V., van der Goot, E., & Halkia, M. … Belyaeva, J. (2010). Sentiment analysis in the news. In *Proceedings of the Seventh Conference on International Language Resources and Evaluation (LREC 2010)*. Valletta, Malta: LREC.

Bao, T., Cao, H., Chen, E., Tian, J., & Xiong, H. (2011). An unsupervised approach to modeling personalized contexts of mobile users. Retrieved from http://dm.ustc.edu.cn/docs/2010/mobileuser.pdf

Barabasi, A., & Albert, R. (1999). Emergence of scaling in random networks. *Science*, 286.

Baralis, E., Garza, P., Quintarelli, E., & Tanca, L. (2007). Answering XML queries by means of data summaries. *ACM Transactions on Information Systems*, *25*(3), 1–33. doi:10.1145/1247715.1247716

Barber, B., & Hamilton, H. J. (2003). Extracting share frequent itemsets with infrequent subsets. *Data Mining and Knowledge Discovery*, *7*(2), 153–185. doi:10.1023/A:1022419032620

Barbosa, E. M., Moro, M. M., Lopes, G. R., & de Oliveira, J. P. M. (2012). VRRC: Web based tool for visualization and recommendation on co-authorship network. In K. S. Candan, Y. Chen, R. Snodgrass, L. Gravano, & A. Fuxman (Eds.), *Proceedings of the 2012 ACM SIGMOD International Conference on Management of Data,* (p. 865). New York, NY: ACM Press.

Barkhuus, L. (2005). Bring your own laptop unless you want to follow the lecture: Alternative communication in the classroom. In *Proceedings of the Internal ACM SIGGROUP Conference on Supporting Group Work (GROUP 2005)*. Sanibel Island, FL: ACM Press.

Basile, P., Gendarmi, D., Lanubile, F., & Semeraro, G. (2007). Recommending smart tags in a social bookmarking system. *Bridging the Gap between Semantic Web and Web, 2*, 22-29.

Baumer, E., & Fisher, D. (2008). Smarter blogroll: An exploration of social topic extraction for manageable blogrolls. In *Proceedings of the Hawaii International Conference on System Sciences*, (pp. 155–155). IEEE.

Bayir, M. A., Demirbas, M., & Eagle, N. (2009). Discovering spatiotemporal mobility profiles of cellphone users. In *Proceedings of the World of Wireless, Mobile and Multimedia Networks & Workshops, 2009, WoWMoM 2009*. IEEE Press.

Becker, I., & Aharonson, V. (2010). Last but definitely not least: On the role of the last sentence in automatic polarity-classification. In *Proceedings of the ACL 2010 Conference Short Papers, ACLShort 2010*, (pp. 331–335). Morristown, NJ: Association for Computational Linguistics.

Begelman, G., Keller, P., & Smadja, F. (2006). Automated tag clustering: Improving search and exploration in the tag space. In *Proceedings of WWW 2006 Collaborative Web Tagging Workshop*, (pp. 15-33). IEEE.

Bender, M., Crecelius, T., Kacimi, M., Michel, S., Neumann, T., & Parreira, J. X. … Weikum, G. (2008). Exploiting social relations for query expansion and result ranking. In *Proceedings of the IEEE 24th International Conference on Data Engineering Workshop,* (vol. 2, pp. 501-506). IEEE Press.

Berger, A., & Lafferty, J. (1999). Information retrieval as statistical translation. [ACM Press.]. *Proceedings of SIGIR, 1999*, 222–229.

Bernardo, A. H., Peter, P., James, E. P., & Rajan, M. L. (1998). Strong regularities in world wide web surfing. *Science, 280*(5360), 95–97. doi:10.1126/science.280.5360.95

Berners-Lee, T. (1999). *Weaving the web: The original design and ultimate destiny of the world wide web by its inventor*. San Francisco, CA: Harper San Francisco.

Berthard, S., Yu, H., Thornton, A., Hativassiloglou, V., & Jurafsky, D. (2004). Automatic extraction of opinion propositions and their holders. In *Proceedings of AAAI Spring Symposium on Exploring Attitude and Affect in Text*. AAAI.

Bhagat, S., Goyal, A., & Lakshmanan, L. V. S. (2012). Maximizing product adoption in social networks. In E. Adar, J. Teevan, E. Agichtein, & Y. Maarek (Eds.), *Proceedings of the Fifth ACM International Conference on Web Search and Data Mining* (*WSDM*), (pp. 603-612). New York, NY: ACM Press.

Bird, S., Klein, E., & Loper, E. (2009). *Natural language processing with Python*. New York, NY: O'Reilly Media.

Blei, D. M., Ng, A. Y., & Jordan, M. I. (2003). Latent dirichlet allocation. *Journal of Machine Learning Research, 3*, 993–1022.

Boldi, P., Codenotti, B., Santini, M., & Vigna, S. (2002). Structural properties of the African web. In *Proceedings of the 11th International Conference on World Wide Web (WWW 2002)*. IEEE.

Borda, J. (1781). Mémoire sur les élections au scrutin. *Comptes rendus de l'Académie des Sciences, 44*, 42–51.

Brandes, U., & Delling, D. (2006). *On modularity-np-completeness and beyond*. Bibliothek.

Brandes, U., Gaertler, M., & Wagner, D. (2003). Experiments on graph clustering algorithms. [ESA.]. *Proceedings of Algorithms-ESA, 2003*, 568–579. doi:10.1007/978-3-540-39658-1_52

Brin, S., & Page, L. (1998). The anatomy of a large-scale hypertextual web search engine. In *Proceedings of the Seventh International Conference on World Wide Web,* (vol. 7, pp. 107-117). IEEE.

Brin, S., Motwani, R., & Silverstein, C. (1997). Beyond market baskets: generalizing association rules to correlations. *SIGMOD Record, 26*(2), 265–276. doi:10.1145/253262.253327

Broder, A. Z., Glassman, S. C., Manasse, M. S., & Zwerg, G. (1997). Syntactic clustering of the web. *Computer Networks and ISDN Systems, 29*(8-13), 1157-1166.

Broder, A., Kumar, R., Maghoul, F., Raghavan, P., Rajagopalan, S., & Stata, R. … Wiener, J. (2000). Graph structure in the web. In *Proceedings of the 9th International Conference on World Wide Web (WWW 2000)*. IEEE.

Bron, C., & Kerbosch, J. (1973). Algorithm 457: Finding all cliques of an undirected graph. *Communications of the ACM, 16*(9), 575–577. doi:10.1145/362342.362367

Brooks, C. H., & Montanez, N. (2006). Improved annotation of the blogosphere via autotagging and hierarchical clustering. *In Proceedings of the 15th International Conference on World Wide Web*, (pp. 625-632). ACM.

Burke, J., Estrin, D., Hansen, M., & Parker, A. (2006). Participatory sensing. In *Proceedings of ACM Sensys World Sensor Web Workshop (WSW 2006)*. Boulder, CO: ACM Press.

Cai, D., Mei, Q., Han, J., & Zhai, C. (2008). Modeling hidden topics on document manifold. In *Proceedings of ACM 17th Conference on Information and Knowledge Management (CIKM 2008)*. ACM Press.

Cameron, J. J., Leung, C. K.-S., & Tanbeer, S. K. (2011). Finding strong groups of friends among friends in social networks. In J. Chen, S. Lawson, & N. Agarwal (Eds.), *Proceedings of the 2011 International Conference on Social Computing and Its Applications* (pp. 824-831). Los Alamitos, CA: IEEE Computer Society.

Campbell, A. T., Eisenman, S. B., Fodor, K., & Lane, N. D. (2008). Transforming the social networking experience with sensing presence from mobile phones. In *Proceedings of the 6th ACM Conference on Embedded Network Sensor Systems (SenSys 2008)*. New York, NY: ACM Press.

Cao, H., Bao, T., Yang, Q., Chen, E., & Tian, J. (2010). An effective approach for mining mobile user habits. In *Proceedings of the 19th ACM International Conference on Information and Knowledge Management*. Toronto, Canada: ACM Press.

Carenini, G., Ng, R. T., & Zhou, X. (2007). Summarizing email conversations with clue words. In *Proceedings of the World Wide Web Conference Series*, (pp. 91–100). IEEE.

Carrasco, J. J., Fain, D. C., Lang, K. J., & Zhukov, L. (2003). Clustering of bipartite advertiser-keyword graph. In *Proceedings of the Third IEEE International Conference on Data Mining, Workshop on Clustering Large Data Sets*. Melbourne, FL: IEEE Press.

Carter, C. L., Hamilton, H. J., & Cercone, N. (1997). Share based measures for itemsets. In J. Komorowski & J. Zytkow (Eds.), *Proceedings of the First European Symposium on Principles of Data Mining and Knowledge Discovery (PKDD)* (pp. 14-24). Berlin, Germany: Springer.

Chakrabarti, S., Dom, B., Gibson, D., Kumar, R., Raghavan, P., Rajagopalan, S., & Tomkins, A. (1998). Experiments in topic distillation. In *Proceedings of the SIGIR Workshop on Hypertext Information Retrieval on the Web*. ACM.

Chang, J., & Blei, D. (2009). Relational topic models for document networks. In *Proceedings of the 12th International Conference on Artificial Intelligence and Statistics (AISTATS 2009)*. IEEE.

Charif, H. N., & McKenna, S. J. (2006). Tracking the activity of participants in a meeting. *Machine Vision and Applications, 17*(2), 83–93. doi:10.1007/s00138-006-0015-5

Chaumartin, F.-R. (2007). Upar7: A knowledge-based system for headline sentiment tagging. In *Proceedings of the 4th International Workshop on Semantic Evaluations, SemEval 2007*, (pp. 422-425). Morristown, NJ: Association for Computational Linguistics.

Cheeger, J. (1970). A lower bound for the smallest eigenvalue of the Laplacian . In *Problems in Analysis* (pp. 195–199). Princeton, NJ: Princeton University Press.

Chen, H., & Dumais, S. (2000). Bringing order to the web: Automatically categorizing search results. In *Proceedings of the SIGCHI Conference on Human Factors in Computing Systems*, (pp. 145-152). ACM Press.

Chen, J., Geyer, W., Dugan, C., Muller, M. J., & Guy, I. (2009). Make new friends, but keep the old: Recommending people on social networking sites. In *Proceedings of CHI*, (pp. 201–210). CHI.

Chen, J., Zaizne, O., & Goebel, R. (2009). Local community identification in social networks. In *Proceedings of the International Conference on Advances in Social Network Analysis and Mining,* (pp. 237-242). IEEE.

Cheng, D., Kannan, R., Vempala, S., & Wang, G. (2006). A divide-and-merge methodology for clustering. *ACM Transactions on Database Systems, 31*(4), 1499–1525. doi:10.1145/1189769.1189779

Chen, Y.-X., Santaía, R., Butz, A., & Therón, R. (2009). Tagclusters: Semantic aggregation of collaborative tags beyond tagclouds . In Butz, A., Fisher, B. D., Christie, M., Krüger, A., Olivier, P., & Therón, R. (Eds.), *Smart Graphics* (pp. 56–67). Berlin, Germany: Springer. doi:10.1007/978-3-642-02115-2_5

Choi, B., Lee, I., Kim, J., & Jeon, Y. (2005). A qualitative cross-national study of cultural influences on mobile data service design. In *Proceedings of the SIGCHI Conference on Human Factors in Computing Systems 2005 (CHI 2005)*, (pp. 661-670). ACM Press.

Choi, Y., Cardie, C., Riloff, E., & Patwardhan, S. (2005). Identifying sources of opinions with conditional random fields and extraction patterns. In *Proceedings of HLT/EMNLP*. HLT/EMNLP.

Chung, F. R. K. (1997). *CBMS conference on recent advances in spectral graph theory, and conference board of the mathematical sciences: Spectral graph theory*. Academic Press.

Chung, F. R. K. (1997). *Spectral graph theory*. New York, NY: AMS Bookstore.

Clauset, A. (2005). Finding local community structure in networks. *Physical Review E: Statistical, Nonlinear, and Soft Matter Physics, 72*(2). doi:10.1103/PhysRevE.72.026132

Clauset, A., Newman, M. E. J., & Moore, C. (2004). Finding community structure in very large networks. *Physical Review E: Statistical, Nonlinear, and Soft Matter Physics, 70*(6). doi:10.1103/PhysRevE.70.066111

Cohn, D., & Chang, H. (2000). Learning to probabilistically identify authoritative documents. In *Proceedings of the International Conference on Machine Learning (ICML 2000)*. ICML.

Cohn, D., & Hofmann, T. (2001). The missing link - A probabilistic model of document content and hypertext connectivity. In *Proceedings of the International Conference on Neural Information Processing Systems (NIPS 2001)*. NIPS.

Condon, A., & Karp, R. M. (2004). Algorithms for graph partitioning on the planted partition model: Randomization, approximation, and combinatorial optimization. *Algorithms and Techniques, 18*(2), 221–232.

Dai, P., Di, H., Dong, L., Tao, L., & Xu, G. (2009). Group interaction analysis in dynamic context. *IEEE Transactions on Systems, Man, and Cybernetics. Part B, Cybernetics, 39*(1), 34–42. doi:10.1109/TSMCB.2008.2009559

Dalvi, N., Kumar, R., & Pang, B. (2012). Object matching in tweets with spatial models. In E. Adar, J. Teevan, E. Agichtein, & Y. Maarek (Eds.), *Proceedings of the Fifth ACM International Conference on Web Search and Data Mining (WSDM)*, (pp. 43-52). New York, NY: ACM Press.

Danon, L., Diaz-Guilera, A., & Arenas, A. (2006). The effect of size heterogeneity on community identification in complex networks. *Journal of Statistical Mechanics*, 11.

Dao, T., & Simpson, T. (2005). Measuring similarity between sentences. Technical Report. *WordNet*. Retrieved from http://wordnetdotnet.googlecode.com/svn/trunk/Projects/Thanh/Paper/WordNetDotNet_Semantic_Similarity.pdf

Dattolo, A., Eynard, D., & Mazzola, L. (2011). An integrated approach to discover tag semantics. In *Proceedings of the 2011 ACM Symposium on Applied Computing SAC 2011*, (pp. 814-820). ACM.

Dave, K., Lawrence, S., & Pennock, D. M. (2003). Mining the peanut gallery: Opinion extraction and semanctic classification of product reviews. In *Proceedings of the 12th International World Wide Web Conference*. Budapest, Hungary: IEEE.

Davidson, B. D. (2000). Topical locality in the web. In *Proceedings of the 23rd Annual International ACM SIGIR Conference on Research and Development in Information Retrieval (SIGIR 2000)*. ACM Press.

de Marneffe, M. C., Maccartney, B., & Manning, C. D. (2006). Generating typed dependency parses from phrase structure parses. In *Proceedings of LREC*. LREC.

Deerwester, S., Dumais, S., Landauer, T., Furnas, G., & Harshman, R. (1990). Indexing by latent semantic analysis. *Journal of the American Society for Information Science and Technology*, *41*(6), 391–407. doi:10.1002/(SICI)1097-4571(199009)41:6<391::AID-ASI1>3.0.CO;2-9

Dempster, A., Laird, N., & Rubin, D. (1977). Maximum likelihood from incomplete data via the EM algorithm. *Journal of the Royal Statistical Society. Series B. Methodological*, *39*, 1–38.

Dhillon, I. S., Mallela, S., & Modha, D. S. (2003). Information-theoretic co-clustering. In *Proceedings of the 9th ACM SIGKDD International Conference on Knowledge Discovery and Data Mining (KDD 2003)*. ACM Press.

Dhillon, I., Guan, Y., & Kullis, B. (2007). Weighted graph cuts without Eigen vectors: A multilevel approach. *IEEE Transactions on Pattern Analysis and Machine Intelligence*, *29*(11), 1944–1957. doi:10.1109/TPAMI.2007.1115

Dietterich, T. G. (1998). Approximate statistical test for comparing supervised classification learning algorithms. *Neural Computation*, *10*(7), 1895–1923. doi:10.1162/089976698300017197

Dietz, L., Bickel, S., & Scheffer, T. (2007). Unsupervised prediction of citation influences. In *Proceedings of the International Conference on Machine Learning (ICML 2007)*. ICML.

Domingue, J., Galizia, S., & Cabral, L. (2005). Choreography in IRS-III- Coping with heterogeneous interaction patterns in web services. *Lecture Notes in Computer Science*, *3729*, 171–185. doi:10.1007/11574620_15

Donath, W. E., & Hoffman, A. J. (1973). Lower bounds for the partitioning of graphs. *IBM Journal of Research and Development*, *17*(5), 420–425. doi:10.1147/rd.175.0420

Dong, F., & Gou, X. (2009). An advanced location based service on mobile social network. In *Proceedings of the 2nd IEEE International Conference on Broadband Network and Multimedia Technology (IC-BNMT 2009)*. Beijing, China: IEEE Press.

Dongen, S. V. (2000). Graph clustering by flow simulation. *Computer Science Review*, *1*(1), 27–64.

Dourisboure, Y., Geraci, F., & Pellegrini, M. (2007). Extraction and classification of dense communities in the web. In *Proceedings of 16th International Conference on World Wide Web*, (pp. 461-470). ACM.

Dragone, M., Duffy, B. R., & O'Hare, G. M. P. (2005). Social interaction between robots, avatars & humans. In *Proceedings of 2005 IEEE International Workshop on Robots and Human Interactive Communication (RO-MAN 2005)*. Nashville, TN: IEEE.

Dredze, M., Wallach, H. M., Puller, D., & Pereira, F. (2008). Generating summary keywords for emails using topics. In *Proceedings of the 13th International Conference on Intelligent User Interfaces*, (pp. 199-206). IEEE.

Dubhashi, D., Laura, L., & Panconesi, A. (2003). Analysis and experimental evaluation of a simple algorithm for collaborative filtering in planted partition models . In *Proceedings of Foundations of Software Technology and Theoretical Computer Science* (pp. 168–182). IEEE. doi:10.1007/978-3-540-24597-1_15

Durao, F., & Dolog, P. (2010). *Extending a hybrid tag-based recommender system with personalization.* Paper presented at the the 25th ACM Symposium on Applied Computing. Sierre, Switzerland.

Eagle, N., Clauset, A., & Quinn, J. A. (2009). *Location segmentation, inference and prediction for anticipatory computing*. Retrieved from http://reality.media.mit.edu/pdfs/anticipatory.pdf

Eagle, N., Pentland, A., & Lazer, D. (2009). Inferring friendship network structure by using mobile phone data. *Proceedings of the National Academy of Sciences of the United States of America*, *106*(36), 15274–15278. doi:10.1073/pnas.0900282106

Edwin, S. (2008). *Clustering tags in enterprise and web folksonomies.* Paper presented at the International Conference on Weblogs and Social Media. Washington, DC.

Ehrig, M., Staab, S., & Sure, Y. (2005). Bootstrapping ontology alignment methods with APFEL. *Lecture Notes in Computer Science*, *3729*, 186–200. doi:10.1007/11574620_16

Erosheva, E., Fienberg, S., & Lafferty, J. (2004). Mixed membership models of scientific publications. *Proceedings of the National Academy of Sciences of the United States of America*, *101*, 5220–5227. doi:10.1073/pnas.0307760101

Facebook. (2011). *Anna Hazare - Movement against corruption*. Retrieved 17 September, 2011 from http://www.facebook.com/AnnaHazareMovement

Fan, R. E., Chang, K.-W., Hsieh, C.-J., Wang, X.-R., & Lin, C.-J. (2008). Liblinear: A library for large linear classification. *Journal of Machine Learning Research*, *9*, 1871–1874.

Felipe, S. M., Juan, A. P. O., & Mikel, L. F. (2006). Target-language-driven agglomerative part-of-speech tagging clustering for machine translation. *Advances in Artificial Intelligence*, *4239*, 844–854.

Fiedler, M. (1973). Algebraic connectivity of graphs. *Czechoslovak Mathematical Journal*, *23*(2), 298–305.

Filatova, E., & Hatzivassiloglou, V. (2004). A formal model for information selection in multi-sentence text extraction. In *Proceedings of the 20th International Conference on Computational Linguistics*, (p. 397). IEEE.

Flake, G. W., Lawrence, S., & Giles, C. L. (2000). Efficient in social and biological networks. In *Proceedings of the Sixth ACM SIGKDD International Conference on Knowledge Discovery and Data Mining*, (p. 160). ACM Press.

Flake, G. W., Lawrence, S., et al. (2000). Efficient identification of web communities. In *Proceedings of the 6th ACM SIGKDD International Conference on Knowledge Discovery and Data Mining*, (pp. 150-160). ACM Press.

Flake, G. W., & Lawrence, S. (2002). Self-organization and Identification of web communities. *Computer*, *35*, 66–70. doi:10.1109/2.989932

Flake, G. W., Lawrence, S., Giles, C. L., & Coetzee, F. M. (2002). Self-organization and identification of web communities. *Computer*, *35*(3), 66–70. doi:10.1109/2.989932

Flake, G. W., Tarjan, R. E., & Tsioutsiouliklis, K. (2004). Graph clustering and minimum cut trees. *Internet Mathematics*, *1*(4), 385–408. doi:10.1080/15427951.2004.10129093

Ford, L. R., & Fulkerson, D. R. (2008). Maximal flow through a network. *Canadian Journal of Mathematics*, *8*(3), 399–404.

Frawley, W. J., Piatetsky-Shapiro, G., & Matheus, C. J. (1991). Knowledge discovery in databases: An overview. In Piatetsky-Shapiro, G., & Frawley, W. J. (Eds.), *Knowledge Discovery in Databases* (pp. 1–30). Cambridge, MA: The MIT Press.

Freeman, L. C. (1977). A set of measures of centrality based on betweenness. *Sociometry*, *40*(1), 35–41. doi:10.2307/3033543

Gabrielsson, S., & Gabrielsson, S. (2006). *The use of self-organizing maps in recommender systems*. (Master's Thesis). Uppsala University. Uppsala, Sweden.

Gallo, G., Grigoriadis, M. D., & Tarjan, R. E. (1989). A fast parametric maximum flow algorithm and applications. *SIAM Journal on Computing*, *18*(1), 30–55. doi:10.1137/0218003

Gaonkar, S., Li, J., & Choudhury, R. R. (2008). Micro-blog: Sharing and querying content through mobile phones and social participation. In *Proceedings of MobiSys 2008*. Breckenridge, CO: MobiSys.

Garcia-Silva, A., Corcho, O., Alani, H., & Gomez-Perez, A. (2011). Review of the state of the art: Discovering and associating semantics to tags in folksonomies. *The Knowledge Engineering Review*, *26*(4).

Garey, M. R., & Johnson, D. S. (1979). *Computers and intractability: A guide to the theory of NP-completeness*. New York, NY: W. H. Freeman.

Gemmell, J., Shepitsen, A., Mobasher, M., & Burke, R. (2008). *Personalization in folksonomies based on tagging clustering*. Paper presented at the the 6th Workshop on Intelligent Techniques for Web Personalization and Recommender Systems. Chicago, IL.

Giannakidou, E., Koutsonikola, V., Vakali, A., & Kompatsiaris, Y. (2008). *Co-clustering tags and social data sources*. Paper presented at the the 9th International Conference on Web-Age Information Management. Zhangjiajie,China.

Gibson, D., Kleinberg, J., & Raghavan, J. (1998). Inferring web communities from link topology. In *Proceedings of the 9th ACM Conference on Hypertext and Hypermedia: Links, Objects, Time and Space*, (pp. 225-234). New York, NY: ACM.

Gibson, D., Kumar, R., & Tomkins, A. (2005). Discovering large dense sub-graphs in massive graphs. In *Proceedings of 31st International Conference on very Large Data Bases*, (pp. 721-732). ACM.

Gibson, D., Kumar, R., McCurley, K. S., & Tomkins, A. (2006). Dense subgraph extraction . In *Mining Graph Data* (pp. 411–441). New York, NY: Wiley. doi:10.1002/9780470073049.ch16

Girvan, M., & Newman, M. E. J. (2002). Community structure in social and biological networks. *Proceedings of the National Academy of Sciences of the United States of America*, *99*(12). doi:10.1073/pnas.122653799

Gkantsidis, C., Mihail, M., & Saberi, A. (2003). Conductance and congestion in power law graphs. In *Proceedings of the 2003 ACM SIGMETRICS International Conference on Measurement and Modeling of Computer Systems*, (pp. 148–159). ACM Press.

Golder, S. A., & Huberman, B. A. (2006). The structure of collaborative tagging systems. *Journal of Information Science*, *32*(2), 198–208. doi:10.1177/0165551506062337

Goldstein, J., Mittal, V., Carbonell, J., & Kantrowitz, M. (2000). Multi-document summarization by sentence extraction. In *Proceedings of the ANLP/NAACL Workshop on Automatic Summarization*, (pp. 40-48). ANLP.

Gong, S., Qu, Y., & Tian, S. (2010). Summarization using Wikipedia. In *Proceedings of Text Analysis Conference*. IEEE.

Górecki, J., Slaninová, K., & Snášel, V. (2011). Visual investigation of similarities in global terrorism database by means of synthetic social networks. In A. Abraham, E. Corchado, R. Alhaj, & V. Snášel (Eds.), *Proceedings of the 2011 International Conference on Computational Aspects of Social Networks (CASoN)*, (pp. 255-260). Los Alamitos, CA: IEEE Computer Society.

Gou, L., Kim, J. H., & Chen, H. H. (2009). MobiSNA: A mobile video social network application . In *Proceedings of MobiDE 2009*. Providence, RI: MobiDE.

Granovetter, M. S. (1973). The strength of weak ties. *American Journal of Sociology*, *78*(6), 1360–1380. doi:10.1086/225469

Griffiths, T. L., & Steyvers, M. (2002). A probabilistic approach to semantic representation. In *Proceedings of the 24th Annual Conference of the Cognitive Science Society*. IEEE.

Griffiths, T. L., & Steyvers, M. (2003). Prediction and semantic association . In *Neural Information Processing Systems*. Cambridge, MA: MIT Press.

Griffiths, T. L., & Steyvers, M. (2004). Finding scientific topics. *Proceedings of the National Academy of Sciences of the United States of America*, *101*, 5228–5235. doi:10.1073/pnas.0307752101

Grisword, W. G., Shanahan, P., & Brown, S. W. (2004). ActivieCampus: Experiments in community-oriented ubiquitous computing. *IEEE Computer*, *37*(10), 73–81. doi:10.1109/MC.2004.149

Groh, G. (2007). Recommendations in taste related domains: Collaborative filtering vs. social filtering. In *Proceedings of ACM Group 2007*, (pp. 127-136). ACM Press.

Gruber, A., Rosen-Zvi, M., & Weiss, Y. (2008). Latent topic models for hypertext. In *Proceedings of the 24th Conference on Uncertainty in Artificial Intelligence (UAI 2008)*. UAI.

Gruhl, D., Guha, R., Kumar, R., Novak, J., & Tomkins, A. (2005). The predictive power of online chatter. In *Proceedings of the Eleventh ACM SIGKDD International Conference on Knowledge Discovery in Data Mining*, (pp. 78–87). ACM Press.

Guan, Z. Y., Bu, J. J., Mei, Q. Z., Chen, C., & Wang, Q. (2009). *Personalized tag recommendation using graph-based ranking on multi-type interrelated objects*. Paper presented at the the 32nd International ACM SIGIR Conference on Research and Development in Information Retrieval. New York, NY.

Guan, Z., Wang, C., Bu, J., Chen, C., Yang, K., Cai, D., et al. (2010). Document recommendation in social tagging services. In *Proceedings of WWW*, (pp. 391-400). IEEE.

Guattery, S., & Miller, G. L. (1998). On the quality of spectral separators. *SIAM Journal on Matrix Analysis and Applications*, *19*(3), 701–719. doi:10.1137/S0895479896312262

Guha, S., Mishra, N., Motwani, R., & O'Callaghan, L. (2000). Clustering data streams. In *Proceedings of the 41st Annual Symposium on Foundations of Computer Science*, (p. 359). IEEE.

Gupta, M., Li, R., Yin, Z., & Han, J. (2010). Survey on social tagging techniques. *SIGKDD Explorations*, *12*(1), 58–72. doi:10.1145/1882471.1882480

Halawani, T., & Mohands, M. (2003). Smart card for smart campus: KFUPM case study. In *Proceedings of 10th IEEE Internal Conference on Electronics, Circuits and Systems (ICECS 2003)*. Sharjah, United Arab Emirates: ICECS.

Han, B., & Baldwin, T. (2011). Lexical normalisation of short text messages: Maknsens a #twitter. In *Proceedings of ACL*, (pp. 368–378). ACL.

Han, J., Pei, J., & Yin, Y. (2000). Mining frequent patterns without candidate generation. In W. Chen, J. Naughton, & P. A. Bernstein (Eds.), *Proceedings of the 2000 ACM SIGMOD International Conference on Management of Data,* (pp. 1-12). New York, NY: ACM Press.

Harb, A., Planitié, M., Poncelet, P., Roche, M., & Trousset, F. (2008). *Détection d'opinions: Apprenons les bons adjectifs*. Paper presented at Actes de l'Atelier Fouille des Données d'Opinions, conjointement Conférence INFORSID 08. Fontainebleau, France.

Hartuv, E., & Shamir, R. (2000). A clustering algorithm based on graph connectivity. *Information Processing Letters*, *76*(4-6), 175–181. doi:10.1016/S0020-0190(00)00142-3

Hayes, C., & Avesani, P. (2007). Using tags and clustering to identify topic-relevant blogs. In *Proceedings of the International Conference on Weblogs and Social Media*. IEEE.

Heep, M., Bachlechner, D., & Siopaes, K. (2006). *Harvesting wiki consensus - Using Wikipedia entries as ontology elements.* Paper presented at the SemWiki2006. Budva, Montenegro.

Heinrich, G. (2009). *Parameter estimation for text analysis. Technical Report*. Darmstadt, Germany: Fraunhofer IGD.

Herlocker, J. L., Konstan, J. A., Terveen, L. G., & Riedl, J. T. (2004). Evaluating collaborative filtering recommender systems. *ACM Transactions on Information Systems*, *22*(1), 5–53. doi:10.1145/963770.963772

Hermes, L., & Buhmann, J. M. (2003). A minimum entropy approach to adaptive image polygonization. *IEEE Transactions on Image Processing*, *12*(10), 1243–1258. doi:10.1109/TIP.2003.817240

Herring, S. C., Kouper, I., Paolillo, J. C., Scheidt, L. A., Tyworth, M., Welsch, P., et al. (2005). Conversations in the blogosphere: An analysis. In *Proceedings of the Thirty-Eighth Hawai'i International Conference on System Sciences (HICSS)*. IEEE.

He, X., & Zha, H. (2002). Web document clustering using hyperlink structures. *Computational Statistics & Data Analysis*, *41*(1), 19–45. doi:10.1016/S0167-9473(02)00070-1

Heymann, P., Ramage, D., & Garcia-Molina, H. (2008). Social tag prediction. In *Proceedings of the 31st Annual International ACM SIGIR Conference on Research and Development in Information Retrieval*, (pp. 531-538). ACM Press.

Hodgson, M. (2003). Classtalk system for predicing and visualizing speech in noise in classrooms. *Canadian Acoustics*, *31*(3), 62–63.

Hoffman, M., Blei, D., & Bach, F. (2010). On-line learning for latent dirichlet allocation. *Neural Information Processing Systems*. Retrieved from http://www.cs.princeton.edu/~blei/papers/HoffmanBleiBach2010b.pdf

Hofmann, T. (1999). Probabilistic latent semantic analysis. In *Proceedings of the Fifteenth Conference on Uncertainty in Artificial Intelligence*. IEEE.

Hofmann, T. (2001). Unsupervised learning by probabilistic latent semantic analysis. *Machine Learning Journal*, *42*(1), 177–196. doi:10.1023/A:1007617005950

Hofmann, T., Puzicha, J., & Jordan, M. I. (1999). Advances in Neural Information Processing Systems: *Vol. 11*. *Unsupervised learning from dyadic data*. Cambridge, MA: MIT Press.

Hong, L., Ahmed, A., Gurumurthy, S., Smola, A. J., & Tsioutsiouliklis, K. (2012). Discovering geographical topics in the twitter stream. In A. Mille, F. Gandon, J. Misselis, M. Rabinovich, & S. Staab (Eds.), *Proceedings of the 21st International World Wide Web Conference (WWW)*, (pp. 769-778). New York, NY: ACM Press.

Hotho, A., Jschke, R., Schmitz, C., & Stumme, G. (2006). Folkrank: A ranking algorithm for folksonomies. In *Proceedings of FGIR 2006*. FGIR.

Hu, D. H., Zhang, X. X., Yin, J., Zheng, V. W., & Yang, Q. (2010). Abnormal activity recognition based on HDP-HMM models. Retrived from http://www.cse.ust.hk/~vincentz/ijcai09_abnormalAR.pdf

Hu, M., & Liu, B. (2004). Mining and summarizing customer reviews. In *Proceedings of the Tenth ACM SIGKDD International Conference on Knowledge Discovery and Data Mining,* (pp. 168–177). ACM Press.

Hummel, H. G., Berg, B. V., Berlanga, A. J., Drachsler, H., Janssen, J., & Nadolski, R. (2007). Combining social-based and information-based approaches for personalized recommendation on sequencing learning activities. *International Journal of Learning Technology*, *3*(2), 152–168. doi:10.1504/IJLT.2007.014842

Ino, H., Kudo, M., & Nakamura, A. (2005). Partitioning of web graphs by community topology. In *Proceedings of the 14th International Conference on World Wide Web (WWW 2005)*. IEEE.

Internet World Stats, U., & Statistics, P. (2011). *Website.* Retrieved 17 September, 2011 from http://www.internetworldstats.com/stats.htm

James, C. L., & Reischel, K. M. (2001). Text input for mobile devices: Comparing model prediction to actual performance. In *Proceedings of CHI 2001,* (pp. 365-371). ACM.

Jansen, B. J., Zhang, M., Sobel, K., & Chowdury, A. (2009). Twitter power: Tweets as electronic word of mouth. *Journal of the American Society for Information Science and Technology*, *60*, 2169–2188. doi:10.1002/asi.21149

Jaroszewicz, S., & Simovici, D. A. (2004). Interestingness of frequent itemsets using Bayesian networks as background knowledge. In *Proceedings of the Tenth ACM SIGKDD International Conference on Knowledge Discovery and Data Mining*, (pp. 178-186). ACM Press.

Jaschk, R., Marinho, L., Hotho, A., Schmidt-Thiem, L., & Stumme, G. (2007). *Tag recommendations in folksonomies.* Paper presented at the the 11th European Conference on Principles and Practice of Knowledge Discovery in Databases. Athens, Greece.

Jiang, Y. X., Tang, C. J., Xu, K. K., et al. (2009). *Core-tagging clustering for Web2.0 based on multi-similarity measurements.* Paper presented at the APWeb/WAIM Workshops. New York, NY.

Kambhampati, S. (1997, Summer). Refinement planning as a unifying framework for plan synthesis. *AI Magazine*, 67–97.

Kamps, J., Marx, M., Mokken, R. J., & de Rijke, M. (2004). Using wordnet to measure semantic orientation of adjectives. [). LREC.]. *Proceedings of LREC*, *4*, 174–181.

Kamvar, M., & Baluja, S. (2006). A large scale study of wireless search behavior: Google mobile search. In *Proceedings of CHI 2006,* (pp. 701-709). ACM Press.

Kannan, R., Vempala, S., & Vetta, A. (2004). On clustering: Good bad and spectral. *Journal of the ACM*, *51*(3), 497–515. doi:10.1145/990308.990313

Kaplan, A. M., & Haenlein, M. (2010). Users of the world, unite! The challenges and opportunities of social media. *Business Horizons*, *53*(1), 59–68. doi:10.1016/j.bushor.2009.09.003

Karydis, I., Nanopoulos, A., Gabriel, H. H., & Spiliopoulou, M. (2009). *Tag-aware spectral clustering of music items.* Paper presented at the the 10th International Society for Music Information Retrieval Conference. Kobe, Japan.

Karypis, G., & Kumar, V. (1998). A fast and high quality multilevel scheme for partitioning irregular graphs. *SIAM Journal on Scientific Computing*, *20*(1), 359–392. doi:10.1137/S1064827595287997

Karypis, G., & Kumar, V. (1999). Parallel multilevel k-way partitioning for irregular graphs. *SIAM Review*, *41*(2), 278–300. doi:10.1137/S0036144598334138

Katharina, S., & Martin, H. (2007). *Folksontology: An integrated approach for turning folksonomies into ontologies.* Paper presented at the Bridging the Gap between Semantic Web and Web 2.0 SemNet 2007. Innsbruck, Austria.

Kelsey, T. (2010). *Social networking spaces: From Facebook to Twitter and everything in between.* New York, NY: Apress.

Kernighan, B. W., & Lin, S. (1970). An efficient heuristic procedure for partitioning graphs. *The Bell System Technical Journal, 49*(2), 291–307.

Khandekar, R., Rao, S., & Vazirani, U. (2009). Graph partitioning using single commodity flows. *Journal of the ACM, 56*(4), 1–15. doi:10.1145/1538902.1538903

Kim, E., Plummer, M., Hiltz, S. R., & Jones, Q. (2007). Perceived benefits and concerns of prospective users of the SmartCampus location-aware community system test-bed. In *Proceedings of the 40th Annual Hawaii International Conference on System Sciences*. Big Island, HI: IEEE.

Kim, S.-M., & Hovy, E. (2004). Determining the sentiment of opinions. In *Proceedings of the 20th International Conference on Computational Linguistics*, (p. 1367). Morristown, NJ: Association for Computational Linguistics.

Kirkpatrick, S., Gelatt, Jr., et al. (1983). Optimization by simulated annealing. *Science, 220*(4598), 671–679. doi:10.1126/science.220.4598.671

Kleinberg, J. M. (1999). Authoritative sources in a hyperlinked environment. *Journal of the ACM, 46*(5), 604–632. doi:10.1145/324133.324140

Kohonen, T. (1990). The self-organizing map. *Proceedings of the IEEE, 78*(9), 1464–1480. doi:10.1109/5.58325

Konstas, I., Stathopoulos, V., & Jose, J. M. (2009). On social networks and collaborative recommendation. In *Proceedings of the 32nd International ACM SIGIR Conference on Research and Development in Information Retrieval (SIGIR 2009)*, (pp. 195-202). ACM Press.

Kontonasios, K., & De Bie, T. (2010). An information-theoretic approach to finding informative noisy tiles in binary databases. In *Proceedings of the SIAM International Conference on Data Mining*. SIAM.

Kritikopoulos, A., Sideri, M., & Varlamis, I. (2006). BlogRank: Ranking weblogs based on connectivity and similarity features. In *Proceedings of the 2nd International Workshop on Advanced Architectures and Algorithms for Internet Delivery and Applications,* (p. 8). IEEE.

Krovetz, R. (1993). Viewing morphology as an inference process. [ACM Press.]. *Proceedings of SIGIR, 1993,* 191–202.

Kumar, R., Raghavan, P., Rajagopalan, S., &Tomkins, A. (1999). Trawling the web for emerging cyber-communities. *Computer Networks, 31*(11-16), 1481-1493.

Kumar, R., Novak, J., Raghavan, P., & Tomkins, A. (2005). On the bursty evolution of blogspace. *World Wide Web (Bussum), 8*(2), 159–178. doi:10.1007/s11280-004-4872-4

Kuter, U., Sirin, E., Nau, D., Parsia, B., & Hendler, J. (2005). Information gathering during planning for web service composition. *Journal of Web Semantics, 3*(2-3), 183–205. doi:10.1016/j.websem.2005.07.001

Kwak, H., Lee, C., Park, H., & Moon, S. B. (2010). What is twitter, a social network or a news media? In *Proceedings of WWW*, (pp. 591–600). IEEE.

Laasonen, K. (2005). *Route prediction from cellular data.* Paper presented at the Workshop on Context-Awareness for Proactive Systems (CAPS). New York, NY.

Laasonen, K. (2005). Clustering and prediction of mobile user routes from cellular data. *Knowledge Discovery in Databases, 3721*, 569–576.

Laasonen, K., Raento, M., & Toivonen, H. (2004). Adaptive on-device location recognition. *Pervasive Computing, 3001*, 287–304. doi:10.1007/978-3-540-24646-6_21

Lacy, S. (2008). *The stories of Facebook, YouTube & MySpace: The people, the hype and the deals behind the giants of web 2.0.* Richmond, UK: Crimson Publishing.

Lang, K., & Rao, S. (2004). A flow-based method for improving the expansion or conductance of graph cuts. *Integer Programming and Combinatorial Optimization,* 383–400.

Langville, A. N., & Meyer, C. D. (2006). *Google's PageRank and beyond: The science of search engine rankings.* Princeton, NJ: Princeton University Press.

Lappas, T., Arai, B., Platakis, M., Kotsakos, D., & Gunopulos, D. (2009). On burstiness-aware search for document sequences. In *Proceedings of the 15th ACM SIGKDD International Conference on Knowledge Discovery and Data Mining*, (pp. 477-486). ACM Press.

Lavrenko, V., & Croft, W. B. (2001). Relevance based language models. [ACM Press.]. *Proceedings of SIGIR, 2001*, 120–127.

Lee, C., Nordstedt, D., & Helal, S. (2003). Enabling smart spaces with OSGi. *IEEE Pervasive Computing / IEEE Computer Society [and] IEEE Communications Society*, *2*(3), 89–95. doi:10.1109/MPRV.2003.1228530

Lehwark, P., Risi, S., & Ultsch, A. (2007). Visualization and clustering of tagged music data. In *Proceedings 31st Annual Conference of the German Classification Society*, (pp. 673-680). Berlin, Germany: Springer.

Lesk, M. (1986). Automatic sense disambiguation using machine readable dictionaries: How to tell a pine code from an ice cream cone. In *Proceedings of the 5th Annual International Conference on Systems Documentation*, (pp. 24-26). New York, NY: ACM Press.

Leskovec, J., Lang, K. J., & Mahoney, M. W. (2010). Empirical comparison of algorithms for network community detection. In *Proceedings of the 19th International Conference on World Wide Web (WWW 2010)*. IEEE.

Leung, C. K.-S., & Carmichael, C. L. (2010). Exploring social networks: A frequent pattern visualization approach. In J. Zhan (Ed.), *Proceedings of the Second IEEE International Conference on Social Computing (SocialCom)*, (pp. 419-424). Los Alamitos, CA: IEEE Computer Society.

Leung, C. K.-S., & Carmichael, C. L. (2011). iVAS: An interactive visual analytic system for frequent set mining. In Q. Zhang, R. S. Segall, & M. Cao (Eds.), *Visual Analytics and Interactive Technologies: Data, Text and Web Mining Applications*, (pp. 213-231). Hershey, PA: IGI Global.

Leung, C. K.-S., & Tanbeer, S. K. (2012). Mining social networks for significant friend groups. In P.-N. Tan, S. Chawla, C. K. Ho, & J. Bailey (Eds.), *Proceedings of the Third International Workshop on Social Networks and Social Web Mining* (pp. 180-192). Berlin, Germany: Springer.

Leung, C. K.-S., Carmichael, C. L., & Teh, E. W. (2011). Visual analytics of social networks: Mining and visualizing co-authorship networks. In D. Schmorrow & C. M. Fidopiastis (Eds.), *Proceedings of the Sixth International Conference of Foundations of Augmented Cognition* (pp. 335-345). Berlin, Germany: Springer.

Leung, C. K.-S. (2009). Constraint-based association rule mining . In Wang, J. (Ed.), *Encyclopedia of Data Warehousing and Mining* (2nd ed., pp. 307–312). Hershey, PA: IGI Global.

Levenshtein, V. I. (1966). Binary codes capable of correction deletions, insertions, and reversals. *Soviet Physics, Doklady*, *10*(8), 707–710.

Li, B., & Zhu, Q. (2008). The determination of semantic dimension in social tagging system based on some model. In *Proceedings of the Second International Symposium on Intelligent Information Technology Application, IITA 2008*, (pp. 909-913). IEEE.

Li, K. A., Sohn, T. Y., Huang, S., & Griswold, G. (2008). PeopleTones: A system for the detection and notification of buddy proximity on mobile phones. In *Proceedings of the 6th International Conference on Mobile System, Application and Services, (MobiSys 2008)*. New York, NY: MobiSys.

Li, L., Xiao, H., & Xu, G. (2012). Finding related microblogs based on WordNet. In P.-N. Tan, S. Chawla, C. K. Ho, & J. Bailey (Eds.), *Proceedings of the Third International Workshop on Social Networks and Social Web Mining* (pp. 115-122). Berlin, Germany: Springer.

Li, L., Yang, Z., Liu, L., & Kitsuregawa, M. (2008). Query-url bipartite based approach to personalized query recommendation. In *Proceedings of AAAI*, (pp. 1189-1194). AAAI.

Li, Q., Zheng, Y., Xie, X., Chen, Y., Liu, W., & Ma, W. Y. (2008). Mining user similarity based on location history. In *Proceedings of the 16th ACM SIGSPATIAL International Conference on Advances in Geographic Information Systems*. Irvine, CA: ACM Press.

Li, X., Guo, L., & Zhao, E. (2008). *Tag-based social interest discovery*. Paper presented at the the 17th International Conference on World Wide Web. Beijing,China.

Li, L., Otsuka, S., & Kitsuregawa, M. (2010). Finding related search engine queries by web community based query enrichment. *World Wide Web (Bussum)*, *13*(1-2), 121–142. doi:10.1007/s11280-009-0077-1

Lin, C., & Hovy, E. (2003). Automatic evaluation of summaries using N-gram co-occurrence statistics. In *Proceedings of the Conference of the North American Chapter of the Association for Computational Linguistics on Human Language Technology*, (vol. 1, pp. 71-78). ACL.

Li, Q., Wang, J., Chen, Y. P., & Lin, Z. (2010). User comments for news recommendation in forum-based social media. *Information Sciences*, *180*(24), 4929–4939. doi:10.1016/j.ins.2010.08.044

Liu, H., & Maes, P. (2005). *Interestmap: Harvesting social network profiles for recommendations*. Paper presented at Beyond Personalization. Los Angeles, CA.

Liu, H., Lafferty, J., & Wasserman, L. (2007). Sparse nonparametric density estimation in high dimensions using the rodeo. *Journal of Machine Learning Research*. Retrieved from http://www.cs.cmu.edu/~lafferty/pub/drodeo_aistats.pdf

Liu, B. (2007). *Web data mining: Exploring hyperlinks, contents, and usage data*. Berlin, Germany: Springer.

Liu, G., & Wong, L. (2008). Effective pruning techniques for mining quasi-cliques. *Lecture Notes in Computer Science*, *5222*, 33–49. doi:10.1007/978-3-540-87481-2_3

Longstaff, B., Reddy, S., & Estrin, D. (2010). *Improving activity classification for health applications on mobile devices using active and semi-supervised learning*. Paper presented at the Pervasive Computing Technologies for Healthcare (PervasiveHealth), 2010 4th International Conference on NO PERMISSIONS. New York, NY.

Luo, F., Wang, J. Z., & Promislow, E. (2008). Exploring local community structures in large networks. *Web Intelligence and Agent Systems*, *6*(4), 387–400.

Macdonald, C., & Ounis, I. (2006). *The TREC Blog06 collection: Creating and analysing a blog test collection. Technical Report*. Glasgow, UK: University of Glasgow.

Makino, K., & Uno, T. (2004). New algorithms for enumerating all maximal cliques. *Algorithm Theory – SWAT 2004*, (pp. 260-272). Retrieved from http://research.nii.ac.jp/~uno/papers/04swat.pdf

Malone, T. W., Crowston, K., & Herman, G. A. (2003). *Organizing business knowledge - The MIT process hand book*. Cambridge, MA: MIT Press.

Mampaey, M., Tatti, N., & Vreeken, J. (2011). Tell me what I need to know: Succinctly summarizing data with itemsets. In *Proceedings of the 17th ACM SIGKDD Conference on Knowledge Discovery and Data Mining*. ACM Press.

Marco, L. S., & Edwin, S. (2009). *Tagging clustering with self organizing maps.*. Retrieved from.http://www.hpl.hp.com/techreports/2009/HPL-2009-338.pdf

Marples, D., & Kriens, P. (2001). The open services gateway initiative: An introductory review. *IEEE Communications Magazine*, *39*(12), 110–114. doi:10.1109/35.968820

Martineau, J., & Finin, T. (2009). Delta TFIDF: An improved feature space for sentiment analysis. In *Proceedings of the Third AAAI Internatonal Conference on Weblogs and Social Media*. San Jose, CA: AAAI Press.

Martin, J. R., & White, P. R. R. (2005). *The language of evaluation: Appraisal in English*. New York, NY: Palgrave Macmillan.

Masr, B. (2009). *Stop, look, what's that sound – The death of Egyptian activism*. Retrieved February 8, 2009 from http://bikyamasr.wordpress.com/2009/08/02/bm-opinion-stop-look-whats-that-sound-the-death-of-egyptian-activism/

Masuoka, R., Parsia, B., & Labrou, Y. (2003). Task computing - The semantic web meets pervasive computing. *Lecture Notes in Computer Science*, *2870*, 866–881. doi:10.1007/978-3-540-39718-2_55

Matera, M. (2009). Social applications . In Liu, L., & Özsu, M. T. (Eds.), *Encyclopedia of Database Systems* (p. 2667). New York, NY: Springer.

Mathioudakis, M., & Koudas, N. (2010). TwitterMonitor: Trend detection over the twitter stream. In *Proceedings of the 2010 International Conference on Management of Data*, (pp. 1155-1158). IEEE.

Matsumoto, S., Takamura, H., & Okumura, M. (2005). Sentiment classification using word subsequences and dependency sub-trees. In *Proceedings of the Pacific-Asia Conference on Knowledge Discovery and Data Mining*. IEEE.

Meena, A., & Prabhakar, T. V. (2007). Sentence level sentiment analysis in the presence of conjuncts using linguistic analysis. *Lecture Notes in Computer Science*, *4425*, 573–580. doi:10.1007/978-3-540-71496-5_53

Mei, Q., Deng, C., Zhang, D., & Zhai, C. (2008). Topic modeling with network regularization. In *Proceedings of the 17th International Conference on World Wide Web (WWW 2008)*. IEEE.

Meneses, F., & Moreira, A. (2006) *Using GSM CellID positioning for place discovering*. Paper presented at the Pervasive Health Conference and Workshops. New York, NY.

Metzler, D., Dumais, S. T., & Meek, C. (2007). Similarity measures for short segments of text. In *Proceedings of ECIR*, (pp. 16–27). ECIR.

Miao, G., Tatemura, J., Hsiung, W., Sawires, A., & Moser, L. (2009). *Extracting data records from the web using tag path clustering*. Paper presented at the the 18th International Conference on World Wide Web. New York, NY.

Miao, Y., & Li, C. (2010). WikiSummarizer - A Wikipedia-based summarization system. In *Proceedings of Text Analysis Conference*. IEEE.

Mihail, M., Gkantisidis, C., & Saberi, A. (2002). *On the semantics of internet topologies*. Atlanta, GA: Georgia Institute of Technology.

Mika, P. (2007). Ontologies are us: A unified model of social networks and semantics. *Journal of Web Semantics*, *5*(1), 5–15. doi:10.1016/j.websem.2006.11.002

Milgram, S. (1967). The small world problem. *Psychology Today*, *2*(1), 60–67.

Miluzzo, E. (2008). Supporting personal sensing presence in social networks: Implementation, evaluation and user experience of the CenceMe application using mobile phones. In *Proceedings of the 6th ACM Conference on Embedded Network Sensor Systems (SenSys 2008)*. New York, NY: ACM Press.

Mirisaee, S. H., Noorzaden, S., & Sami, A. (2010). Mining friendship from cell phone switch data. In *Proceedings of the 3rd International Conference on Human-Centric Computing (HumanCom 2010)*. HumanCom.

Mishne, G., & de Rijke, M. (2006). Deriving wishlists from blogs show us your blog, and we'll tell you what books to buy. In *Proceedings of the 15th International Conference on World Wide Web*, (pp. 925–926). IEEE.

Motahari-Nexhad, H. R., Martens, A., Curbera, F., & Casati, F. (2007). Semi-automated adaptation of service interactions. [IEEE.]. *Proceedings of WWW*, *2007*, 993–1002.

Naganuma, T., & Kurakake, S. (2005). Task knowledge based retrieval for services relevant to mobile user's activity. [ISWC.]. *Proceedings of the ISWC*, *2005*, 959–973.

Nakagawa, T., Inui, K., & Kurohashi, S. (2010). Dependency tree-based sentiment classification using crfs with hidden variables. In *Proceedings of the Human Language Technologies: The 2010 Annual Conference of the North American Chapter of the Association for Computational Linguistics, HLT 2010*, (pp. 786-794). Morristown, NJ: Association for Computational Linguistics.

Nakanishi, H. (2004). FreeWalk: A social interaction platform for group behaviour in a virtual space. *International Journal of Human-Computer Studies*, *60*(4), 421–454. doi:10.1016/j.ijhcs.2003.11.003

Nallapati, R. M., Ahmed, A., Xing, E. P., & Cohen, W. W. (2008). Joint latent topic models for text and citations. In *Proceedings of The 14th ACM SIGKDD International Conference on Knowledge Discovery and Data Mining (KDD 2008)*. ACM Press.

Nastase, V. (2008). Topic-driven multi-document summarization with encyclopedic knowledge and spreading activation. In *Proceedings of Conference on Empirical Methods on Natural Language Processing*, (pp. 763-772). IEEE.

Nepali Blogger. (2011). *Bloggers meet up in Kathmandu*. Retrieved 18 September, 2011 from http://nepaliblogger.com/bloggers/bloggers-meet-up-in-kathmandu/1174/

Newman, M. E. J. (2003). Mixing patterns in networks. *Physical Review E: Statistical, Nonlinear, and Soft Matter Physics*, *67*(2). doi:10.1103/PhysRevE.67.026126

Newman, M. E. J. (2004b). Fast algorithm for detecting community structure in networks. *Physical Review E: Statistical, Nonlinear, and Soft Matter Physics*, *69*(6). doi:10.1103/PhysRevE.69.066133

Newman, M. E. J. (2004c). Analysis of weighted networks. *Physical Review E: Statistical, Nonlinear, and Soft Matter Physics*, *70*(5). doi:10.1103/PhysRevE.70.056131

Newman, M. E. J. (2006). Modularity and community structure in networks. *Proceedings of the National Academy of Sciences of the United States of America*, *103*(23). doi:10.1073/pnas.0601602103

Newman, M. E. J., & Girvan, M. (2004a). Finding and evaluating community structure in networks. *Physical Review E: Statistical, Nonlinear, and Soft Matter Physics*, *69*(2). doi:10.1103/PhysRevE.69.026113

Nham, B., Siangliulue, S., & Yeung, S. (2009). *Predicting mode of transport from iPhone accelerometer data*. Palo Alto, CA: Stanford University.

Nicosia, V., Mangioni, G., Carchiolo, V., & Malgeri, M. (2009). Extending the definition of modularity to directed graphs with overlapping communities. *Journal of Statistical Mechanics: Theory and Experiment*. Retrieved from arxiv.org

Nigam, K., McCallum, A. K., Thrun, S., & Mitchell, T. (2000). Text classification from labeled and unlabeled documents using EM. *Machine Learning*, *39*(2), 103–134. doi:10.1023/A:1007692713085

Noy, N. F., & McGuiness, D. (2001). *Ontology development 101: A guide to creating your first ontology*. New York, NY: Knowledge Systems Laboratory.

Omelayenko, B. (2003). RDFT: A mapping meta-ontology for web service integration . In Omelayenko, B., & Klein, M. (Eds.), *Knowledge Transformation for the Semantic Web* (pp. 137–153). Boca Raton, FL: IOS Press.

Orponen, P., Schaeffer, S. E., & Gaytan, V. A. (2008). *Locally computable approximations for spectral clustering and apsorption times of random walks*. Retrieved from arxiv.org

Orponen, P., & Schaeffer, S. E. (2005). Local clustering of large graphs by approximate Fiedler vectors. *Experimental and Efficient Algorithms*, *3503*, 524–533. doi:10.1007/11427186_45

Otsuka, S., Toyoda, M., Hirai, J., & Kitsuregawa, M. (2004). Extracting user behavior by web communities technology on global web logs. In *Proceedings of the 15th International Conference on Database and Expert Systems Applications (DEXA 2004)*. DEXA.

Pak, A., & Paroubek, P. (2010). Twitter as a corpus for sentiment analysis and opinion mining. In *Proceedings of LREC*. LREC.

Paltoglou, G., & Thelwall, M. (2010). A study of information retrieval weighting schemes for sentiment analysis. In *Proceedings of the 48th Annual Meeting of the Association for Computational Linguistics, ACL 2010*. Morristown, NJ: Association for Computational Linguistics.

Pang, B., & Lee, L. (2004). A sentimental education: Sentiment analysis using subjectivity summarization based on minimum cuts. In *Proceedings of the ACL 2004*, (pp. 271–278). ACL.

Pang, B., Lee, L., & Vaithyanathan, S. (2002). Thumbs up? Sentiment classification using machine learning techniques. In *Proceedings of the ACL 2002 Conference on Empirical Methods in Natural Language Processing, EMNLP 2002*, (vol. 10, pp. 79-86). Morristown, NJ: Association for Computational Linguistics.

Pang, B., & Lee, L. (2008). Opinion mining and sentiment analysis. *Foundations and Trends in Information Retrieval*, *2*(1-2), 1–135. doi:10.1561/1500000011

Paroubek, P., Pak, A., & Mostefa, D. (2010). Annotations for opinion mining evaluation in the industrial context of the DOXA project. In *Proceedings of LREC*. LREC.

Pennacchiotti, M., & Popescu, A.-M. (2011). Democrats, republicans and Starbucks afficionados: User classification in twitter. In C. Apte, J. Ghosh, & P. Smyth (Eds.), *Proceedings of the 17th ACM SIGKDD International Conference on Knowledge Discovery and Data Mining* (pp. 430-438). New York, NY: ACM Press.

Perelomov, I., Azcarraga, A. P., Tan, J., & Chua, T. S. (2002). Using structured self-organizing maps in news integration websites. In *Proceedings of the 11th International World Wide Web Conference*. ACM.

Phelan, O., McCarthy, K., & Smyth, B. (2009). Using Twitter to recommend real-time topical news. In *Proceedings of the Third ACM Conference on Recommender Systems*, (pp. 385-388). ACM Press.

Pierrakos, D., Paliouras, G., Papatheodorou, C., Karkaletsis, V., & Dikaiakos, M. D. (2003). Web community directories: A new approach to web personalization. In *Proceedings of the 1st European Web Mining Forum (EWMF 2003)*. EWMF.

Pietilainen, A. K., Oliver, E., & LeBrun, J. (2009). MobiClique: Middleware for mobile social networking. In *Proceedings of WOSN 2009*. Barcelona, Spain: WOSN.

Porter, M. F. (1980). An algorithm for suffix stripping. *Program*, *14*(3), 130–137. doi:10.1108/eb046814

Pothen, A., Simon, H. D., & Liou, K. (1990). Partitioning sparse matrices with eigenvectors of graphs. *SLAM Journal on Metrix Analysis and Applications*, *11*(3).

Pulse, B. (2011). *Stats 2011*. Retrieved 17 September, 2011 from http://www.blogpulse.com/

Qiu, H., & Hancock, E. R. (2006). Graph matching and clustering using spectral partitions. *Pattern Recognition*, *39*(1), 22–34. doi:10.1016/j.patcog.2005.06.014

Quirk, P. W. (2009). *Iran's Twitter revolution: Foreign policy in focus*. Retrieved June 17, 2009 from http://www.fpif.org/articles/irans_twitter_revolution

Quirk, R., Leech, G., & Startvik, J. (1985). *A comprehensive grammar of the English language*. New York, NY: Longman.

Radev, D. R. (2004). Lexrank: Graph-based lexical centrality as salience in text summarization. *Journal of Artificial Intelligence Research*, 22.

Radev, D. R., Jing, H., Sty, M., & Tam, D. (2004). Centroid-based summarization of multiple documents. *Information Processing & Management*, *40*(6), 919–938. doi:10.1016/j.ipm.2003.10.006

Ralphs, T. K., & Güzelsoy, M. (2005) The symphony callable library for mixed-integer linear programming. In *Proceedings of the Ninth INFORMS Computing Society Conference*, (pp. 61-76). INFORMS.

Ramage, D., Heymann, P., Manning, C. D., et al. (2009). *Clustering the tagged web*. Paper presented at the the Second ACM International Conference on Web Search and Data Mining. Barcelona, Spain.

Randell, C., & Muller, H. (2000). *Context awareness by analyzing accelerometer data*. Paper presented at the Fourth International Symposium on the Wearable Computers. New York, NY.

Rauber, A., Merkl, D., & Dittenbach, M. (2002). The growing hierarchical self-organizing map: Exploratory analysis of high-dimensional data. *IEEE Transactions on Neural Networks*, *13*(6), 1331–1341. doi:10.1109/TNN.2002.804221

Ravi, N. D. N., Mysore, P., & Littman, M. L. (2005). *Activity recognition from accelerometer data*. Paper presented at the Proceedings of the IAAI 2005. New York, NY.

Reddy, P. K., & Kitsuregawa, M. (2001). An approach to relate the web communities through bipartite graphs. In *Proceedings of WISE* (p. 301). IEEE Press.

Reddy, S., Burke, J., Estrin, D., Hansen, M., & Srivastava, M. (2008). Determining transportation mode on mobile phones. In *Proceedings of the Wearable Computers, 2008*. IEEE Press.

Resnik, P., & Hardisty, E. (2010). *Gibbs sampling for the uninitiated*. University Park, MD: University of Maryland.

Richardson, M., & Domingos, P. (2002). Mining knowledge-sharing sites for viral marketing. In *Proceedings of the Eighth ACM SIGKDD International Conference on Knowledge Discovery and Data Mining*, (pp. 61–70). ACM Press.

Riloff, E., Patwardhan, S., & Wiebe, J. (2006). Feature subsumption for opinion analysis. In *Proceedings of EMNLP*. EMNLP.

Riloff, E., Wiebe, J., & Wilson, T. (2003). Learning subjective noun using extraction pattern bootstrapping. In *Proceedings of the 7th Conference on Natural Language Learning*, (pp. 25–32). Edmonton, Canada: ACL.

Rocchio, J. J. (1971). *Relevance feedback in information retrieval*. Upper Saddle River, NJ: Prentice-Hall.

Rohs, M., & Bohn, R. (2003). Entry points into a smart campus environment – Overview of the ETHOC system. In *Proceedings of the 23rd Internal Conference on Distributed Computing Systems (ICDCSW 2003)*. ISDCSW.

Roh, T. H., Oh, K. J., & Han, I. (2003). The collaborative filtering recommendation based on SOM cluster-indexing CBR. *Expert Systems with Applications*, *25*(3), 413–423. doi:10.1016/S0957-4174(03)00067-8

Rotem, N. (2006). *Open text summarizer (OTS)*. Retrieved in July 2011 from http://libots.sourceforge.net

Russell, B., Yin, H., & Allinson, N. M. (2002). Document clustering using the 1 + 1 dimensional self-organising map. In *Proceedings of the Intelligent Data Engineering and Automated Learning—IDEAL 2002*, (pp. 167-174). Berlin, Germany: Springer.

Rymaszewski, M., Au, W. J., Wallace, M., Winters, C., Ondrejka, C., & Batstone-Cunningham, B. (2007). *Second life: The official guide*. Indianapolis, IN: Wiley Publishing.

Sasajima, M., Kitamura, Y., Naganuma, T., Fujii, K., Kurakake, S., & Mizoguchi, R. (2008). Obstacles reveal the needs of mobile internet services -OOPS: Ontology-based obstacle, prevention, and solution modeling framework. *Journal of Web Engineering*, *7*(2), 133–157.

Sasajima, M., Kitamura, Y., Naganuma, T., Kurakake, S., & Mizoguchi, R. (2006). Task ontology-based framework for modeling users' activities for mobile service navigation. [ESWC.]. *Proceedings of Posters and Demos of the ESWC*, *2006*, 71–72.

Sbodio, M. L., & Simpson, E. (2009). *Tag clustering with self organizing maps. HP Labs Technical Reports*. New York, NY: HP Labs.

Schaeffer, S. E. (2005). Stochastic local clustering for massive graphs. *Lecture Notes in Computer Science*, *3518*, 413–424. doi:10.1007/11430919_42

Schreiber, G., Akkermans, H., Anjewierden, A., de Hoog, R., Shadbolt, N. V., de Velde, W., & Wielinga, B. (2000). *Knowledge engineering and management - The CommonKADS methodology*. Cambridge, MA: MIT Press.

Schwagereit, F., & Staab, S. (2009). Social networks . In Liu, L., & Özsu, M. T. (Eds.), *Encyclopedia of Database Systems* (pp. 2667–2672). New York, NY: Springer.

Schwarz, G. (1978). Estimating the dimension of a model. *Annals of Statistics*, *6*(2), 461–464. doi:10.1214/aos/1176344136

Scoble, R., Israel, S., & Corporation, E. (2006). *Naked conversations: How blogs are changing the way businesses talk with customers*. New York, NY: John Wiley.

Sen, S., Lam, S. K., Rashid, A. M., Cosley, D., Frankowski, D., & Osterhouse, J. … Riedl, J. (2006). Tagging, communities, vocabulary, evolution. In *Proceedings of the 2006 20th Anniversary Conference on Computer Supported Cooperative Work*, (pp. 181-190). ACM.

Sharan, A., & Gupta, S. L. (2009). Identification of web communities through link based approaches. In *Proceedings of the International Conference on Information Management and Engineering*, (pp. 703-708). IEEE.

Sharifi, B., Hutton, M. A., & Kalita, J. (2010). Automatic summarization of Twitter topics. In *Proceedings of the National Workshop on Design and Analysis of Algorithms*. IEEE.

Shepitsen, A., Gemmell, J., Mobasher, B., & Burke, R. (2008). *Personalized recommendation in social tagging systems using hierarchical clustering*. Paper presented at the the 2008 ACM Conference on Recommender Systems. Bilbao, Spain.

Shi, J., & Malik, J. (2002). Normalized cuts and image segmentation. *IEEE Transactions on Pattern Analysis and Machine Intelligence*, *22*(8), 888–905.

Shi, Y., Xie, W., & Xu, G. (2003). The smart classroom: merging technologies for seamless tele-education. *IEEE Pervasive Computing / IEEE Computer Society [and] IEEE Communications Society*, *2*(2), 47–55. doi:10.1109/MPRV.2003.1203753

Sima, J., & Schaeffer, S. (2006). On the NP-completeness of some graph cluster measures. In *Proceedings of the SOFSEM 2006: Theory and Practice of Computer Science*, (pp. 530-537). SOFSEM.

Simpson, E. (2008). *Clustering tags in enterprise and web folksonomies. HP Labs Technical Reports*. New York, NY: HP Labs.

Singh, V., Mahata, D., & Adhikari, R. (2010). Mining the blogosphere from a socio-political perspective. In *Proceedings of the 6th International Conference on Next Generation Web Services Practices*, (pp. 365-370). Gwalior, India: IEEE Press.

Smith, G. (2008). *Tagging: People-powered metadata for the social web*. Berkeley, CA: New Riders.

Somasundaran, S., Ruppenhofer, J., & Wiebe, J. (2008). Discourse level opinion relations: An annotation study. In *Proceedings of the 9th SIGdial Workshop on Discourse and Dialogue*, (pp. 129–137). Association for Computational Linguistics.

Somboonviwat, K. (2008). *Research on language specific crawling and building of Thai web archive*. (Unpublished Doctoral Dissertation). University of Tokyo. Tokyo, Japan.

Song, Y., Zhang, L., & Giles, C. L. (2011). Automatic tag recommendation algorithms for social recommender systems. *ACM Transactions on the Web, 5*(1), 4:1-4:31.

Specia, L., & Motta, E. (2007). Integrating folksonomies with the semantic web . In Franconi, E., Kifer, M., & May, W. (Eds.), *The Semantic Web: Research and Applications* (pp. 624–639). Berlin, Germany: Springer. doi:10.1007/978-3-540-72667-8_44

Spertus, E., Sahami, M., & Buyukkokten, O. (2005). Evaluating similarity measures: A large-scale study in the orkut social network. In *Proceedings of KDD*, (pp. 678-684). KDD.

Spielman, D. A., & Teng, S. H. (2004). Nearly-linear time algorithms for graph partitioning, graph sparsification, and solving linear systems. In *Proceedings of the Thirty-Sixth Annual ACM Symposium on Theory of Computing*, (pp. 81–90). ACM Press.

Statistics, F. (2012). *Webpage*. Retrieved 27 January, 2012 from http://www.facebook.com/press/info.php?statistics

Stoyanov, V., Cardie, C., Littman, D., & Wiebe, J. (2004). Evaluating an opinion annotation scheme using a new multiperspective question and answer corpus. In *Proceedings of the AAAI Spring Symposium on Exploring Attitude and Affect in Text*. AAAI.

Suaris, P. R., & Kedem, G. (2002). An algorithm for quadrisection and its application to standard cell placement. *IEEE Transactions on Circuits and Systems, 35*(3), 294–303. doi:10.1109/31.1742

Sun, J., Qu, H., Chakrabarti, D., & Faloutsos, C. (2005). Neighborhood formation and anomaly detection in bipartite graphs. In *Proceedings of ICDM*, (pp. 418-425). ICDM.

Sun, T., Shi, X. Y., & Shen, Y. M. (2010). iLife: A novel mobile social network services on mobile phones. In *Proceedings of the 10th International Conference on Computer and Information Technology (CIT)*. Bradford, UK: CIT.

Sun, Y., Han, J., Gao, J., & Yu, Y. (2009). iTopicModel: Information network-integrated topic modeling. In *Proceedings of 2009 International Conference on Data Mining (ICDM 2009)*. ICDM.

Suo, Y., Miyata, N., & Morikawa, H. (2009). Open smart classroom: extensible and scalable learning system in smart space using web service technology. *IEEE Transactions on Knowledge and Data Engineering, 21*(6), 814–828. doi:10.1109/TKDE.2008.117

Takamura, H., & Okumura, M. (2009a). Text summarization model based on maximum coverage problem and its variant. In *Proceedings of the 12th Conference of the European Chapter of the Association for Computational Linguistics*, (pp. 781-789). ACL.

Takamura, H., & Okumura, M. (2009b). Text summarization model based on the budgeted median problem. In *Proceeding of the 18th ACM Conference on Information and Knowledge Management*, (pp. 1589-1592). ACM Press.

Tamura, T., Somboonviwat, K., & Kitsuregawa, M. (2007). A method for language-specific web crawling and its evaluation. *Systems and Computers in Japan, 38*(2), 10–20. doi:10.1002/scj.20693

Tang, J., Yao, L., & Chen, D. (2009). Multi-topic based query-oriented summarization. In *Proceedings of the SIAM International Conference Data Mining*. SIAM.

Tan, P. N., Steinbach, M., & Kumar, V. (2006). *Introduction to data mining*. Boston, MA: Pearson Addison Wesley.

Tarrow, S., & Tollefson. (1994). *Power in movement: Social movements, collective action and politics*. Cambridge, UK: Cambridge University Press.

Tatti, N., & Heikinheimo, H. (2008). Decomposable families of itemsets. In *Proceedings of the Machine Learning and Knowledge Discovery in Databases*, (pp. 472-487). IEEE.

Tatti, N., & Mampaey, M. (2010). Using background knowledge to rank itemsets. *ACM Data Mining and Knowledge Discovery*, *21*(2), 293–309. doi:10.1007/s10618-010-0188-4

Technorati's State of the Blogosphere. (2011). *Website*. Retrieved 17 September, 2011 from http://technorati.com/state-of-the-blogosphere/

Tengfei, B., Happia, C., Enhong, C., Jilei, T., & Hui, X. (2010). An unsupervised approach to modeling personalized contexts of mobile users. In *Proceedings of the Data Mining (ICDM), 2010*. IEEE Press.

TexLexAn. (2011). *Texlexan: An open-source text summarizer*. Retrieved March 15, 2011, from http://texlexan.sourceforge.net/

Thakkar, K., Dharaskar, R., & Chandak, M. (2010). Graph-based algorithms for text summarization. In *Proceedings of the Third International Conference on Emerging Trends in Engineering and Technology*, (pp. 516–519). IEEE.

Thelwall, M. (2006). Bloggers under the London attacks: Top information sources and topics. In *Proceedings of the 3rd Annual Workshop on Webloging Ecosystem: Aggregation, Analysis and Dynamics*. IEEE.

Toprak, C., Jakob, N., & Gurevych, I. (2010). Sentence and expression level annotation of opinions in user-generated discourse. In *Proceedings of the 48th Annual Meeting of the Association for Computational Linguistics, ACL 2010*. Stroudsburg, PA: Association for Computational Linguistics.

Toyoda, M., & Kitsuregawa, M. (2001). Creating a web community chart for navigating related communities. In *Proceedings of the 12th ACM Conference on Hypertext and Hypermedia (HT 2001)*. ACM Press.

Toyoda, M., & Kitsuregawa, M. (2003). Extracting evolution of web communities from a series of web archives. In *Proceedings of the 14th ACM Conference on Hypertext and Hypermedia (HT 2003)*. ACM Press.

Tso-Sutter, K. H. L., Marinho, L. B., & Schmidt-Thieme, L. (2008). *Tag-aware recommender systems by fusion of collaborative filtering algorithms*. Paper presented at the the 23rd Annual ACM Symposium on Applied Computing. Ceará, Brazil.

Turney, P. D. (2002). Thumbs up or thumbs down? Semantic orientation applied to unsupervised classification of reviews. In *Proceedings of the 40th Annual Meeting on Association for Computational Linguistics*, (pp. 417–424). ACL.

van Dam, J., Vandic, D., Hogenboom, F., & Frasincar, F. (2010). *Searching and browsing tag spaces using the semantic tagging clustering search framework*. Paper presented at the the 4th IEEE International Conference on Semantic Computing. Pittsburgh, PA.

Vandic, D., van Dam, J.-W., Hogenboom, F., & Frasincar, F. (2011). A semantic clustering-based approach for searching and browsing tag spaces. In *Proceedings of the 2011 ACM Symposium on Applied Computing, SAC 2011*, (pp. 1693-1699). ACM Press.

Vesanto, J., & Alhoniemi, E. (2000). Clustering of the self-organizing map. *IEEE Transactions on Neural Networks*, *11*(3), 586–600. doi:10.1109/72.846731

Wan, X., & Yang, J. (2006). Improved affinity graph based multi-document summarization. In *Proceedings of the Human Language Technology Conference of the NAACL*, (pp. 181-184). NAACL.

Wang, D., & Li, T. (2010). Document update summarization using incremental hierarchical clustering. In *Proceedings of the 19th ACM International Conference on Information and Knowledge Management*, (pp. 279–288). ACM Press.

Wang, J., Li, Q., & Chen, Y. P. (2010). User comments for news recommendation in social media. In *Proceeding of the 33rd International ACM SIGIR Conference on Research and Development in Information Retrieval*, (pp. 881-882). ACM Press.

Wang, D., Zhu, S., Li, T., Chi, Y., & Gong, Y. (2011). Integrating document clustering and multidocument summarization. *ACM Transactions on Knowledge Discovery from Data*, *5*(3), 14. doi:10.1145/1993077.1993078

Wasserman, S., & Faust, K. (1994). *Social network analysis: Methods and applications*. Cambridge, UK: Cambridge University Press. doi:10.1017/CBO9780511815478

Watts, D. J., & Strogatz, S. H. (1998). Collective dynamics of "small-world" networks. *Nature, 393*(6684), 440–442. doi:10.1038/30918

White, S., & Smyth, P. (2005). A spectral clustering approach to finding communities in graphs. In *Proceedings of the Fifth SIAM International Conference on Data Mining*, (p. 274). Society for Industrial Mathematics.

Whitelaw, C., Garg, N., & Argamon, S. (2005). Using appraisal groups for sentiment analysis. In *Proceedings of the 14th ACM International Conference on Information and Knowledge Management, CIKM 2005*, (pp. 625–631). New York, NY: ACM.

Wiebe, J., Wilson, T., & Cardie, C. (2005). *Annotating expressions of opinions and emotions in language*. Dordrecht, The Netherlands: Kluwer Academic Publishers. doi:10.1007/s10579-005-7880-9

Xi, L., Bin, Y., & Aarts, R. M. (2009). *Single-accelerometer-based daily physical activity classification. In Proceedings of the Engineering in Medicine and Biology Society, 2009*. IEEE Press.

Xu, J. (2009). Identifying cohesive local community structures in networks. In *Proceedings of ICIS 2009*, (p. 112). ICIS.

Yang, T., Chi, Y., Zhu, S., & Jin, R. (2010). Directed network community detection: A popularity and productivity link model. In *Proceedings of the 2010 SIAM International Conference on Data Mining (SDM 2010)*. SIAM.

Yang, T., Jin, R., Chi, Y., & Zhu, S. (2009). Combining link and content for community detection: A discriminative approach. In *Proceedings of The 15th ACM SIGKDD International Conference on Knowledge Discovery and Data Mining (KDD 2009)*. ACM Press.

Yang, Z., Cai, K., Tang, J., Zhang, L., Su, Z., & Li, J. (2011). Social context summarization. In *Proceedings of the International ACM SIGIR Conference on Research and Development in Information Retrieval*. ACM Press.

Yang, G. (2009). Discovering significant places from mobile phones: A mass market solution. *Mobile Entity Localization and Tracking in GPS-less Environments, 5801*, 34–49. doi:10.1007/978-3-642-04385-7_3

Yang, Z., Yu, J. X., Liu, Z., & Kitsuregawa, M. (2010). Fires on the web: Towards efficient exploring historical web graphs. *Lecture Notes in Computer Science, 5981*, 612–626. doi:10.1007/978-3-642-12026-8_46

Yannik-Mathieu, Y. (1991). *Les verbes de sentiment – De l'analyse linguistique au traitement automatique*. Paris, France: CNRS Editions.

Yau, S. S., Gupta, S. K. S., & Fariaz, K. (2003). Smart classroom: Enhancing collaborative learning using pervasive computing technology. In *Proceedings of Annual Conference and Exposition: Staying in Tune with Engineering Education (ASEE)*. Nashville, TN: ASEE.

Yi, J., Nasukawa, T., Bunescu, R., & Niblack, W. (2003). Sentiment analyzer: Extracting sentiments about a given topic using natural language processing techniques. In *Proceedings of the third IEEE International Conference on Data Mining, 2003*, (pp. 427–434). IEEE Press.

Yin, Z., Li, R., Mei, Q., & Han, J. (2009). Exploring social tagging graph for web object classification. In *Proceedings of the 15th ACM SIGKDD International Conference on Knowledge Discovery and Data Mining*, (pp. 957-966). ACM Press.

Yin, D., Hong, L., Xiong, X., & Davison, B. D. (2011). Link formation analysis in microblogs. [Beijing, China: ACM Press.]. *Proceedings of SIGIR, 2001*, 24–28.

Yu, H., & Hatzivassiloglou, V. (2003). Towards answering opinion questions: Separating facts from opinions and identifying the polarity of opinion sentences. In *Proceedings of EMNLP*, (pp. 129–136). Sapporo, Japan: EMNLP.

Yu, S., Moor, B. D., & Moreau, Y. (2009). Clustering by heterogeneous data fusion: Framework and applications. In *Proceedings of the NIPS Workshop*. NIPS.

Yu, Z. W., Yu, Z. Y., Aoyama, H., Ozeki, M., & Nakamura, Y. (2010). Capture, recognition, and visualization of human semantic interactions in meetings. In *Proceedings of the 8th IEEE International Conference on Pervasive Computing and Communications (PerCom 2010)*. Mannheim, Germany: IEEE Press.

Zhai, C., & Lafferty, J. (2001). Model-based feedback in the language modeling approach to information retrieval. [CIKM.]. *Proceedings of CIKM*, *2001*, 403–410.

Zhang, X., Li, Y., & Liang, W. (2010b). C&C: An effective algorithm for extracting web community cores . In *Database Systems for Advanced Applications* (pp. 316–326). Berlin, Germany: Springer. doi:10.1007/978-3-642-14589-6_32

Zhang, X., Wang, L., Li, Y., & Liang, W. (2011). Extracting local community structure from local cores. [Springer.]. *Proceedings of Database Systems for Advanced Applications: SNSMW*, *2011*, 287–298.

Zhang, X., Xu, W., & Liang, W. (2010a). Extracting local web communities using lexical similarity . In *Database Systems for Advanced Applications* (pp. 327–337). Berlin, Germany: Springer. doi:10.1007/978-3-642-14589-6_33

Zhang, Y., Xu Yu, J., & Hou, J. (2006). *Web communities: Analysis and construction*. Berlin, Germany: Springer.

Zhao, S., Du, N., & Nauerz, A. (2008). Improved recommendation based on collaborative tagging behaviors. In *Proceedings of the 13th International Conference on Intelligent User Interfaces*. New York, NY: IEEE.

Zheng, Y., Liu, L., Wang, L., & Xie, X. (2008). Learning transportation mode from raw GPS data for geographic applications on the web. In *Proceedings of the 17th International Conference on World Wide Web*. Beijing, China. IEEE.

Zheng, Y., Zhang, L., Xie, X., & Ma, W.-Y. (2009). Mining interesting locations and travel sequences from GPS trajectories. In *Proceedings of the 18th International Conference on World Wide Web*. Madrid, Spain: IEEE.

Zhu, J., Wang, C., He, X., Bu, J., Chen, C., & Shang, S. ... Lu, G. (2009). Tag-oriented document summarization. In *Proceedings of the 18th International ACM Conference on World Wide Web,* (pp. 1195-1196). ACM Press.

Zhu, S., Yu, K., Chi, Y., & Gong, Y. (2007). Combining content and link for classification using matrix factorization. In *Proceedings of the 30th Annual International ACM SIGIR Conference on Research and Development in Information Retrieval (SIGIR 2007)*. ACM Press.

Zhuang, L., Jing, F., & Zhu, X.-Y. (2006). Movie review mining and summarization. In *Proceedings of the 15th ACM International Conference on Information and Knowledge Management, CIKM 2006*, (pp. 43–50). New York, NY: ACM.

Zubiaga, A., García-Plaza, A. P., Fresno, V., & Martínez, R. (2009). Content-based clustering for tag cloud visualization. In *Proceedings of the 2009 International Conference on Advances in Social Network Analysis and Mining*, (pp. 316-319). IEEE Computer Society.

About the Contributors

Guandong Xu received his PhD degree in Computer Science from Victoria University, Australia, in 2008. Prior to this, he received B.Eng and M.Eng degree in Computer Science and Computer Engineering from Zhejiang University in 1989 and 1992, respectively. He is now a Lecturer in the Advanced Analytics Institute at the University of Technology Sydney after being a Research Fellow in the Centre for Applied Informatics at Victoria University. During 2008-2009, he was awarded a prestigious Endeavour Research Fellowships by Australian Federal Government and working in the University of Tokyo. His research interests cover Web data management and analysis, data mining, machine learning, Internet computing and e-commerce, advanced Web applications such as Web communities, Web recommendation and personalization, as well as health informatics. He has extensively published papers in international journals and conference proceedings in the areas of Web data mining, Web recommendation, and health informatics.

Lin Li is an Associate Professor with School of Computer Science and Technology, Wuhan University of Technology, China. She received a Ph.D degree in Information Science and Technology from the University of Tokyo in 2009. Her research interests cover Web mining, machine learning, natural language processing, information retrieval, and so on. She has extensively published papers in international journals and conference proceedings in the areas of information retrieval, Web personalization, recommendation, and social network mining.

* * *

Nitin Agarwal is an Assistant Professor in the Information Science Department at the University of Arkansas at Little Rock. Agarwal has a Ph.D. in Computer Science from Arizona State University with outstanding dissertation recognition. He studies the computational aspects of social media and the underlying social network processes including behavioral modeling, knowledge extraction, prediction, and evaluating phenomena such as influence, trust, collective wisdom, collective action, crowd dynamics, and community extraction. His research leverages fundamentals of data mining, graph mining, content analysis, and large-scale data management. His research is sponsored by the US National Science Foundation (NSF) and the US Office of Naval Research (ONR). He was recently nominated among the top 20 influentials in their 20s by *Arkansas Business Magazine*, a statewide business publication. He has published his work in over 50 leading journals and conferences, which are highly cited and include best paper awards. He has authored two books, edited several special issues for leading journals, and delivered several well-received talks and tutorials in industry and academia. He is currently serving as program chair and on program committees of several prestigious conferences. More details can be found at www.ualr.edu/nxagarwal/.

Tengfei Bao received a B.E. degree from University of Science and Technology of China in 2007. He is currently a Ph.D. student at the school of Computer Science, University of Science and Technology of China. His research interests include mobile context modeling and mobile user geo-trajectory mining. During his Ph.D. candidate career, he won the Chinese Academic Science President Award.

Luca Cagliero is a Post-Doc Fellow at the Dipartimento di Automatica e Informatica of the Politecnico di Torino since March 2012. He holds a Master degree in Computer and Communication Networks and a PhD in Computer Engineering from Politecnico di Torino. His current research interests are in the fields of data mining and database systems. In particular, he has worked on structured and unstructured data mining by means of itemset and association rule mining algorithms.

Huanhuan Cao is currently a Senior Researcher of Nokia Research Center, GEL Lab. He received a B.E. degree and a Ph.D. degree from University of Science and Technology of China, China, in 2005 and 2009, respectively. His major research interests include mobile context mining and mobile user behavior analysis. In this field, he applied more than twenty invention patents and published several high quality research papers. Moreover, during his Ph.D. candidate career, he did a series of studies of leveraging large-scale search logs for supporting context-aware search. Due to his academic achievement in this field, he won the Microsoft Fellow, Chinese Academic Science President Award, and CCF best PhD thesis award.

Enhong Chen is a Professor at the School of Computer Science and Technology, University of Science and Technology of China. He received his Doctor degree in USTC in 1996. He serves as a Councilor of Machine Learning Society of Chinese Association for Artificial Intelligence, and a Councilor of Artificial Intelligence and Pattern Recognition Society of Chinese Computer Federation. His research interests include machine learning and data mining, personalized recommendation systems, social network. He has published over 100 technical papers in peer-reviewed journals and conference proceedings. He received the Best Application Paper Award in KDD 2008, and Best Research Paper in ICDM 2011. Prof. Chen has been actively involved in the research community by serving as a PC member for a number of international conferences and workshops. He is a senior member of the IEEE, and a member of the ACM.

Alessandro Fiori received the European Ph.D. degree from Politecnico di Torino, Italy. He is a Project Manager at the Institute for Cancer Research and Treatment (IRCC) of Candiolo, Italy, since January 2012, and a Researcher at the DBDMG group of Politenico di Torino. His research interests are in the field of data mining, in particular bioinformatics and text mining. His activity is focused on the development of information systems and analysis frameworks oriented to the management and integration of biological and molecular data. His research activities are also devoted to text summarization and social network analysis.

Víctor Fresno is Assistant Professor at UNED. He received the Bachelor of Theoretical Physics from Universidad Autónoma de Madrid in 1999, the M.Sc. in Telecommunication Engineering from Universidad Politécnica de Madrid in 2004, and the Ph.D. in Computer Science from Universidad Rey Juan Carlos in 2006. His research interests include Web page characterization for classification/clustering and information retrieval, fuzzy logic, and natural language processing tools for text mining. He has published more than 40 papers in national/international journals, conferences, and workshops. He has been involved in projects funded by Spanish Government with academic and industrial partners.

Alberto Pérez García-Plaza earned a B.S. and M.S. degree in Computer Engineering from Universidad Rey Juan Carlos, Madrid (Spain) in 2003 and 2006, respectively. He is currently working as Teaching Assistant at the Department of Computer Systems and Languages of the UNED University in Madrid, Spain, where he is also Ph.D. candidate and member of the NLP&IR Group. His main research interests are Web page clustering, document representation, fuzzy logic, and social media mining.

Bin Guo is now an Associate Professor in the School of Computer Science, Northwestern Polytechnical University, China. He received his Ph.D. degree from Keio University, Japan, in 2009. He was previous a Post-Doctoral Researcher at Institute TELECOM SudParis, France. His research interest includes ubiquitous computing, mobile social networking, and social intelligence.

Xiao Huifan is a postgraduate student of School of Computer Science and Technology, Wuhan University of Technology, China. She received her Bachelor of Engineering degree in Computer Science and Technology from Wuhan University of Technology(WHUT) and Bachelor of Arts degree in English from Huazhong University of Science and Technology(HUST) both in 2011. Data mining and information retrieval are her main research interests.

Yoshinobu Kitamura is an Associate Professor, the Institute of Scientific and Industrial Research, Osaka University. He received his Bachelor, Master, and Doctoral degrees of Engineering in 1991, 1993, and 2000 from Osaka University, respectively. He was a Visiting Associate Professor at Stanford University, CA, from 2007 to 2008. His research interest includes ontological engineering mainly in engineering domain. He has served as an Associate Editor of the ASME JCISE, a member of the Editorial Board of *Advanced Engineering Informatics* and that of the Board of Directors of the Japanese Society for Artificial Intelligence (JSAI).

Carson K.-S. Leung received his B.Sc. (Hons.), M.Sc., and Ph.D. degrees, all in Computer Science, from the University of British Columbia, Canada. Currently, he is an Associate Professor in Department of Computer Science at University of Manitoba, Canada. His research interests include the areas of databases, data mining, social media mining, social network analysis, and social computing. His work has been published in refereed international journals and conferences such as *ACM Transactions on Database Systems*, IEEE International Conference on Data Engineering (ICDE), IEEE International Conference on Data Mining (ICDM), and Pacific-Asia Conference on Knowledge Discovery and Data Mining (PAKDD). In the past few years, he has served as a Program Chair for International C* Conference on Computer Science and Software Engineering (C^3S^2E) 2009-2010, an organizing committee member of ACM SIGMOD 2008 and IEEE ICDM 2011, and a PC member of numerous international conferences including ACM KDD, CIKM, and ECML/PKDD.

Yueting Li is currently a System Engineer at Baidu, Inc. She received her B.E. and M.E. degrees in Computer Science from Dalian University of Technology, China, in 2008 and 2011, respectively. Her main research interests include Web community extraction and machine learning.

Wenxin Liang is an Associate Professor at Dalian University of Technology, China. He received his B.E. and M.E. degrees from Xi'an Jiaotong University, China, in 1998 and 2001, respectively. He received his Ph.D. degree in Computer Science from Tokyo Institute of Technology in 2006. He was a Postdoc Research Fellow, CREST of Japan Science and Technology Agency (JST), and a Guest Research Associate, GSIC of Tokyo Institute of Technology from Oct. 2006 to Mar. 2009. His main research interests include Web-based IR, knowledge discovery and management, XML data processing and management, etc. He is a senior member of China Computer Federation (CCF), and a member of IEEE, ACM, ACM SIGMOD Japan Chapter, and Database Society of Japan (DBSJ).

Yunji Liang is a Ph.D. candidate in the School of Computer Science, Northwestern Polytechnical University, Shaanxi, China. He received the B.S. and M.S. degree in Computer Science and Technology in 2007 and 2010, respectively. His research interests include context-aware systems and social computing.

Raquel Martínez Unanue, Ph.D. in Computer Science and Bs. D. in Computer Science, is Profesor Titular (Associate Professor) at the Dept. of Lenguajes y Sistemas Informáticos (LSI) at UNED (National Distance Learning University), in Madrid, Spain. Her research lines include: multilingual document clustering: document representation and clustering algorithms, person name disambiguation, and data mining in social media.

Debanjan Mahata is a Doctoral student in Information Science Department at the University of Arkansas at Little Rock. He has a Masters in Computer Science from Banaras Hindu University, India. He studies the computational aspects of social media and is mainly interested in analyzing real-life events from the sources available from social media. He leverages the fundamentals of data mining, text mining, content analysis, information retrieval, and artificial intelligence for his research.

Irish J. M. Medina is pursuing her B.C.Sc. (Hons.) degree in Computer Science with a minor in Mathematics. During her current undergraduate degree studies, she has received several awards and honours including a few institutional scholarships as well as recognitions to be on the University 1 Honour List and the Dean's Honour List. Moreover, she has also received the University of Manitoba Undergraduate Summer Research Award for her summer academic research in the Database and Data Mining Lab, led by Prof. Carson K.-S. Leung, in Department of Computer Science at University of Manitoba, Canada. Ms. Medina is interested in databases and data mining.

Riichiro Mizoguchi is currently Professor of the Institute of Scientific and Industrial Research, Osaka University. His research interests include non-parametric data analyses, knowledge-based systems, ontological engineering, and intelligent learning support systems. Dr. Mizoguchi was President of International AI in Education Society and Asia-Pacific Society for Computers in Education from 2001 to 2003 and President of Japanese Society for Artificial Intelligence (JSAI) from 2005 to 2007, Vice-President of SWSA (Semantic Web Science Association) from 2005 - 2009 and Co-Editor-in-Chief of *Journal of Web Semantics* from 2007 to 2011. He is currently associate editor of IEEE TLT and ACM TiiS.

Alexander Pak is a PhD Scholar at University Paris-Sud, lab. LIMSI-CNRS, France. He holds a Master degree in Computer Science from Advanced University of Science and Technology, Taejon, Republic of Korea. His current research involves multilingual and multidomain sentiment analysis and opinion mining.

Patrick Paroubek is a Research Engineer in the Language, Information, and Representation group of the Human-Machine Communication Department at LIMSI-CNRS, France. He received PhD degree in Computer Science from Université Pierre and Marie Curie, Paris, France. His current research focus is on basic natural language processing (POS tagging, parsing), particularly in the context of evaluation, but also addressing machine understanding, opinion mining, and sentiment analysis.

Munehiko Sasajima is an Assistant Professor of the Department of Knowledge Systems, the Institute of Scientific and Industrial Research (ISIR),Osaka University. He received his Ph.D. from Graduate School of Engineering Science, Osaka University, 1997. He worked at Multimedia Labs, Corporate R&D Center, Toshiba Corporation, from 1997 to 2004. His research interests include knowledge-based systems, ontological engineering, and Web intelligence and interaction. He is currently a Chief Examiner of the Steering Committee of IPSJ (Information Processing Society of Japan) Journal.

Kulwadee Somboonviwat received a Ph.D. degree from the University of Tokyo in 2008. She is currently a Lecturer in Software Engineering Program at the International College, KMITL, Bangkok Thailand. Her current research interests include database engineering, Web mining, and social media mining.

Syed K. Tanbeer received his B.S. degree in Applied Physics and Electronics in 1996 and his M.S. degree in Computer Science in 1998, both from University of Dhaka, Bangladesh. He then worked as a faculty member in Department of Computer Science and Information Technology at IUT-OIC, Dhaka, Bangladesh. He received his Ph.D. degree in Computer Engineering in 2010 from Kyung Hee University, South Korea. Currently, he is a Post-Doctoral Fellow in the Database and Data Mining Lab, led by Prof. Carson K.-S. Leung, in Department of Computer Science at University of Manitoba, Canada. Dr. Tanbeer's research areas are data mining, parallel and distributed mining, and knowledge discovery from social networks.

Liang Wang is currently a system engineer at Baidu, Inc. He received his B.E. and M.E. degrees in Computer Science from Dalian University of Technology, China, in 2009 and 2012, respectively. His main research interests include Web community extraction and data mining.

Yue Yang is currently a Master student in the School of Computer Science, Northwestern Polytechnical University, China. She received her B.S. degree in Computer Science and Technology in 2010. Her research interests include participatory sensing, social computing, and link prediction.

Xun Yi is an Associate Professor with School of Engineering and Science, Victoria University, Australia. Before joining Victoria University, he was an Assistant Professor with School of Information Science, Japan Advanced Institute of Science and Technology (JAIST), Japan, and an Assistant Professor with School of Electrical and Electronic Engineering, Nanyang Technological University (NTU), Singapore. He obtained a PhD degree in Engineering from Xidian University in 1995. His research interests include applied cryptography, privacy-preserving data mining, computer and network security, mobile and wireless communication security, and intelligent agent technology. He has published about 100 research papers in international journals, such as *IEEE Transactions on Knowledge and Data Engineering, IEEE Transactions on Wireless Communication, IEEE Transactions on Dependable and Secure Computing, IEEE Transactions on Circuit and Systems, IEEE Transactions on Vehicular Technologies, IEEE Communication Letters*, and conference proceedings. He is a member of Editorial Reviewer Board for *International Journal of Electronic Commerce in Organization*. He has undertaken program committee membership for about 20 international conferences. He is leading an Australia Research Council (ARC) Discovery Project on Privacy Protection in Distributed Data Mining.

Zhiwen Yu is a Professor and Vice Dean in the School of Computer Science, Northwestern Polytechnical University, China. He received his B.Eng, M.Eng, and Ph.D degree of Engineering in Computer Science and Technology in 2000, 2003, and 2005, respectively. He has worked as an Alexander Von Humboldt Fellow at Mannheim University, Germany, from Nov. 2009 to Oct. 2010, a research fellow at Kyoto University, Japan, in 2007-2009, and a Post-Doctoral Researcher at Nagoya University, Japan, in 2006-2007. Dr. Yu has published over 100 scientific papers in refereed journals and conferences, e.g., *ACM Computing Surveys, IEEE Pervasive Computing*, IEEE TKDE, PerCom, etc. His research interests cover pervasive computing, context-aware systems, human-computer interaction, and personalization. He is a senior member of IEEE and IEEE Computer Society.

Xianchao Zhang is a Full Professor at Dalian University of Technology, China. He received his B.S. degree in Applied Mathematics and M.S. degree in Computational Mathematics from National University of Defense Technology in 1994 and 1998, respectively. He received his Ph.D. in Computer Theory and Software from University of Science and Technology of China in 2000. He joined Dalian University of Technology in 2003 after 2 years of industrial working experience at international companies. He worked as a Visiting Scholar at the Australian National University and the City University of Hong Kong in 2005 and 2009, respectively. His research interests include algorithms, machine learning, data mining, and information retrieval.

Yu Zong is an Associate Professor of the Department of Information and Engineering, West Anhui University, and he is also a Postdoc of University of Science and Technology of China. Dr Zong is a Joint PhD of Dalian University of Technology, China, and Victory University, Australia, and he obtained a PhD degree in Computer Science from the Dalian University of Technology, China, in 2010. Prior to this, he received M.Eng degree in Software Engineering from Dalian University of Technology, China, and he also received B.Eng in Computer Application from Agriculture University of Anhui, China, in 2005 and 1999, respectively. Prof. Zong's research interests include data mining, social network mining, recommendation, intelligent algorithm, as well as computer applications. He has extensively published papers in international journals and conference proceedings in the areas of data mining, Web recommendation, and social network mining.

Arkaitz Zubiaga is a Post-Doctoral Researcher at the Queens College and Graduate Center of the City University of New York. Arkaitz earned his Ph.D. in Computer Science in July 2011 under the supervision of Prof. Raquel Martínez and Prof. Víctor Fresno at the NLP&IR Group of the UNED University in Madrid, Spain. His main research interests lie in the fields of social media mining, machine learning, and data mining. In his post-doctoral research, he has continued with his research in mining social media. Specifically, he has explored the power of real-time social streams as a means to characterize and discover information about current events and trends in social media.

Index

A

B

C

D

E

F

G

H

I

K

L

M

N

O

P

Q

R

S

T

U

V

W